Extreme Value and Related Models with Applications in Engineering and Science

Extreme Value and Related Models with Applications in Engineering and Science

Enrique Castillo
University of Cantabria
and University of Castilla La Mancha

Ali S. Hadi
The American University in Cairo
and Cornell University

N. Balakrishnan
McMaster University

José María Sarabia
University of Cantabria

WILEY-INTERSCIENCE

A JOHN WILEY & SONS, INC., PUBLICATION

Published by John Wiley & Sons, Inc., Hoboken, New Jersey.
Published simultaneously in Canada.

Library of Congress Cataloging-in-Publication Data:

Extreme value and related models with applications in engineering and science / Enrique Castillo . . . [et al.].
p. cm.
Includes bibliographical references and index.
ISBN 0-471-67172-X
1. Extreme value theory. 2. Random variables. 3. Mathematical statistics. I. Castillo, Enrique, 1946–

QA273.6.E968 2005
519.2'4—dc22 2004049165

10 9 8 7 6 5 4 3 2 1

DEDICATION

To Juan Ramón Ruiz Tolosa on the occasion of his 70th birthday, E. C.

To my friend Enrique Castillo and his family with much appreciation, A. S. H.

To my sister, Chitra Ramachandran, for showering her love and affection on me all these years, and hoping to receive the same for many years to come, N. B.

To Corito, José María and Belén, J. M. S.

Contents

Preface

The field of extremes, maxima and minima of random variables, has attracted the attention of engineers, scientists, probabilists, and statisticians for many years. The fact that engineering works need to be designed for extreme conditions forces one to pay special attention to singular values more than to regular (or mean) values. The statistical theory for dealing with mean values is very different from that required for extremes, so that one cannot solve the above indicated problems without a knowledge of statistical theory for extremes.

In 1988, the first author published the book *Extreme Value Theory in Engineering* (Academic Press), after spending a sabbatical year at Temple University with Prof. Janos Galambos. This book had an intentional practical orientation, though some lemmas, theorems, and corollaries made life a little difficult for practicing engineers, and a need arose to make the theoretical discoveries accessible to practitioners. Today, many years later, important new material have become available. Consequently, we decided to write a book which is more practically oriented than the previous one and intended for engineers, mathematicians, statisticians, and scientists in general who wish to learn about extreme values and use that knowledge to solve practical problems in their own fields.

The book is structured in five parts. Part I is an introduction to the problem of extremes and includes the description of a wide variety of engineering problems where extreme value theory is of direct importance. These applications include ocean, structural and hydraulics engineering, meteorology, and the study of material strength, traffic, corrosion, pollution, and so on. It also includes descriptions of the sets of data that are used as examples and/or exercises in the subsequent chapters of the book.

Part II is devoted to a description of the probabilistic models that are useful in extreme value problems. They include discrete, continuous, univariate, and multivariate models. Some examples relevant to extremes are given to illustrate the concepts and the presented models.

Part III is dedicated to model estimation, selection, and validation. Though this topic is valid to general statistics, some special methods are given for extremes. The main tools for model selection and validation are probability paper plots (P-P and Q-Q plots), which are described in detail and are illustrated with a wide selection of examples.

Part IV deals with models for order statistics and extremes. Important concepts such as order statistics, return period, exceedances, and shortfalls are explained. Detailed derivations of the exact distributions of these statistics are presented and illustrated by many examples and graphs. One chapter is dedicated to point processes and exact models, where the reader can discover some important ways of modeling engineering problems. Applications of these models are also illustrated by some examples.

Part V is devoted to the important problem of asymptotic models, which are among the most common models in practice. The limit distributions of maxima, minima, and other order statistics of different types, for the cases of independent as well as dependent observations are presented. The important cases of exceedances and shortfalls are treated in a separate chapter, where the prominent generalized Pareto model is discussed. Finally, the multivariate case is analyzed in the last chapter of the book.

In addition to the theory and methods described in this book, we strongly feel that it is also important for readers to have access to a package of computer programs that will enable them to apply all these methods in practice. Though not part of this book, it is our intention to prepare such a package and make it available to the readers at: http://personales.unican.es/castie/extremes. This will assist the readers to (a) apply the methods presented in this book to problems in their own fields, (b) solve some of the exercises that require computations, and (c) reproduce and/or augment the examples included in this book, and possibly even correct some errors that may have occurred in our calculations for these examples. The computer programs will include a wide collection of univariate and multivariate methods such as:

1. Plots of all types (probability papers, P-P and Q-Q plots, plots of order statistics).

2. Determination of domains of attraction based on probability papers, the curvature method, the characterization theorem, etc.

3. Estimation of the parameters and quantiles of the generalized extreme value and generalized Pareto distributions by various methods such as the maximum likelihood, the elemental percentile method, the probability weighted moments, and the least squares.

4. Estimation and plot of multivariate models.

5. Tests of hypotheses.

We are grateful to the University of Cantabria, the University of Castilla-La Mancha, the Dirección General de Investigación Clientífica y Técirca (projects PB98-0421 and DPI2002-04172-C04-02), and the American University in Cairo for partial support.

Enrique Castillo
Ali S. Hadi
N. Balakrishnan
Jose M. Sarabia

Part I

Data, Introduction, and Motivation

Chapter 1

Introduction and Motivation

1.1 What Are Extreme Values?

Often, when natural calamities of great magnitude happen, we are left wondering about their occurrence and frequency, and whether anything could have been done either to avert them or at least to have been better prepared for them. These could include, for example, the extraordinary dry spell in the western regions of the United States and Canada during the summer of 2003 (and numerous forest fires that resulted from this dry spell), the devastating earthquake that destroyed almost the entire historic Iranian city of Bam in 2003, and the massive snowfall in the eastern regions of the United States and Canada during February 2004 (which shut down many cities for several days at a stretch). The same is true for destructive hurricanes and devastating floods that affect many parts of the world. For this reason, an architect in Japan may be quite interested in constructing a high-rise building that could withstand an earthquake of great magnitude, maybe a "100-year earthquake"; or, an engineer building a bridge across the mighty Mississippi river may be interested in fixing its height so that the water may be expected to go over the bridge once in 200 years, say. It is evident that the characteristics of interest in all these cases are *extremes* in that they correspond to either minimum (e.g., minimum amount of precipitation) or maximum (e.g., maximum amount of water flow) values.

Even though the examples listed above are only few and are all connected with natural phenomena, there are many other practical situations wherein we will be primarily concerned with extremes. These include: maximum wind velocity during a tropical storm (which is, in fact, used to categorize the storm), minimum stress at which a component breaks, maximum number of vehicles passing through an intersection at a peak hour (which would facilitate better planning of the traffic flow), minimum weight at which a structure develops a crack, minimum strength of materials, maximum speed of vehicles on a certain section of a highway (which could be used for employing patrol cars), maximum height of waves at a waterfront location, and so on.

Since the primary issues of interest in all the above examples concern the occurrence of such events and their frequency, a careful statistical analysis would require the availability of data on such extremes (preferably of a large size, for making predictions accurately) and an appropriate statistical model for those extremes (which would lead to correct predictions).

1.2 Why Are Extreme Value Models Important?

In many statistical applications, the interest is centered on estimating some population central characteristics (e.g., the average rainfall, the average temperature, the median income, etc.) based on random samples taken from a population under study. However, in some other areas of applications, we are not interested in estimating the average but rather in estimating the maximum or the minimum (see Weibull (1951, 1952), Galambos (1987), Castillo (1994)). For example, in designing a dam, engineers, in addition to being interested in the average flood, which gives the total amount of water to be stored, are also interested in the maximum flood, the maximum earthquake intensity or the minimum strength of the concrete used in building the dam.

It is well known to engineers that design values of engineering works (e.g., dams, buildings, bridges, etc.) are obtained based on a compromise between safety and cost, that is, between guaranteeing that they survive when subject to extreme operating conditions and reasonable costs. Estimating extreme capacities or operating conditions is very difficult because of the lack of available data. The use of safety factors has been a classical solution to the problem, but now it is known that it is not completely satisfactory in terms of safety and cost, because high probabilities of failure can be obtained on one hand, and large and unnecessary waste of money, on the other. Consequently, the safety factor approach is not an optimal solution to the engineering design problem.

The knowledge of the distributions of the maxima and minima of the relevant phenomena is important in obtaining good solutions to engineering design problems.

Note that engineering design must be based on extremes, because largest values, such as loads, earthquakes, winds, floods, waves, etc., and smallest values such as strength, stress, etc. are the key parameters leading to failure of engineering works.

There are many areas where extreme value theory plays an important role; see, for example, Castillo (1988), Coles (2001), Galambos (1994, 1998, 2000), Galambos and Macri (2000), Kotz and Nadarajah (2000), and Nadarajah (2003).

1.3 Examples of Applications

In this section examples of some of the fields where common engineering problems involve extremes or other order statistics are given.[1]

1.3.1 Ocean Engineering

In the area of ocean engineering, it is known that wave height is the main factor to be considered for design purposes. Thus, the designs of offshore platforms, breakwaters, dikes, and other harbor works rely upon the knowledge of the probability distribution of the highest waves. Another problem of crucial interest in this area is to find the joint distribution of the heights and periods of the sea waves. More precisely, the engineer is interested in the periods associated with the largest waves. This is clearly a problem, which in the extreme value field is known as the *concomitants* of order statistics. Some of the publications dealing with these problems are found in Arena (2002), Battjes (1977), Borgman (1963, 1970, 1973), Bretchneider (1959), Bryant (1983), Castillo and Sarabia (1992, 1994), Cavanie, Arhan, and Ezraty (1976), Chakrabarti and Cooley (1977), Court (1953), Draper (1963), Earle, Effermeyer, and Evans (1974), Goodknight and Russel (1963), Günbak (1978), Hasofer (1979), Houmb and Overvik (1977), Longuet-Higgins (1952, 1975), Tiago de Oliveira (1979), Onorato, Osborne, and Serio (2002), Putz (1952), Sellars (1975), Sjö (2000, 2001), Thom (1968a,b, 1969, 1971, 1973), Thrasher and Aagard (1970), Tucker (1963), Wiegel (1964), Wilson (1966), and Yang, Tayfun, and Hsiao (1974).

1.3.2 Structural Engineering

Modern building codes and standards provide information on: (a) extreme winds in the form of wind speeds corresponding to various specified mean recurrence intervals, (b) design loads, and (c) seismic incidence in the form of areas of equal risk. Wind speeds are estimates of extreme winds that can occur at the place where the building or engineering work is to be located and have a large influence on their design characteristics and final costs. Design loads are also closely related to the largest loads acting on the structure during its lifetime. Small design loads can lead to collapse of the structure and associated damages. On the other hand, large design loads lead to a waste of money. A correct design is possible only if the statistical properties of largest loads are well known. For a complete analysis of this problem, the reader is referred to Ang (1973), Court (1953), Davenport (1968a,b, 1972, 1978), Grigoriu (1984), Hasofer (1972), Hasofer and Sharpe (1969), Lévi (1949), Mistéth (1973), Moses (1974), Murzewski (1972), Prot (1949a,b, 1950), Sachs (1972), Simiu, Biétry, and Filliben (1978), Simiu, Changery, and Filliben (1979), Simiu and Filliben (1975, 1976), Simiu, Filliben, and Shaver (1982), Simiu and Scanlan (1977), Thom (1967, 1968a,b), Wilson (1966), and Zidek, Navin, and Lockhart (1979).

[1]Some of these examples are reprinted from the book *Extreme Value Theory in Engineering*, by E. Castillo, Copyright © Academic Press (1988), with permission from Elsevier.

A building or engineering work will survive if it is designed to withstand the most severe earthquake occurring during its design period. Thus, the maximum earthquake intensity plays a central role in design. The probabilistic risk assessment of seismic events is especially important in nuclear power plants where the losses are due not only to material damage of the structures involved but also to the very dangerous collateral damages that follow due to nuclear contamination. Precise estimation of the probabilities of occurrence of extreme winds, loads, earthquakes is required in order to allow for realistic safety margins in structural design, on one hand, and for economical solutions, on the other. Design engineers also need to extrapolate from small laboratory specimens to the actual lengths of structures such as cable-stayed or suspended bridges. In order for this extrapolation to be made with reasonable reliability, extra knowledge is required. Some material related to this problem can be found in Bogdanoff and Schiff (1972).

1.3.3 Hydraulics Engineering

Knowledge of the recurrence intervals of long hydrologic events is important in reservoir storage-yield investigations, drought studies, and operation analysis. It has been usual to base the estimate of the required capacity of a headwater storage on a critical historical drought sequence. It is desirable that the recurrence interval of such an event be known.

There is a continuing need to determine the probability of rare floods for their inclusion in risk assessment studies. Stream discharge and flood flow have long been measured and used by engineers in the design of hydraulic structures (dams, canals, etc.), flood protection works, and in planning for floodplain use. Riverine flooding and dams overtopping are very common problems of concern. A flood frequency analysis is the basis for the engineering design of many projects and the economic analysis of flood-control projects. High losses in human lives and property due to damages caused by floods have recently emphasized the need for precise estimates of probabilities and return periods of these extreme events. However, hydraulic structures and flood protection works are affected not only by the intensity of floods but also by their frequency, as occurs with a levee, for example. Thus, we can conclude that quantifying uncertainty in flood magnitude estimators is an important problem in floodplain development, including risk assessment for floodplain management, risk-based design of hydraulic structures and estimation of expected annual flood damages. Some works related to these problems are found in Beard (1962), Benson (1968), Chow (1951, 1964), Embrechts, Klüppelberg, and Mikosch (1997), Gumbel and Goldstein (1964), Gupta, Duckstein, and Peebles (1976), Hershfield (1962), Karr (1976), Kirby (1969), Matalas and Wallis (1973), Mistéth (1974), Morrison and Smith (2001), Mustafi (1963), North (1980), Shane and Lynn (1964), Todorovic (1978, 1979), and Zelenhasic (1970).

1.3.4 Meteorology

Extreme meteorological conditions are known to influence many aspects of human life such as in the flourishing of agriculture and animals, the behavior of some machines, and the lifetime of certain materials. In all these cases the engineers, instead of centering interest on the mean values (temperature, rainfall, etc.), are concerned only with the occurrence of extreme events (very high or very low temperature, rainfall, etc.). Accurate prediction of the probabilities of those rare events thus becomes the aim of the analysis. For related discussions, the reader can refer to Ferro and Segers (2003), Galambos and Macri (2002), Leadbetter, Lindgren, and Rootzén (1983), and Sneyers (1984).

1.3.5 Material Strength

One interesting application of extreme value theory to material strength is the analysis of size effect. In many engineering problems, the strength of actual structures has to be inferred from the strength of small elements of reduced size samples, prototype or models, which are tested under laboratory conditions. In such cases, extrapolation from small to much larger sizes is needed. In this context, extreme value theory becomes very useful in order to analyze the size effect and to make extrapolations not only possible but also reliable. If the strength of a piece is determined or largely affected by the strength of its weakest (real or imaginary) subpiece into which the piece can be subdivided, as it usually occurs, the minimum strength of the weakest subpiece determines the strength of the entire piece. Thus, large pieces are statistically weaker than small pieces. For a complete list of references before 1978, the reader is referred to Harter (1977, 1978a,b).

1.3.6 Fatigue Strength

Modern fracture mechanics theory reveals that fatigue failure is due to propagation of cracks when elements are under the action of repetitive loads. The fatigue strength of a piece is governed by the largest crack in the piece. If the size and shape of the crack were known, the lifetime, measured in number of cycles to failure, could be deterministically obtained. However, the presence of cracks in pieces is random in number, size, and shape, and, thus, resulting in a random character of fatigue strength. Assume a longitudinal piece hypothetically subdivided into subpieces of the same length and being subjected to a fatigue test. Then all the pieces are subjected to the same loads and the lifetime of the piece is that of the weakest subpiece. Thus, the minimum lifetime of the subpieces determines the lifetime of the piece. Some key references related to fatigue are Anderson and Coles (2002), Andrä and Saul (1974, 1979), Arnold, Castillo, and Sarabia (1996), Batdorf (1982), Batdorf and Ghaffanian (1982), Birnbaum and Saunders (1958), Bühler and Schreiber (1957), Castillo, Ascorbe, and Fernández-Canteli (1983a), Castillo et al. (1983b, 1984a), Castillo et al. (1985), Castillo et al. (1990), Castillo and

Hadi (1995b), Castillo et al. (1987), Castillo et al. (1984b), Coleman (1956, 1957a,b, 1958a,b,c), Dengel (1971), Duebelbeiss (1979), Epstein (1954), Epstein and Sobel (1954), Fernández-Canteli (1982), Fernández-Canteli, Esslinger, and Thurlimann (1984), Freudenthal (1975), Gabriel (1979), Grover (1966), Hajdin (1976), Helgason and Hanson (1976), Lindgren and Rootzén (1987), Maennig (1967, 1970), Mann, Schafer, and Singpurwalla (1974), Mendenhall (1958), Phoenix (1978), Phoenix and Smith (1983), Phoenix and Tierney (1983), Phoenix and Wu (1983), Rychlik (1996), Smith (1980, 1981), Spindel, Board, and Haibach (1979), Takahashi and Sibuya (2002), Tide and van Horn (1966), Tierney (1982), Tilly and Moss (1982), Warner and Hulsbos (1966), Weibull (1959), and Yang, Tayfun, and Hsiao (1974).

1.3.7 Electrical Strength of Materials

The lifetime of some electrical devices depends not only on their random quality but also on the random voltage levels acting on them. The device survives a given period if the maximum voltage level does not surpass a critical value. Thus, the maximum voltage in the period is one of the governing variables in this problem. For some related discussions, the reader may refer to Endicott and Weber (1956, 1957), Hill and Schmidt (1948), Lawless (2003), Nelson (2004), and Weber and Endicott (1956, 1957).

1.3.8 Highway Traffic

Due to economic considerations, many highways are designed in such a manner that traffic collapse is assumed to take place a limited number (say k) of times during a given period of time. Thus, the design traffic is that associated not with the maximum but with the kth largest traffic intensity during that period. Obtaining accurate estimates of the probability distribution of the kth order statistic pertains to the theory of extreme order statistics and allows a reliable design to be made. Some pertinent references are Glynn and Whitt (1995), Gómez-Corral (2001), Kang and Serfozo (1997), and McCormick and Park (1992).

1.3.9 Corrosion Resistance

Corrosion failure takes place by the progressive size increase and penetration of initially small pits through the thickness of an element, due to the action of chemical agents. It is clear that the corrosion resistance of an element is determined by the largest pits and largest concentrations of chemical agents and that small and intermediate pits and concentrations do not have any effect on the corrosion strength of the element. Some references related to this area are Aziz (1956), Eldredge (1957), Logan (1936), Logan and Grodsky (1931), Reiss and Thomas (2001), and Thiruvengadam (1972).

A similar model explains the leakage failure of batteries, which gives another example where extremes are the design values.

1.3.10 Pollution Studies

With the existence of large concentrations of people (producing smoke, human wastes, etc.) or the appearance of new industries (chemical, nuclear, etc.), the pollution of air, rivers, and coasts has become a common problem for many countries. The pollutant concentration, expressed as the amount of pollutant per unit volume (of air or water), is forced, by government regulations, to remain below a given critical level. Thus, the regulations are satisfied if, and only if, the largest pollution concentration during the period of interest is less than the critical level. Here then, the largest value plays the fundamental role in design. For some relevant discussions, the interested reader may refer to Barlow (1972), Barlow and Singpurwalla (1974), Larsen (1969), Leadbetter (1995), Leadbetter, Lindgren, and Rootzén (1983), Midlarsky (1989), Roberts (1979a,b), and Singpurwalla (1972).

1.4 Univariate Data Sets

To illustrate the different methods to be described in this book, several sets of data with relevance to extreme values have been selected. In this section a detailed description of the data are given with the aim of facilitating model selection. Data should not be statistically treated unless a previous understanding of the physical meaning behind them is known and the aim of the analysis is clearly established. In fact, the decision of the importance of upper or lower order statistics, maxima or minima cannot be done without this knowledge. This knowledge is especially important when extrapolation is needed and predictions are to be made for important decision-making.

1.4.1 Wind Data

The yearly maximum wind speed, in miles per hour, registered at a given location during a period of 50 years is presented in Table 1.1. We assume that this data will be used to determine a design wind speed for structural building purposes. Important facts to be taken into consideration for these data are its nonnegative character and, perhaps, the existence of a not clearly defined finite upper end (the maximum conceivable wind speed is bounded). Some important references for wind problems are de Haan and de Ronde (1998), Lighthill (1999), and Walshaw (2000).

1.4.2 Flood Data

The yearly maximum flow discharge, in cubic meters, measured at a given location of a river during 60 years is shown in Table 1.2. The aim of the data analysis is supposed to help in the design of a flood protection device at that location. Similar characteristics such as those for the wind data appear here: a lower end clearly defined (zero) and an obscure upper end.

Table 1.1: Yearly Maxima Wind Data.

22.64	22.80	23.75	24.01	24.04
24.24	24.74	25.45	25.55	25.66
25.99	26.63	26.69	26.88	26.89
27.12	27.43	27.69	27.71	28.12
28.58	28.88	29.12	29.45	29.48
30.18	31.31	31.55	31.57	32.54
32.98	33.83	33.86	34.64	35.21
36.82	37.23	38.09	38.26	38.82
38.90	38.96	42.99	43.66	44.61
45.24	47.91	54.75	69.40	98.16

Table 1.2: Flood Data: Maxima Yearly Floods in a Given Section of a River.

24.21	26.46	29.48	30.32	31.60	32.88
33.03	33.63	35.14	35.23	35.59	35.89
35.95	36.07	36.49	36.50	37.13	37.48
38.01	38.21	38.53	38.91	39.26	39.45
40.32	40.36	40.49	40.69	41.03	41.05
41.54	42.62	42.82	42.91	43.05	43.31
43.34	43.42	43.65	43.87	44.71	45.04
45.58	46.00	48.29	48.76	49.28	49.43
50.17	50.45	50.73	51.90	52.54	52.94
54.01	57.84	60.10	61.95	67.76	75.70

1.4.3 Wave Data

The yearly maximum wave heights, in feet, observed at a given location over 50 years are shown in Table 1.3. Data have been obtained in shallow water, and will be used for designing a breakwater. The wave height is, by definition, a nonnegative random variable, which is bounded from above. In addition, this end is clear for shallow water, but unclear for open sea.

1.4.4 Oldest Age at Death in Sweden Data

The yearly oldest ages at death in Sweden during the period from 1905 to 1958 for women and men, respectively, are given in Tables 1.4 and 1.5. The analysis is needed to forecast oldest ages at death in the future.

1.4.5 Houmb's Data

The yearly maximum significant wave height measured in Miken-Skomvaer (Norway) and published by Houmb and Overvik (1977) is shown in Table 1.6. The data can be used for the design of sea structures.

Table 1.3: Wave Data: Annual Maximum Wave Heights in a Given Location.

2.91	3.74	4.09	5.88	6.42
6.93	7.21	7.92	8.26	8.79
9.17	9.50	9.62	10.00	10.14
10.28	10.45	10.77	11.65	11.65
11.82	12.27	12.68	13.28	13.46
13.88	13.98	14.32	14.38	14.46
14.86	15.03	15.30	16.07	16.23
17.36	18.68	18.72	19.44	20.09
21.06	21.13	21.53	21.80	23.15
24.75	25.45	28.13	29.95	37.19

Table 1.4: Oldest Ages at Death in Sweden Data (Women).

101.50	101.69	102.12	102.32	102.54	102.72
102.78	102.92	103.14	103.31	103.38	103.41
103.46	103.53	103.56	103.56	103.77	103.83
103.86	103.94	103.97	104.01	104.12	104.27
104.33	104.37	104.40	104.42	104.46	104.52
104.71	104.85	104.87	105.01	105.01	105.02
105.03	105.19	105.32	105.45	105.64	105.71
105.83	105.86	105.87	105.88	105.98	106.13
106.15	106.15	106.52	107.49	107.89	107.90

Table 1.5: Oldest Ages at Death in Sweden Data (Men).

100.08	100.49	100.82	100.88	100.90	101.17
101.26	101.41	101.63	101.66	101.67	101.70
101.76	102.41	102.52	102.54	102.55	102.57
102.57	102.61	102.63	102.69	102.78	102.88
102.94	103.00	103.06	103.15	103.17	103.24
103.25	103.36	103.40	103.43	103.47	103.57
103.80	103.98	104.01	104.22	104.65	104.88
104.92	105.00	105.12	105.12	105.12	105.48
105.55	105.72	105.83	106.09	106.48	106.50

1.4.6 Telephone Calls Data

The times between 41 (in seconds) and 48 (in minutes) consecutive telephone calls to a company's switchboard are shown in Tables 1.7 and 1.8. The aim of the analysis is to determine the ability of the company's computer to handle very close, consecutive calls because of a limited response time. A clear lower bound (zero) can be estimated from physical considerations.

Table 1.6: Houmb's Data: The Yearly Maximum Significant Wave Height.

5.60	6.55	6.65	7.35	7.80	7.90	8.00	8.50
9.05	9.15	9.40	9.60	9.80	9.90	10.85	10.90
11.10	11.30	11.30	11.55	11.75	12.85	12.90	13.40

Table 1.7: Telephone Data 1: Times Between 35 Consecutive Telephone Calls (in Seconds).

0.60	0.74	1.12	2.00	2.21	2.36	2.85
2.98	3.37	3.74	3.89	4.16	4.87	5.59
6.32	6.45	8.13	9.60	11.00	11.30	11.70
12.80	13.00	13.50	14.20	15.60	15.90	17.20
22.10	23.80	24.80	26.40	33.90	34.80	52.60

Table 1.8: Telephone Data 2: Times (in Minutes) Between 48 Consecutive Calls.

1.34	0.14	0.33	1.68	1.86	1.31	0.83	0.33
2.20	0.62	3.20	1.38	0.96	0.28	0.44	0.59
0.25	0.51	1.61	1.85	0.47	0.41	1.46	0.09
2.18	0.07	0.02	0.64	0.28	0.68	1.07	3.25
0.59	2.39	0.27	0.34	2.18	0.41	1.08	0.57
0.35	0.69	0.25	0.57	1.90	0.56	0.09	0.28

1.4.7 Epicenter Data

The distances, in miles, from a nuclear power plant to the epicenters of the most recent 60 earthquakes and intensity above a given threshold value are shown in Table 1.9. Data are needed to evaluate the risks associated with earthquakes occurring close to the central site. In addition, geological reports indicate that a fault, which is 50 km away from the plant, is the main cause of earthquakes in the area.

1.4.8 Chain Strength Data

A set of 20 chain links have been tested for strength and the results are given in Table 1.10. The data are used for quality control, and minimum strength characteristics are needed.

1.4.9 Electrical Insulation Data

The lifetimes of 30 electrical insulation elements are shown in Table 1.11. The data are used for quality control, and minimum lifetime characteristics are needed.

Table 1.9: Epicenter Data: Distances from Epicenters to a Nuclear Power Plant.

58.2	58.2	59.5	61.8	65.8	67.8
68.5	70.9	73.7	77.0	80.8	83.7
84.3	89.0	97.6	98.3	99.6	101.4
105.1	105.8	106.7	119.1	119.5	119.9
121.9	125.7	128.4	146.1	153.9	154.6
155.8	157.4	157.7	163.7	172.7	173.9
174.2	175.1	176.0	178.7	179.1	179.5
180.7	182.1	182.7	186.7	187.5	191.0
192.6	193.0	199.4	211.6	212.1	216.8
222.9	227.3	229.4	234.5	236.8	238.9

Table 1.10: Strengths (in kg) for 20 Chains.

51.1	57.1	65.1	69.2	71.2
73.1	76.1	81.9	84.1	86.6
88.1	92.6	94.9	96.6	97.0
101.4	103.4	103.8	105.2	119.1

Table 1.11: Lifetime (in Days) of 30 Electric Insulators.

744	822	847	885	920	948
968	985	1010	1018	1019	1028
1029	1031	1040	1047	1071	1074
1097	1134	1147	1170	1174	1209
1251	1273	1320	1383	1388	1462

1.4.10 Fatigue Data

Thirty five specimens of wire were tested for fatigue strength to failure and the results are shown in Table 1.12. The aim of the study is to determine a design fatigue stress.

1.4.11 Precipitation Data

The yearly total precipitation in Philadelphia for the last 40 years, measured in inches, is shown in Table 1.13. The aim of the study is related to drought risk determination.

1.4.12 Bilbao Wave Heights Data

The Zero-crossing hourly mean periods (in seconds) of the sea waves measured in a Bilbao buoy in January 1997 are given in Table 1.14. Only periods above

Table 1.12: Fatigue Data: Number of Million Cycles Until the Occurrence of Fatigue.

0.039611	0.044132	0.044209	0.045898	0.050139
0.054625	0.058970	0.064703	0.064950	0.066508
0.070208	0.072098	0.075001	0.080393	0.081868
0.082202	0.082447	0.089268	0.090021	0.096136
0.096723	0.101610	0.101833	0.106055	0.112833
0.119154	0.122366	0.134511	0.135220	0.136395
0.138378	0.153790	0.184916	0.216370	0.240316

Table 1.13: Precipitation Data.

29.34	29.88	32.20	33.03	33.27	34.04	34.95	35.15
35.45	36.77	37.78	38.35	38.37	39.14	39.52	40.00
40.47	40.48	41.05	41.15	41.75	42.62	43.05	43.36
44.46	44.82	44.85	45.40	45.84	45.95	46.00	46.06
46.62	47.79	47.87	48.13	49.06	49.42	49.63	52.13

7 seconds are listed.

Table 1.14: The Bilbao Waves Heights Data: The Zero-Crossing Hourly Mean Periods, Above Seven Seconds, of the Sea Waves Measured in a Bilbao Buoy in January 1997.

7.05	7.12	7.15	7.18	7.19	7.20	7.20	7.20	7.20	7.25
7.26	7.27	7.28	7.30	7.31	7.31	7.32	7.33	7.37	7.40
7.46	7.46	7.47	7.48	7.48	7.52	7.54	7.55	7.55	7.58
7.59	7.59	7.61	7.63	7.65	7.66	7.66	7.67	7.67	7.68
7.69	7.72	7.72	7.72	7.72	7.72	7.77	7.77	7.79	7.79
7.82	7.83	7.83	7.83	7.84	7.85	7.85	7.88	7.88	7.90
7.90	7.91	7.93	7.93	7.93	7.94	7.95	7.95	7.97	7.97
7.97	7.99	8.00	8.03	8.03	8.05	8.06	8.06	8.07	8.10
8.11	8.12	8.15	8.15	8.15	8.18	8.18	8.18	8.19	8.20
8.21	8.23	8.23	8.30	8.30	8.31	8.31	8.32	8.32	8.33
8.40	8.41	8.42	8.43	8.43	8.45	8.48	8.49	8.50	8.50
8.51	8.52	8.53	8.54	8.56	8.58	8.59	8.59	8.60	8.65
8.69	8.71	8.72	8.74	8.74	8.74	8.74	8.79	8.81	8.84
8.85	8.86	8.88	8.88	8.94	8.98	8.98	8.99	9.01	9.03
9.06	9.12	9.16	9.17	9.17	9.18	9.18	9.18	9.21	9.22
9.23	9.24	9.27	9.29	9.30	9.32	9.33	9.36	9.38	9.43
9.46	9.47	9.59	9.59	9.60	9.61	9.62	9.63	9.66	9.74
9.75	9.78	9.79	9.79	9.80	9.84	9.85	9.89	9.90	

Table 1.15: Yearly Maximum Floods of the Ocmulgee River Data Downstream at Macon (q_1) and Upstream at Hawkinsville (q_2) from 1910 to 1949.

q_1	q_2	q_1	q_2	q_1	q_2	q_1	q_2	q_1	q_2
28.8	18.8	8.5	5.9	44.8	44.4	51.0	52.0	4.8	5.9
19.1	20.1	47.8	40.4	25.4	27.0	14.3	14.3	31.0	40.0
66.2	45.2	37.0	30.0	48.6	44.0	28.3	30.3	21.0	15.2
72.5	79.0	28.3	19.3	7.9	7.6	47.1	42.4	73.4	70.5
64.4	50.0	10.7	12.2	19.6	16.2	19.0	19.9	16.9	17.4
22.7	13.5	65.3	61.0	33.3	25.8	31.0	33.0	33.9	37.9
14.2	13.3	7.3	6.9	73.4	57.0	44.8	41.6	50.2	46.8
40.4	26.2	57.6	35.4	32.6	34.8	24.0	28.2	84.0	68.0

1.5 Multivariate Data Sets

Multivariate data are encountered when several magnitudes are measured instead of a single one. Some multivariate data sets are given below.

1.5.1 Ocmulgee River Data

The yearly maximum water discharge of the Ocmulgee River, measured at two different locations, Macon and Hawkinsville, between 1910 and 1949, and published by Gumbel (1964) are given in Table 1.15. The aim of the analysis is to help in the designs of the flood protection structures.

1.5.2 The Yearly Maximum Wind Data

The bivariate data (V_1, V_2) in Table 1.16 correspond to the yearly maximum wind speeds (in km/hour) at two close locations. An analysis is needed to forecast yearly maximum wind speeds in the future at these locations, and also to study their association characteristics. If there is little or no association between the two, then the data from each location could be analyzed separately as a univariate data (that is, not as a bivariate data).

1.5.3 The Maximum Car Speed Data

The bivariate data (V_1, V_2) in Table 1.17 correspond to the maximum weekend car speeds registered at two given roads 1 and 2, a highway and a mountain road, respectively, corresponding to 200 dry weeks and the first 1000 cars passing through two given locations. The data will be used to predict future maximum weekend car speeds.

Table 1.16: Yearly Maximum Wind Data at Two Close Locations.

v_1	v_2	v_1	v_2	v_1	v_2	v_1	v_2	v_1	v_2
101.2	96.8	150.6	154.9	101.7	115.0	123.9	112.5	86.2	87.1
94.7	89.0	114.7	114.4	82.2	78.6	96.5	90.7	98.0	92.0
70.1	76.6	100.9	93.1	108.2	112.4	101.2	93.1	135.7	152.1
113.4	120.5	80.2	86.3	99.4	89.3	88.3	91.1	125.5	136.6
74.3	96.0	80.9	84.3	116.7	107.6	87.1	91.9	92.8	95.7
81.8	89.6	83.4	79.9	91.2	92.7	80.9	76.2	105.2	105.7
90.8	93.8	83.4	82.3	102.0	112.2	79.8	79.2	89.8	108.8
99.8	92.1	123.0	112.1	96.2	89.1	91.7	71.8	111.2	105.8
108.6	105.4	129.6	113.4	76.3	76.7	120.0	115.5	104.9	102.2
110.3	107.9	84.6	79.6	98.9	100.9	85.3	91.1	99.8	101.5
70.8	86.1	81.8	73.8	77.1	81.0	123.6	105.5	102.1	96.7
141.0	142.9	93.6	96.8	109.4	111.7	86.2	83.4	90.2	90.7
98.2	98.0	84.8	85.9	119.1	112.8	99.6	92.4	113.5	101.9
87.7	87.2	111.1	121.0	106.5	112.4	78.9	84.2	97.3	115.3
97.5	101.7	79.9	76.0	119.4	130.1	133.1	127.8	96.3	102.
139.4	125.8	110.5	105.5	73.2	73.2	120.5	118.9	174.6	170.4
88.0	91.1	86.7	89.3	100.1	93.5	101.2	110.1	109.3	98.1
73.0	68.9	107.6	118.2	71.6	78.3	133.7	162.4	107.5	109.3
134.8	106.5	87.9	96.8	114.7	108.8	82.5	86.9	91.6	87.2
81.7	86.6	136.8	131.7	85.0	83.0	122.0	119.4	83.4	86.0
82.0	71.8	82.9	71.8	67.5	83.7	83.5	78.7	118.7	114.4
126.1	122.3	91.8	96.0	99.7	94.5	77.7	76.0	119.5	113.7
104.1	109.1	98.1	98.9	80.7	78.4	79.7	84.0	107.1	111.7
84.8	89.4	88.5	103.4	132.9	147.8	88.0	88.8	75.1	64.3
90.3	92.2	105.3	106.9	92.6	80.7	117.9	114.5	82.5	84.9
95.3	97.1	92.1	83.2	102.2	96.0	88.5	82.7	99.2	92.3
66.7	69.4	73.4	91.3	84.1	95.6	106.3	114.7	88.5	90.5
95.1	92.7	75.3	80.4	98.9	92.9	100.4	106.1	85.0	89.2
107.9	110.2	101.7	95.9	86.7	83.7	81.9	86.4	89.8	80.7
130.4	130.7	117.9	110.2	88.2	84.2	81.5	83.3	92.9	122.
92.8	84.7	104.5	81.2	116.3	111.2	117.9	122.5	92.5	97.2
101.8	86.1	100.8	98.4	80.0	91.3	77.2	73.4	80.1	84.3
97.4	102.3	98.5	100.9	95.0	91.8	123.2	136.1	113.0	99.3
130.5	126.3	83.0	74.1	106.4	100.7	93.0	115.5	58.1	69.6
89.1	96.0	91.0	77.5	86.0	86.2	98.3	103.4	125.3	127.1
105.9	87.3	92.6	84.4	105.8	125.2	94.7	106.8	75.6	96.8
113.9	116.4	98.3	87.2	90.7	96.5	129.0	123.6	86.1	98.4
173.0	169.0	123.2	120.0	71.5	73.1	94.1	101.6	74.6	65.9
105.1	101.6	87.9	79.6	70.5	81.4	74.6	73.2	89.2	90.6
128.2	112.1	109.0	105.3	95.5	94.2	92.9	101.4	83.6	73.9

Table 1.17: Data Corresponding to the Maximum Weekend Car Speeds Registered at Two Given Locations 1 and 2 in 200 Dry Weeks.

V_1	V_2	V_1	V_2	V_1	V_2	V_1	V_2	V_1	V_2
68.9	51.6	88.2	56.6	108.2	63.3	97.7	61.1	152.3	66.8
81.7	54.5	98.4	60.9	88.6	56.9	87.5	46.3	83.4	51.0
67.7	59.5	124.4	64.8	124.8	65.7	83.2	56.1	114.1	63.4
109.3	67.0	165.8	69.0	85.6	60.2	122.9	64.5	93.9	57.4
118.9	62.6	84.9	59.3	75.1	53.3	106.0	61.2	117.0	65.3
127.7	64.8	88.2	61.6	80.9	59.0	118.7	65.1	108.5	65.6
109.4	60.6	115.0	62.6	98.0	61.3	97.1	62.1	108.5	61.1
128.0	67.8	102.6	54.9	88.4	54.4	129.2	67.4	81.2	63.2
102.9	60.0	75.0	53.1	71.8	51.8	105.6	62.6	111.8	60.1
114.3	65.0	141.3	68.9	164.5	68.2	132.7	66.1	123.6	64.7
102.3	57.1	115.5	63.3	76.2	57.9	80.5	55.7	121.3	66.9
65.1	42.0	96.6	55.4	100.9	63.3	79.6	55.7	76.2	48.7
120.4	61.6	86.7	57.6	83.1	58.5	126.6	64.0	83.8	59.0
83.9	49.5	81.5	52.4	83.5	57.3	92.6	59.4	91.4	56.7
83.0	53.4	83.8	55.0	94.9	55.0	95.6	64.4	98.2	56.2
90.8	65.0	91.5	57.0	89.3	56.6	104.1	59.5	90.6	62.7
120.9	63.1	107.7	62.9	97.3	61.9	87.3	59.7	64.1	43.2
93.2	52.0	88.7	48.1	90.7	58.2	111.1	62.3	84.0	64.2
81.1	58.8	81.5	61.8	89.4	61.7	114.6	64.3	138.3	66.8
134.3	66.6	119.0	65.4	88.1	59.6	91.4	56.2	98.2	60.5
93.8	57.0	142.2	66.7	107.0	55.9	81.8	57.8	99.1	55.0
84.4	58.8	82.0	53.3	120.2	67.6	98.9	65.2	111.5	64.0
117.0	62.7	95.1	58.8	112.0	63.3	98.3	61.7	104.5	61.7
88.0	57.8	93.0	60.8	108.0	57.2	106.8	56.8	90.1	47.4
98.9	51.4	92.1	57.1	88.1	60.2	113.3	63.1	78.5	57.3
93.4	52.2	99.9	62.4	101.7	63.4	97.1	60.8	83.6	54.4
91.9	57.5	109.2	65.0	138.2	67.4	92.4	60.8	98.5	55.7
85.8	55.4	81.3	50.0	118.8	59.9	121.3	65.2	133.8	67.7
107.1	61.6	78.9	64.2	81.0	55.9	99.0	63.5	80.4	54.3
105.7	62.7	116.2	65.7	110.	61.8	71.7	55.9	85.3	57.3
84.8	49.7	79.6	56.2	95.4	54.4	72.4	48.1	115.1	64.7
90.4	61.0	97.1	59.2	100.7	60.5	85.6	57.3	78.2	53.1
90.1	55.8	108.6	62.1	90.4	64.4	81.1	56.0	102.3	64.5
101.3	60.5	83.6	55.2	99.5	62.2	95.2	58.8	127.5	67.3
90.4	59.3	114.2	62.4	93.8	61.2	103.5	59.5	79.5	55.5
104.5	63.4	139.4	67.4	101.4	56.1	87.4	61.9	87.4	64.8
92.4	54.7	68.5	53.2	72.8	50.3	82.6	56.9	74.3	59.3
85.8	52.1	103.4	60.1	71.0	42.6	91.9	68.0	102.2	62.6
93.3	64.5	93.0	57.1	127.3	65.8	106.7	64.7	75.3	51.3
101.2	51.9	93.8	62.3	73.9	53.8	130.8	67.2	131.3	65.2

Part II

Probabilistic Models Useful for Extremes

Chapter 2

Discrete Probabilistic Models

When we talk about a random variable, it is helpful to think of an associated *random experiment* or *trial*. A random experiment or trial can be thought of as any activity that will result in one and only one of several well-defined outcomes, but one does not know in advance which one will occur. The set of all possible outcomes of a random experiment E, denoted by $S(E)$, is called the *sample space* of the random experiment E.

Suppose that the structural condition of a concrete structure (e.g., a bridge) can be classified into one of three categories: poor, fair, or good. An engineer examines one such structure to assess its condition. This is a random experiment and its sample space, $S(E) = \{poor, fair, good\}$, has three elements.

Definition 2.1 (Random variable). *A random variable can be defined as a real-valued function defined over a sample space of a random experiment. That is, the function assigns a real value to every element in the sample space of a random experiment. The set of all possible values of a random variable X, denoted by $S(X)$, is called the support or range of the random variable X.*

Example 2.1 (Concrete structure). In the previous concrete example, let X be $-1, 0$, or 1, depending on whether the structure is poor, fair, or good, respectively. Then X is a random variable with support $S(X) = \{-1, 0, 1\}$. The condition of the structure can also be assessed using a continuous scale, say, from 0 to 10, to measure the concrete quality, with 0 indicating the worst possible condition and 10 indicating the best. Let Y be the assessed condition of the structure. Then Y is a random variable with support $S(Y) = \{y : 0 \leq y \leq 10\}$. ■

We consistently use the customary notation of denoting random variables by uppercase letters such as X, Y, and Z or $X_1, X_2, \ldots, X_n$, where n is the number of random variables under consideration. Realizations of random variables (that is, the actual values they may take) are denoted by the corresponding lowercase letters such as x, y, and z or $x_1, x_2, \ldots, x_n$.

A random variable is said to be *discrete* if it can assume only a finite or countably infinite number of distinct values. Otherwise, it is said to be *continuous*. Thus, a continuous random variable can take an uncountable set of real values. The random variable X in Example 2.1 is discrete, whereas the random variable Y is continuous.

When we deal with a single random quantity, we have a *univariate* random variable. When we deal with two or more random quantities simultaneously, we have a *multivariate* random variable. Section 2.1 presents some functions that are useful to all univariate discrete random variables. Section 2.2 presents some of the commonly encountered univariate discrete random variables. Multivariate random variables (both discrete and continuous) are treated in Chapter 4.

2.1 Univariate Discrete Random Variables

To specify a random variable we need to know (a) its range or support, $S(X)$, which is the set of all possible values of the random variable, and (b) a tool by which we can obtain the probability associated with every subset in its support, $S(X)$. These tools are some functions such as the *probability mass function* (pmf), the *cumulative distribution function* (cdf), or the characteristic function. The pmf, cdf, and the so-called *moments* of random variables are described in this section.

2.1.1 Probability Mass Function

Every discrete random variable has a *probability mass function* (pmf). The pmf of a discrete random variable X is a function that assigns to each real value x the probability of X taking on the value x. That is, $P_X(x) = \Pr(X = x)$. For notational simplicity we sometimes use $P(x)$ instead of $P_X(x)$. Every pmf $P(x)$ must satisfy the following conditions:

$$P(x) > 0 \quad \text{for all } x \in S(X), \qquad \text{and} \qquad \sum_{x \in S(X)} P(x) = 1. \tag{2.1}$$

Example 2.2 (Concrete structure). Suppose in Example 2.1 that 20% of all concrete structures we are interested in are in poor condition, 30% are in fair condition, and the remaining 50% are in good condition. Then if one such structure is selected at random, the probability that the selected structure is in poor condition is $P(-1) = 0.2$, the probability that it is in fair condition is $P(0) = 0.3$, and the probability that it is in good condition is $P(1) = 0.5$. ■

The pmf of a random variable X can be displayed in a table known as a *probability distribution table*. For example, Table 2.1 is the probability distribution table for the random variable X in Example 2.2. The first column in a probability distribution table is a list of the values of $x \in S(X)$, that is, only

Table 2.1: The Probability Mass Function (pmf) of a Random Variable X.

x	$P(x)$
−1	0.2
0	0.3
1	0.5
Total	1.0

Table 2.2: The Probability Mass Function (pmf) and the Cumulative Distribution Function (cdf) of a Random Variable X.

x	$P(x)$	$F(x)$
−1	0.2	0.2
0	0.3	0.5
1	0.5	1.0

the values of x for which $P(x) > 0$. The second column displays $P(x)$. It is understood that $P(x) = 0$ for every $x \notin S(X)$.

2.1.2 Cumulative Distribution Function

Every random variable also has a *cumulative distribution function* (cdf). The cdf of a random variable X, denoted by $F(x)$, is a function that assigns to each real value x the probability of X taking on values less than or equal to x, that is,

$$F(x) = \Pr(X \leq x) = \sum_{a \leq x} P(a).$$

Accordingly, the cdf can be obtained from the pmf and vice versa. For example, the cdf in the last column of Table 2.2 is computed from the pmf in Table 2.1 by accumulating $P(x)$ in the second column.

The pmf and cdf of any discrete random variable X can be displayed in probability distribution tables, such as Table 2.2, or they can be displayed graphically. For example, the graphs of the pmf and cdf in Table 2.2 are shown in Figure 2.1. In the graph of pmf, the height of a line on top of x is $P(x)$. The graph of the cdf for discrete random variable is a step function. The height of the step function is $F(x)$.

The cdf has the following properties as a direct consequence of the definitions of cdf and probability (see, for example, Fig. 2.1):

1. $F(\infty) = 1$ and $F(-\infty) = 0$.

2. $F(x)$ is nondecreasing and right continuous.

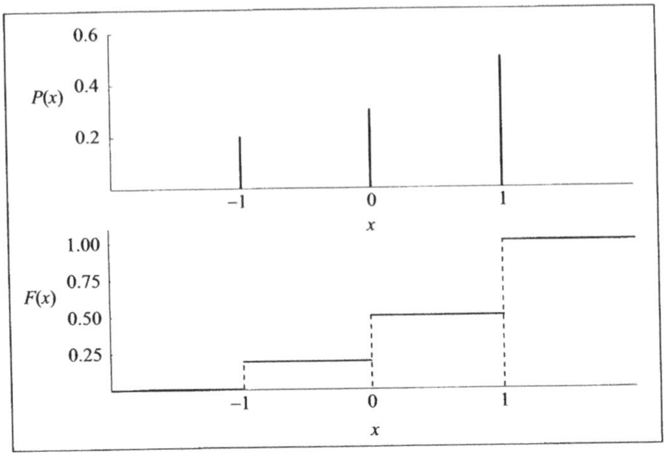

Figure 2.1: Graphs of the pmf and cdf of the random variable in Table 2.2.

3. $P(x)$ is the jump of the cdf at x.
4. $\Pr(a < X \leq b) = F(b) - F(a)$.

2.1.3 Moments

Let $g(X)$ be a function of a discrete random variable X. The *expected value* of $g(X)$ is defined by

$$E[g(X)] = \sum_{x \in S(X)} g(x)P(x). \tag{2.2}$$

For example, letting $g(X) = X^r$, we obtain the so-called rth *moment* of the discrete random variable X, with respect to the origin

$$E(X^r) = \sum_{x \in S(X)} x^r P(x). \tag{2.3}$$

When $r = 1$, we obtain the *mean*, μ, of the discrete random variable X,

$$\mu = E(X) = \sum_{x \in S(X)} xP(x). \tag{2.4}$$

Thus, the mean, μ, is the first moment of X with respect to the origin.

Letting $g(X) = (X - \mu)^r$, we obtain the rth *central moment*,

$$E[(X - \mu)^r] = \sum_{x \in S(X)} (x - \mu)^r P(x). \tag{2.5}$$

When $r = 1$, it can be shown that the *first* central moment of any random variable X is zero, that is,

$$E(X - \mu) = 0. \tag{2.6}$$

Table 2.3: Calculations of the Mean and Variance of a Random Variable X.

x	$P(x)$	$xP(x)$	x^2	$x^2P(x)$
–1	0.2	–0.2	1	0.2
0	0.3	0.0	0	0.0
1	0.5	0.5	1	0.5
Total	1.0	0.3		0.7

When $r = 2$, we obtain the second central moment, better known as the *variance*, σ^2, of the discrete random variable X, that is,

$$\sigma^2 = E[(X - \mu)^2] = \sum_{x \in S(X)} (x - \mu)^2 P(x). \tag{2.7}$$

The *standard deviation*, σ, of the random variable X is the positive square root of its variance. The mean can be thought of as a measure of *center* and the standard deviation (or, equivalently, the variance) as a measure of *spread* or *variability*. It can be shown that the variance can also be expressed as

$$\sigma^2 = E(X^2) - \mu^2, \tag{2.8}$$

where

$$E(X^2) = \sum_{x \in S(X)} x^2 P(x)$$

is the second moment of X with respect to the origin. For example, the calculations of the mean and variance of the random variable X are shown in Table 2.3. Accordingly, the mean and variance of X are $\mu = 0.3$ and $\sigma^2 = 0.7 - 0.3^2 = 0.61$, respectively.

The expected value, defined in (2.2), can be thought of as an *operator*, which has the following properties:

1. $E(c) = c$, for any constant c, that is, the expected value of a constant (a degenerate random variable) is the constant.

2. $E[cg(X)] = cE[g(X)]$.

3. $E[g(X) + h(X)] = E[g(X)] + E[h(X)]$, for any functions $g(X)$ and $h(X)$.

For example, $E(c + X) = E(c) + E(X) = c + \mu$. In other words, the mean of a constant plus a random variable is the constant plus the mean of the random variable. As another example,

$$\sigma^2 = E[(X - \mu)^2] = E[X^2 - 2\mu X + \mu^2] = E(X^2) - 2\mu E(X) + \mu^2 = E(X^2) - \mu^2.$$

This is actually the proof of the identity in (2.8).

2.2 Common Univariate Discrete Models

In this section we present several important discrete random variables that often arise in extreme value problems. For a more detailed description and some additional discrete random variables, see, for example, the books by Balakrishnan and Nevzorov (2003), Thoft-Christensen, Galambos (1995), Johnson, Kotz, and Kemp (1992), Ross (1992), and Wackerly, Mendenhall, and Scheaffer (2001).

2.2.1 Discrete Uniform Distribution

When a random variable X can have one of n possible values and they are all equally likely, then X is said to have a *discrete uniform distribution*. Since the possible values are equally likely, the probability for each one of them is equal to $1/n$. Without loss of generality, let us assume these values are $1, \ldots, n$. Then the pmf of X is

$$P(x) = \frac{1}{n}, \quad x = 1, 2, \ldots, n. \tag{2.9}$$

This discrete uniform distribution is denoted by $U(n)$. The mean and variance of $U(n)$ are

$$\mu = (n+1)/2 \quad \text{and} \quad \sigma^2 = (n^2 - 1)/12, \tag{2.10}$$

respectively.

Example 2.3 (Failure types). A computer system has four possible types of failure. Let $X = i$ if the system results in a failure of type i, with $i = 1, 2, 3, 4$. If these failure types are equally likely to occur, then the distribution of X is $U(4)$ and the pmf is

$$P(x) = \frac{1}{4}, \quad x = 1, 2, 3, 4.$$

The mean and variance can be shown to be 2.5 and 1.25, respectively. ∎

2.2.2 Bernoulli Distribution

The *Bernoulli* random variable arises in situations where we have a random experiment, which has two possible mutually exclusive outcomes: *success* or *failure*. The probability of success is p and the probability of failure is $1 - p$. This random experiment is called a *Bernoulli trial* or *experiment*. Define a random variable X by

$$X = \begin{cases} 0, & \text{if a failure is observed,} \\ 1, & \text{if a success is observed.} \end{cases}$$

This is called a *Bernoulli random variable*. Its distribution is called a *Bernoulli distribution*. The pmf of X is

$$P(x) = p^x (1-p)^{1-x}, \quad x = 0, 1, \tag{2.11}$$

and its cdf is

$$F(x) = \begin{cases} 0, & \text{if } x < 0, \\ 1-p, & \text{if } 0 \le x < 1, \\ 1, & \text{if } x \ge 1. \end{cases} \tag{2.12}$$

The mean and variance of a Bernoulli random variable are

$$\mu = p \quad \text{and} \quad \sigma^2 = p(1-p), \tag{2.13}$$

respectively. Note that if $p = 1$, then X becomes a degenerate random variable (that is, a constant) and the pmf of X is

$$P(x) = \begin{cases} 1, & \text{if } x = 1, \\ 0, & \text{otherwise.} \end{cases} \tag{2.14}$$

This is known as the *Dirac* function.

Example 2.4 (Concrete structure). Suppose we are interested in knowing whether or not a given concrete structure is in poor condition. Then, a random variable X can be defined as

$$X = \begin{cases} 1, & \text{if the condition is poor,} \\ 0, & \text{otherwise.} \end{cases}$$

This is a Bernoulli random variable. From Example 2.2, 20% of structures are in poor condition. Then the pmf is

$$P(x) = 0.2^x (1-0.2)^{1-x}, \quad x = 0, 1.$$

The mean and variance of X are $\mu = p = 0.2$ and $\sigma^2 = p(1-p) = 0.16$. ■

Bernoulli random variables arise frequently while handling extremes. Engineers are often interested in events that cause failure such as exceedances of a random variable over a threshold value.

Definition 2.2 (Exceedances). *Let X be a random variable and u a given threshold value. The event $\{X = x\}$ is said to be an exceedance at the level u if $x > u$.*

For example, waves can destroy a breakwater when their heights exceed a given value, say 9 m. Then it does not matter whether the height of a wave is 9.5, 10, or 12 m because the consequences of these events are the same.

Let X be a random variable representing heights of waves and Y_u be defined as

$$Y_u = \begin{cases} 0, & \text{if no exceedance occurred,} \\ 1, & \text{if an exceedance occurred.} \end{cases}$$

Then Y_u is a Bernoulli random variable with success probability $p_u = \Pr(X > u)$.

Bernoulli random variables arise in many important practical engineering situations, a few of which are described below.

Example 2.5 (Yearly maximum wave height). When designing a breakwater, civil engineers need to define the so-called design wave height, which is a wave height such that, when occurring, the breakwater will be able to withstand it without failure. Then, a natural design wave height would be the maximum wave height reaching the breakwater during its lifetime. However, this value is random and cannot be found. So, the only thing that an engineer can do is to choose this value with a small probability of being exceeded. In order to obtain this probability, it is important to know the probability of exceedances of certain values during a year. Then, if we are concerned with whether the yearly maximum wave height exceeds a given threshold value h_0, we have a Bernoulli experiment. ■

Example 2.6 (Tensile strength). Suspension bridges are supported by long cables. However, long cables are much weaker than short cables, the only ones tested in the laboratory. This is so because of the weakest link principle, which states that the strength of a long piece is the minimum strength of all its constituent pieces. Thus, the engineer has to extrapolate from lab results to real cables. The design of a suspension bridge requires the knowledge of the probability of the strength of the cable to fall below certain values. That is why values below a threshold are important. ■

Example 2.7 (Nuclear power plant). When designing a nuclear power plant, one has to consider the occurrence of earthquakes that can lead to disastrous consequences. Apart from the earthquake intensity, one of the main parameters to be considered is the distance from the earthquake epicenter to the location of the plant. Damage will be more severe for short distances than for long ones. Thus, engineers need to know whether this distance is below a given threshold value. ■

Example 2.8 (Temperature). Temperatures have a great influence on engineering works and can cause problems either for large or small values. Then, values above or below given threshold values are important. ■

Example 2.9 (Water flows). The water circulating through rivers greatly influences the life of humans. If the amount of water exceeds a given level, large areas can be flooded. On the other hand, if the water levels are below given values, the environment can be seriously damaged. ■

2.2.3 Binomial Distribution

Suppose now that n Bernoulli experiments are run such that the following conditions hold:

1. The experiments are identical, that is, the probability of success p is the same for all trials.

2. The experiments are independent, that is, the outcome of an experiment has no influence on the outcomes of the others.

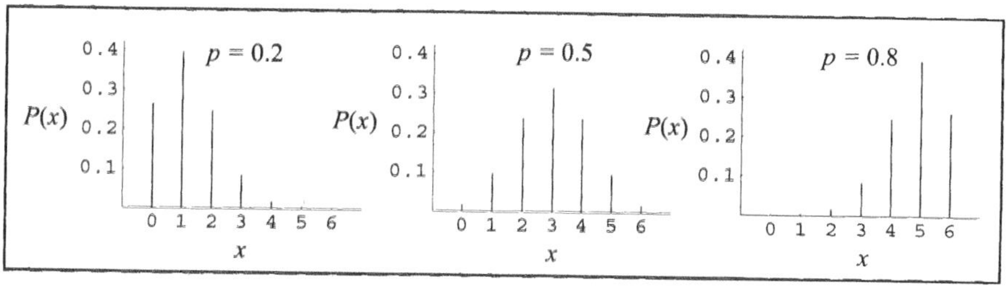

Figure 2.2: Examples of probability mass functions of binomial random variables with $n = 6$ and three values of p.

Let X be the number of successes in these n experiments. Then X is a random variable. To obtain the pmf of X, we first consider the event of obtaining x successes. If we obtained x successes, it also means that we obtained $n - x$ failures. Because the experiments are identical and independent, the probability of obtaining x successes and $n - x$ failures is

$$p^x(1-p)^{n-x}.$$

Note also that the number of possible ways of obtaining x successes (and $n - x$ failures) is obtained using the combinations formula:

$$\binom{n}{x} = \frac{n!}{x!(n-x)!}.$$

Therefore, the pmf of X is

$$P(x) = \binom{n}{x} p^x(1-p)^{n-x}, \quad x = 0, 1, \ldots, n. \tag{2.15}$$

This random variable is known as the *binomial random variable* and is denoted by $X \sim B(n,p)$ and the distribution in (2.15) is called the *binomial distribution*.

The mean and variance of a $B(n,p)$ random variable can be shown to be

$$\mu = np \quad \text{and} \quad \sigma^2 = np(1-p). \tag{2.16}$$

Figure 2.2 shows the graphs of the pmf of three binomial random variables with $n = 6$ and three values of p. From these graphs, it can be seen that when $p = 0.5$, the pmf is symmetric; otherwise, it is skewed.

Since X is the number of successes in these n identical and independent Bernoulli experiments, one may think of X as the sum of n identical and independent Bernoulli random variables, that is, $X = X_1 + X_2 + \ldots + X_n$, where X_i is a Bernoulli random variable with probability of success equal to p. Note that when $n = 1$, then a $B(1,p)$ random variable is a Bernoulli random variable.

Another important property of binomial random variables is *reproductivity* with respect to the parameter n. This means that the sum of two independent binomial random variables with the same p is also a binomial random variable. More precisely, if $X_1 \sim B(n_1, p)$ and $X_2 \sim B(n_2, p)$, then

$$X_1 + X_2 \sim B(n_1 + n_2, p).$$

Example 2.10 (Exceedances). An interesting practical problem consists of determining the probability of r exceedances over a value u in n identical and independent repetitions of the experiment. Since there are only two possible outcomes (exceedance or not exceedance), these are Bernoulli experiments. Consequently, the number of exceedances M_u over the value u of the associated random variable X is a $B(n, p_u)$ random variable with parameters n and p_u, where p_u is the probability of an exceedance over the level u of X. Therefore, the pmf of M_u is

$$\Pr(M_u = r) = \binom{n}{r} (p_u)^r (1 - p_u)^{n-r}, \quad r = 0, 1, \ldots, n. \tag{2.17}$$

Moreover, since p_u can be written as

$$p_u = \Pr(X > u) = 1 - F(u), \tag{2.18}$$

where $F(\cdot)$ is the cdf of X, (2.17) becomes

$$\Pr(M_u = r) = \binom{n}{r} [1 - F(u)]^r [F(u)]^{n-r}, \quad r = 0, 1, \ldots, n. \tag{2.19}$$

■

Example 2.11 (Concrete structures). Suppose that an engineer examined $n = 6$ concrete structures to determine which ones are in poor condition. As in Example 2.4, the probability that a given structure is in poor condition is $p = 0.2$. If X is the number of structures that are in poor condition, then X is a binomial random variable $B(6, 0.2)$. From (2.15), the pmf is

$$P(x) = \binom{6}{x} p^x (1 - p)^{6-x}, \quad x = 0, 1, \ldots, 6. \tag{2.20}$$

The graph of this pmf is given in Figure 2.2. For example, the probability that none of the six structures is found to be in poor condition is

$$P(0) = \binom{6}{0} 0.2^0 \, 0.8^6 = 0.2621,$$

and the probability that only one of the six structures is found to be in poor condition is

$$P(1) = \binom{6}{1} 0.2^1 \, 0.8^5 = 0.3932.$$

■

Example 2.12 (Yearly maximum wave height). Consider a breakwater that is to be designed for a lifetime of 50 years. Assume also that the probability of yearly exceedance of a wave height of 9 m is 0.05. Then, the probability of having 5 years with exceedances during its lifetime is given by

$$P(5) = \binom{50}{5} 0.05^5 (1 - 0.05)^{50-5} = 0.06584.$$

Note that we have admitted the two basic assumptions of the binomial model, that is, identical and independent Bernoulli experiments. In this case, both assumptions are reasonable. Note, however, that if the considered period were one day or one month instead of one year this would not be the case, because the wave heights in consecutive days are not independent events. (Assume both days belong to the same storm, then the maximum wave heights would be both high. On the other hand, if the periods were calm, both would be low.)

It is well known that there are some periodical phenomena ruling the waves that can last for more than one year. For that reason, it would be even better to consider periods of longer duration. ∎

Example 2.13 (Earthquake epicenter). From past experience, the epicenters of 10% of the earthquakes are within 50 km from a nuclear power plant. Now, consider a sequence of 10 such earthquakes and let X be the number of earthquakes whose epicenters are within 50 km from the nuclear power plant. Assume for the moment that the distances associated with different earthquakes are independent random variables and that all the earthquakes have the same probabilities of having their epicenters at distances within 50 km. Then, X is a $B(10, 0.1)$ random variable. Accordingly, the probability that none of the 10 earthquakes will occur within 50 km is

$$P(0) = \binom{10}{0} 0.1^0 (1 - 0.1)^{10-0} = 0.348678.$$

Note that this probability is based on two assumptions:

1. The distances associated with any two earthquakes are independent random variables.

2. The occurrence of an earthquake does not change the possible locations for others.

Both assumptions are not very realistic, because once an earthquake has occurred some others usually occur in the same or nearby location until the accumulated energy is released. Moreover, the probability of occurrence at the same location becomes much smaller, because no energy has been built up yet. ∎

2.2.4 Geometric or Pascal Distribution

Consider again a series of identical and independent Bernoulli experiments, which are repeated until the first success is obtained. Let X be the number

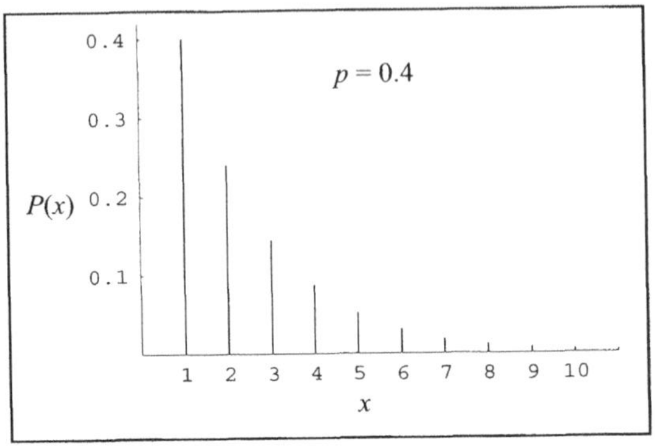

Figure 2.3: Graph of the probability mass function of the geometric random variable with $p = 0.4$.

of trial on which the first success occurs. What is the pmf of the random variable X? Note that if the first success has occurred at the trial number x, then the first $(x-1)$ trials must have been failures. Since the probability of a success is p and the probability of the $(x-1)$ failures is $(1-p)^{x-1}$ (because the trials are identical and independent), the pmf of X is

$$P(x) = p(1-p)^{x-1}, \quad x = 1, 2, \ldots. \tag{2.21}$$

This random variable is called a *geometric* or *Pascal* random variable and is denoted by $G(p)$. The pmf of $G(p)$ random variable is decreasing in x, which means that the largest value of $P(x)$ is at $x = 1$. A graph of the pmf of $G(0.4)$ is shown in Figure 2.3. The mean and variance of $G(p)$ are

$$\mu = \frac{1}{p} \quad \text{and} \quad \sigma^2 = \frac{1-p}{p^2}. \tag{2.22}$$

Example 2.14 (Job interviews). A company has one vacant position to fill. It is known that 80% of job applicants for this position are actually qualified for the job. The company interviews the applicants one at a time as they come in. The interview process stops as soon as one qualified applicant is found. How many interviews will have to be conducted until the first qualified applicant is found? This can be thought of as a series of identical and independent Bernoulli trials each with success probability $p = 0.8$. If X is the number of interviews required to find the first qualified applicant, then X is $G(0.8)$ random variable. For example, the probability that the first qualified applicant is found on the third interview is

$$P(3) = 0.8(1-0.8)^{3-1} = 0.032.$$

Also, the probability that the company will conduct at least three interviews to find the first qualified applicant is

$$\Pr(X \geq 3) = 1 - \Pr(X \leq 2) = 1 - F(2) = 1 - [P(1) + P(2)] = 0.04.$$

If each interview costs the company \$500, then the expected costs of filling the vacant position is

$$500\,E(X) = 500\,(1/0.8) = \$625.$$

■

Assume now that a given event (flood, dam failure, exceedance over a given temperature, etc.) is such that its probability of occurrence during a period of unit duration (normally one year) is a small value p. Assume also that the occurrences of such event in nonoverlapping periods are independent. Then, as time passes, we have a sequence of identical Bernoulli experiments (occurrence or not of the given event). Thus, the time measured in the above units until the first occurrence of this event is the number of experiments until the first occurrence and then it can be considered as a geometric $G(p)$ random variable, whose mean is $1/p$. This suggests the following definition.

Definition 2.3 (Return period). *Let A be an event, and T be the random time between successive occurrences of A. The mean value, μ, of the random variable T is called the return period of A (note that it is the mean time for the return of such an event).*

For the return period to be approximately $1/p$, the following conditions must hold:

1. The probability of one event occurring during a short period of time is small.

2. The probability of more than one event occurring during a short period of time is negligible.

2.2.5 Negative Binomial Distribution

The geometric distribution arises when we are interested in the number of Bernoulli trials that are required until we get the first success. Now suppose that we define the random variable X as the number of identical and independent Bernoulli trials that are required until we get the rth success. For the rth success to occur at the trial number x, we must have $r-1$ successes in the $x-1$ previous trials and one success in the trial number x. The number of possible ways of obtaining $r-1$ successes in $x-1$ trials is obtained using the combinations formula:

$$\binom{x-1}{r-1} = \frac{(x-1)!}{(r-1)!\,(x-r)!}.$$

Therefore, the pmf of X is

$$P(x) = \binom{x-1}{r-1} p^r (1-p)^{x-r}, \quad x = r, r+1, \ldots. \tag{2.23}$$

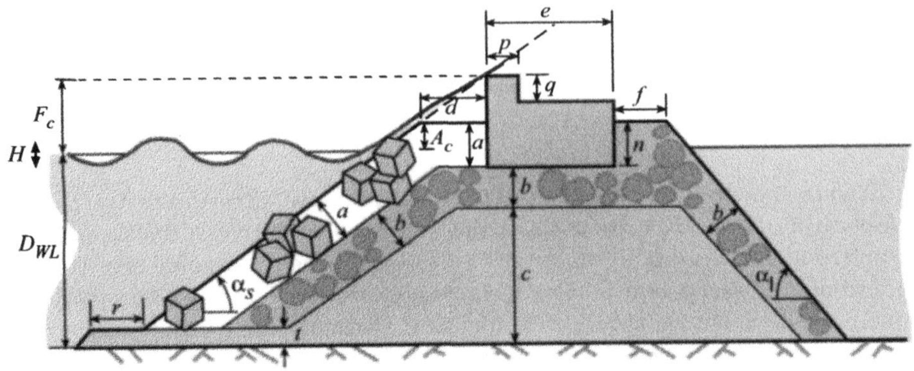

Figure 2.4: Illustration of the parts of a breakwater.

This random variable is called a *negative binomial* random variable and is denoted by $X \sim NB(r, p)$. Note that the geometric distribution is a special case of the negative binomial distribution obtained by setting $(r = 1)$, that is, $G(p) \sim NB(1, p)$. The mean and variance of an $NB(r, p)$ variable are

$$\mu = \frac{r}{p} \quad \text{and} \quad \sigma^2 = \frac{r\,(1-p)}{p^2}. \tag{2.24}$$

Example 2.15 (Job interviews). Suppose that the company in Example 2.14 wishes to fill two vacant positions. Thus, the interview process stops as soon as two qualified applicants are found. If X is the number of interviews needed to fill the two vacant positions, then X is an $NB(2, 0.8)$ random variable. For example, the probability that the second qualified applicant is found on the third interview is

$$P(3) = \binom{3-1}{2-1}\, 0.8^2\, 0.2^{3-2} = 0.128.$$

■

Example 2.16 (Rubble-mound breakwater). A rubble-mound breakwater is made of a supported crownwall on an earthfill that is protected by a mound armor (large pieces of stone to protect the earthfill from the waves) (see Fig. 2.4). The geometrical connectivity and structural stress transmission in the armor occurs by friction and interlocking between units. While failure of rigid breakwaters occurs when a single wave exceeds a given threshold value, a rubble-mound breakwater fails after the occurrence of several waves above a given threshold value. This is because the failure is progressive, that is, the first wave produces some movement of the stone pieces, the second increases the damage, etc.

Then, a failure occurs when the rth Bernoulli event (wave height exceeding the threshold) occurs. Thus, the negative binomial random variable plays a key role. Assume that a wave produces some damage on the armor if its height exceeds 7 m, and that the probability of this event is 0.001. Then, if the rubble-mound breakwater fails after, say, eight such waves, then the number of waves occurring until failure is a negative binomial random variable. Consequently, the pmf of the number of waves until failure is

$$P(x) = \binom{x-1}{8-1} 0.001^8 (1-0.001)^{x-8}, \quad x = 8, 9, \ldots.$$

■

Like the binomial random variable, the negative binomial random variable is also reproductive with respect to parameter r. This means that the sum of independent negative binomial random variables with the same probability of success p is a negative binomial random variable. More precisely, if $X_1 \sim NB(r_1, p)$ and $X_2 \sim NB(r_2, p)$, then

$$X_1 + X_2 \sim NB(r_1 + r_2, p).$$

2.2.6 Hypergeometric Distribution

Suppose we have a finite population consisting of N elements, where each element can be classified into one of two distinct groups. Say, for example, that we have N products of which D products are defective and the remaining $N - D$ are acceptable (nondefective). Suppose further that we wish to draw a random sample of size $n < N$ from this population without replacement. The random variable X, which is the number of defective items in the sample, is called a *hypergeometric* random variable and is denoted by $HG(N, p, n)$, where $p = D/N$ is the proportion of defective items in the population.

It is clear that the number of defective elements, X, cannot exceed either the total number of defective elements D, or the sample size n. Also, it cannot be less than 0 or less than $n - (N - D)$. Therefore, the support of X is

$$\max(0, n - qN) \le X \le \min(n, D),$$

where $q = 1 - p$ is the proportion of acceptable items in the population. The probability mass function of X is

$$P(x) = \frac{\binom{D}{x}\binom{N-D}{n-x}}{\binom{N}{n}}, \quad \max(0, n - qN) \le x \le \min(n, Np). \tag{2.25}$$

The numerator is the number of samples that can be obtained from the population with x defective elements and $n-x$ nondefective elements. The denominator is the total number of possible samples of size n that can be drawn. The mean and variance of $HG(N, p, n)$ are

$$\mu = np \quad \text{and} \quad \sigma^2 = \frac{N-n}{N-1}\, n\, p\, (1-p). \tag{2.26}$$

When N and D both tend to infinity such that D/N tends to p, this distribution tends to the binomial distribution.

Example 2.17 (Urn problem). An urn contains 20 balls, 5 white and 15 black. We draw a sample of size 10 without replacement. What is the probability that the drawn sample contains exactly 2 white balls? Here $N = 20$, $p = 5/20 = 1/4$, and $n = 10$. Letting X be the number of white balls in the sample, then X is $HG(20, 1/4, 10)$. From (2.25), we have

$$P(2) = \frac{\binom{5}{2}\binom{15}{8}}{\binom{20}{10}} = 0.348.$$

■

2.2.7 Poisson Distribution

Suppose we are interested in the number of occurrences of an event over a given interval of time or space. For example, let X be the number of traffic accidents occurring during a time interval t, or the number of vehicles arriving at a given intersection during a time interval of duration t. Then, X is a random variable and we are interested in finding its pmf. The experiment here consists of counting the number of times an event occurs during a given interval of duration t. Note that t does not have to be time; it could be location, area, volume, etc.

To derive the pmf of X we make the following *Poissonian* assumptions:

1. The probability p of the occurrence of a single event in a short interval d is proportional to its duration, that is, $p = \alpha d$, where α is a positive constant, known as the *arrival* or *intensity rate*.

2. The probability of the occurrence of more than one event in the same interval is negligible.

3. The number of occurrences in one interval is independent of the number of occurrences in other nonoverlapping intervals.

4. The number of events occurring in two intervals of the same duration have the same probability distribution.

Now, divide the interval t into n small and equal subintervals of duration $d = t/n$. Then, with the above assumptions, we may think of the n subintervals as n identical and independent Bernoulli trials $X_1, X_2, \ldots, X_n$, with $\Pr(X_i = 1) = p$ for $i = 1, 2, \ldots, n$. Letting $X = \sum_{i=1}^{n} X_i$, then X is a binomial random variable with parameters n and $p = \alpha t/n$. To guarantee that no more than a single event occurs in a subinterval d, the interval t may have to be divided

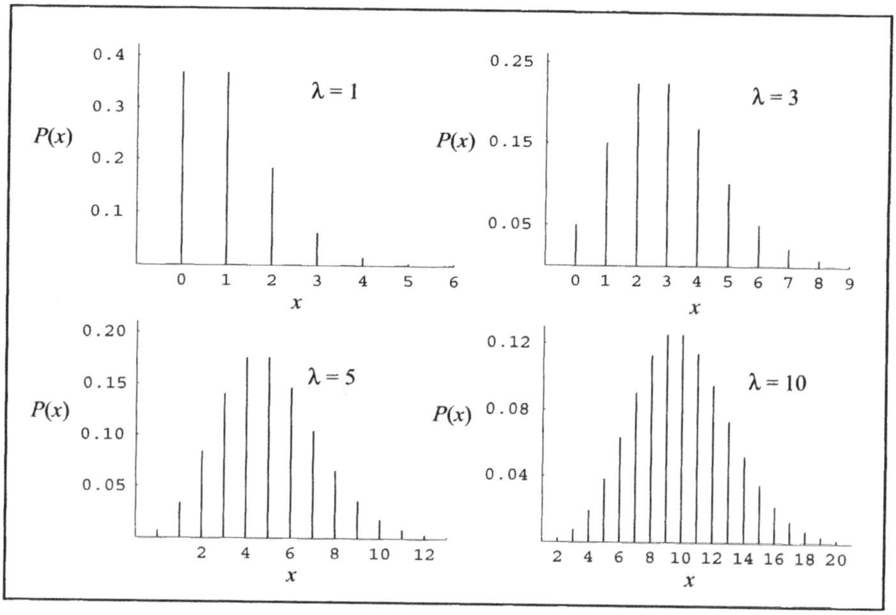

Figure 2.5: Some examples of probability mass functions of the Poisson random variable with four different values of λ.

into a very large number of intervals. So, we are interested in the pmf of X as $n \to \infty$. Under the above assumptions, one can show that

$$P(x) = \lim_{n\to\infty} \binom{n}{x} p^x (1-p)^{n-x} = \frac{e^{-\alpha t}(\alpha t)^x}{x!}, \quad x = 0, 1, \ldots, \lambda > 0. \tag{2.27}$$

Letting $\lambda = \alpha t$, we obtain

$$P(x) = \frac{e^{-\lambda}\lambda^x}{x!}, \quad x = 0, 1, \ldots, \lambda > 0. \tag{2.28}$$

This random variable, which is the number of events occurring in period of a given duration t, is known as a Poisson random variable with parameter $\lambda = \alpha t$ and is denoted by $P(\lambda)$. Note that the parameter λ is equal to the intensity α times the duration t.

Figure 2.5 shows the graphs of the pmf of some Poisson random variables. It can be seen that as the parameter λ gets larger, the pmf becomes more symmetric.

The mean and variance of a $P(\lambda)$ variable are

$$\mu = \lambda \quad \text{and} \quad \sigma^2 = \lambda. \tag{2.29}$$

Like the binomial random variable, the Poisson random variables are also reproductive, that is, if $X_1 \sim P(\lambda_1)$ and $X_2 \sim P(\lambda_2)$ are independent, then

$$X_1 + X_2 \sim P(\lambda_1 + \lambda_2).$$

The Poisson random variable is particularly appropriate for modeling the number of occurrences of rare events such as storms, earthquakes, and floods.

Example 2.18 (Storms). Suppose that storms of a certain level occur once every 50 years on the average. We wish to compute the probability of no such storm will occur during a single year. Assuming that X has a Poisson random variable with parameter λ, then $\lambda = 1/50 = 0.02$ and, using (2.28), we have

$$P(0) = \frac{e^{-1/50}(1/50)^0}{0!} = e^{-0.02} = 0.98.$$

That is, it is highly likely that no storms will occur in a single year.

For this model to be correct we need to check the assumptions of the above Poisson model. The first two assumptions are reasonable, because if several storms occur during a short interval, they could be considered as a single storm. The third and fourth assumptions are not true for close intervals, but they are true for far enough intervals. ∎

Example 2.19 (Parking garage). A parking garage has three car entrances. Assume that the number of cars coming into the garage using different entrances are independent Poisson random variables with parameters λ_1, λ_2, and λ_3. Then, using the reproductivity property of Poisson random variables, the total number of cars entering the garage is a Poisson random variable $P(\lambda_1+\lambda_2+\lambda_3)$.

The definition of reproductivity assumes that the random variables being considered in the sum are independent. Then, the number of cars entering at each entrance must be independent. This assumption must be checked before the above Poisson model is used. ∎

Poisson Approximation of the Binomial Distribution

If X is a $B(n, p)$ random variable, but p is small, say $p \leq 0.01$, and $np \leq 5$, then the pmf of X can be approximated by the pmf of the Poisson random variable with $\lambda = np$, the mean of the binomial random variable, that is,

$$P(x) \approx \frac{e^{-\lambda}\,\lambda^x}{x!}, \quad x = 0, 1, \ldots, n. \tag{2.30}$$

This is why the Poisson process is known as the *rare* events process.

Example 2.20 (Storms). Consider the storms in Example 2.18, and suppose that we are interested in the number of years with storms over a 40-year period. Although X is $B(40, 1/50)$, it can be approximated by a $P(40/50)$ random variable. For example, $P(3)$ can be computed either exactly, using the binomial pmf

$$P(x) = \binom{40}{3}\left(\frac{1}{50}\right)^3\left(1 - \frac{1}{50}\right)^{40-3} = 0.0374293,$$

or approximately, using the Poisson pmf,

$$P(x) \approx \frac{(40/50)^3 e^{-40/50}}{3!} = 0.0383427.$$

Table 2.4: Some Discrete Random Variables that Arise in Engineering Applications, Together with Their Probability Mass Functions, Parameters, and Supports.

Distribution	$P(x)$	Parameters and Support
Bernoulli	$P(x) = p^x(1-p)^{1-x}$	$0 < p < 1$ $x = 0, 1$
Binomial	$\binom{n}{x} p^x (1-p)^{n-x}$	$n = 1, 2, \ldots$ $0 < p < 1$ $x = 0, 1, \ldots, n$
Geometric	$p(1-p)^{x-1}$	$0 < p < 1$ $x = 1, 2, \ldots$
Negative Binomial	$\binom{x-1}{r-1} p^r (1-p)^{x-r}$	$0 < p < 1$ $x = r, r+1, \ldots$
Poisson	$\dfrac{e^{-\lambda}\lambda^x}{x!}$	$\lambda > 0$ $x = 0, 1, \ldots$
Nonzero Poisson	$\dfrac{e^{-\lambda}\lambda^x}{x!(1-e^{-\lambda})}$	$\lambda > 0$ $x = 1, 2, \ldots$

The error of approximation in this case is $0.0374293 - 0.0383427 = -0.0009134$. ∎

2.2.8 Nonzero Poisson Distribution

In certain practical applications we are interested in the number of occurrences of an event over a period of duration t, but we also know that at least one event has to occur during the period. If the Poissonian assumptions hold, then it can be shown that the random variable X has the following pmf

$$P(x) = \frac{e^{-\lambda}\lambda^x}{x!(1-e^{-\lambda})}, \quad x = 1, 2, \ldots, \lambda > 0. \tag{2.31}$$

This distribution is known as the *nonzero* or the *zero-truncated* Poisson distribution and is denoted by $P_0(\lambda)$. The mean and variance of $P_0(\lambda)$ are

$$\mu = \frac{\lambda}{1-e^{-\lambda}} \quad \text{and} \quad \sigma^2 = \frac{e^{\lambda}\lambda(-1+e^{\lambda}-\lambda)}{(e^{\lambda}-1)^2}.$$

A summary of the random variables we discussed in this section is given in Table 2.4.

Exercises

2.1 Show that the mean and variance of a Bernoulli random variable, with success probability p, are $\mu = p$ and $\sigma^2 = p(1-p)$.

2.2 Show that the mean and variance of a $B(n, p)$ random variable are $\mu = np$ and $\sigma^2 = np(1-p)$.

2.3 For the $B(6, 0.2)$ random variable in Example 2.11, whose pmf is given in (2.20), compute the probability of each of the following events:

(a) At least one structure is in poor condition.

(b) At most four structures are in poor conditions.

(c) Between two and four structures are in poor conditions.

2.4 Let $X = X_1 + \ldots + X_n$, where $X_i, i = 1, \ldots, n$, be identical and independent Bernoulli random variables with probability of success equal to p. Use the reproductivity property of the binomial random variable to show that $X \sim B(n, p)$.

2.5 Show that the pmf of any $G(p)$ random variable is decreasing in x.

2.6 Suppose that the company in Example 2.14 wishes to fill two vacant positions.

(a) What is the probability that at least four interviews will be required to fill the two vacant positions?

(b) If an interview costs the company \$500 on the average, what is the expected cost of filling the two vacant positions?

2.7 Derive the expressions of the mean and variance of a hypergeometric random variable in (2.26).

2.8 Prove the result in Equation (2.27).

2.9 Use (4.3) to show that the pmf of the nonzero Poisson random variable is given by (2.31).

2.10 The probability of a river to be flooded at a certain location during a period of 1 year is 0.02.

(a) Find the pmf of the number of floods to be registered during a period of 20 years.

(b) Find the pmf of the number of years required until the first with a flood.

(c) Find the pmf of the number of years required until the fifth with a flood.

2.11 Assume that the occurrence of earthquakes with intensity above a given threshold is Poissonian with rate of two earthquakes per year. Compute:

(a) The probability of having no earthquakes in a period of 6 months (the time a dam is being repaired).

(b) The pmf of the number of earthquakes occurring in a period of 5 years.

(c) Discuss the validity of the assumptions for using the Poisson model.

2.12 It is said that a system (e.g., a dam or a dike) has been designed for the N-year if it withstands floods that occur once in N years, that is, its probability of occurrence is $1/N$ in any year. Assuming that floods in different years are independent, calculate:

(a) The probability of having a flood larger or equal to the N-year flood during a period of 50 years.

(b) The probability of having one or more such floods in 50 years.

(c) If a company designs 20 independent systems (located far enough) for the 500-year flood, what is the cdf of the number of systems that will fail in 50 years?

2.13 The presence of cracks in a long cable has a Poisson process of intensity of two cracks per meter. Twenty percent of them are of critical size (large). Determine:

(a) The pmf of the number of cracks in a piece of 10 m.

(b) The length up to the first crack.

(c) The distance from the origin of the piece to the third crack.

(d) The number of cracks of critical size in the piece of 10 m.

2.14 The number of waves of height above 8 m at a given location and during a storm follows a Poisson process of intensity of three waves per hour. Determine:

(a) The pmf of the number of waves of height above 8 m during a storm of 5 hours.

(b) The time up to the first wave of height above 8 m.

(c) The time between the third and the sixth waves of height above 8 m.

(d) If a rouble-mound breakwater fails after 12 such waves, obtain the probability of failure of the breakwater during a storm of 2 hours duration.

2.15 The number of important floods during the last 50 years at a given location was 20. Determine:

(a) The pmf of the number floods during the next 10 years.

(b) The mean number of yearly floods required for having a probability of 0.9 of no floods in the next 10 years.

2.16 Thirty percent of car accidents involve deaths (or fatal). Determine:

(a) The number of fatal accidents of a set of 20 accidents.

(b) The number of accidents up to a fatal accident.

(c) The number of accidents up to the fifth fatal accident.

Chapter 3

Continuous Probabilistic Models

In Chapter 2 we discussed several commonly used univariate discrete random variables. This chapter deals with univariate continuous random variables. Multivariate random variables (both discrete and continuous) are discussed in Chapter 4.

We start with a discussion in Section 3.1 of some methods for dealing with univariate continuous random variables that include probability density function, cumulative distribution function, and moments of random variables. Then, several commonly used univariate continuous random variables are presented in Section 3.2. They are viewed with an eye on their application to extremes. Section 3.3 is devoted to truncated distributions, which have important applications. Section 3.4 presents four important functions associated with random variables, which are the survival, hazard, moment generating, and characteristic functions.

3.1 Univariate Continuous Random Variables

As in the case of discrete random variables, there are several methods for dealing with continuous random variables, some of which are presented below.

3.1.1 Probability Density Function

As mentioned in Chapter 2, *continuous* random variables take on an uncountable set of real values. Every continuous random variable has a *probability density function* (pdf). The pdf of a continuous random variable X is denoted by $f_X(x)$. For notational simplicity, we sometimes use $f(x)$ instead of $f_X(x)$. Note that $f(x)$ is not $\Pr(X = x)$, as in the discrete case. But it is the height of the density curve at the point x. Also, if we integrate $f(x)$ on a given set A, we obtain $\Pr(X \in A)$.

Every pdf $f(x)$ must satisfy two conditions:

$$f(x) \geq 0 \quad \forall x \tag{3.1}$$

and

$$\int_{x \in S(X)} f(x)dx = 1, \tag{3.2}$$

where $S(X)$ is the support of the random variable X, the set of all values x for which $f(x) > 0$.

3.1.2 Cumulative Distribution Function

Every random variable also has a *cumulative distribution function* (cdf). The cdf of a random variable X, denoted by $F(x)$, is a function that assigns to each real value x the probability of X being less than or equal to x, that is,

$$F(x) = \Pr(X \leq x) = \int_{-\infty}^{x} f(t)dt, \tag{3.3}$$

which implies that

$$f(x) = \frac{dF(x)}{dx}. \tag{3.4}$$

The probability that the random variable X takes values in the interval $(a, b]$, with $a \leq b$, is given by

$$\Pr(a < X \leq b) = \int_{a}^{b} f(x)dx = F(b) - F(a). \tag{3.5}$$

Thus, $\Pr(a < X \leq b)$ is the area under the pdf on top of the interval $(a, b]$, as can be seen in Figure 3.1, which shows the graphs of the pdf and cdf of a continuous random variable X. Note that, while $f(x)$ is the height of the density curve at x, $F(x)$ is the area under the curve to the left of x. From (3.2), the area under the pdf of any continuous random variable is 1.

Note also that

$$\Pr(X = x) = \Pr(x < X \leq x) = F(x) - F(x) = 0, \tag{3.6}$$

that is, while it is possible for a continuous random variable X to take a given value in its support, it is improbable that it will take this exact value. This is due to the fact that there are uncountably many possible values.

The cdf has the following properties as a direct consequence of the definitions of cdf and probability:

1. $F(-\infty) = 0$ and $F(\infty) = 1$.

2. $F(x)$ is nondecreasing and right continuous.

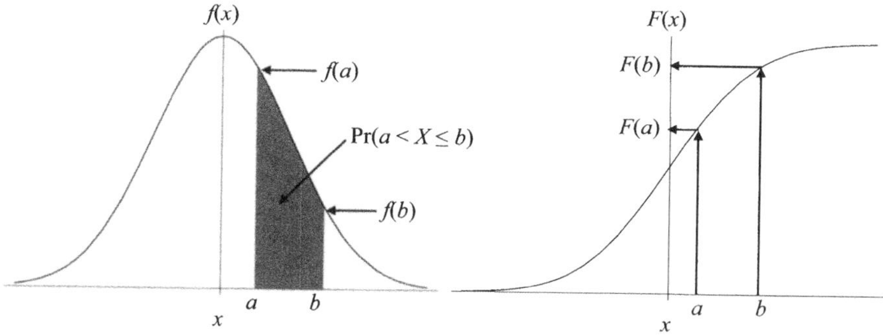

Figure 3.1: Graphs of the pdf and cdf of a continuous random variable X. The pdf, $f(x)$, is the height of the curve at x, and the cdf, $F(x)$, is the area under $f(x)$ to the left of x. Then $\Pr(a < X \leq b) = F(b) - F(a)$ is the area under the pdf on top of the interval $(a, b]$.

3.1.3 Moments

Let $g(X)$ be a function of a continuous random variable X. The *expected value* of $g(X)$ is defined by

$$E[g(X)] = \int_{x \in S(X)} g(x) f(x) dx. \tag{3.7}$$

For example, letting $g(X) = X^r$, we obtain the rth *moment* of the continuous random variable X,

$$E(X^r) = \int_{x \in S(X)} x^r f(x) dx. \tag{3.8}$$

When $r = 1$, we obtain the *mean*, μ, of the continuous random variable X,

$$\mu = E(X) = \int_{x \in S(X)} x \, f(x) dx. \tag{3.9}$$

Letting $g(X) = (X - \mu)^r$, we obtain the rth *central moment*,

$$E[(X - \mu)^r] = \int_{x \in S(X)} (x - \mu)^r f(x) dx. \tag{3.10}$$

When $r = 2$, we obtain the second central moment of the continuous random variable X, that is,

$$\sigma^2 = E[(X - \mu)^2] = \int_{x \in S(X)} (x - \mu)^2 f(x) dx, \tag{3.11}$$

which is known as the *variance*. The *standard deviation*, σ, of the random variable X is the positive square root of its variance. The variance can also be expressed as

$$\sigma^2 = E(X^2) - \mu^2, \tag{3.12}$$

where

$$E(X^2) = \int_{x \in S(X)} x^2 f(x) dx.$$

The expected value operator in the continuous case has the same properties that it has in the discrete case (see page 25).

3.2 Common Univariate Continuous Models

In this section, we present several important continuous random variables that often arise in extreme value applications. For more detailed descriptions as well as additional models, see, for example, the books by Balakrishnan and Nevzorov (2003) Johnson, Kotz, and Balakrishnan (1994, 1995), Ross (1992), Thoft-Christensen, and Wackerly, Mendenhall, and Scheaffer (2001).

3.2.1 Continuous Uniform Distribution

The continuous uniform random variable on the interval $[\alpha, \beta]$, denoted by $U(\alpha, \beta)$, has the following pdf

$$f(x) = \frac{1}{\beta - \alpha}, \quad \alpha \leq x \leq \beta, \tag{3.13}$$

from which it follows that the cdf can be written as

$$F(x) = \begin{cases} 0, & \text{if } x < \alpha \\ \dfrac{x - \alpha}{\beta - \alpha}, & \text{if } \alpha \leq x < \beta, \\ 1, & \text{if } x \geq \beta. \end{cases}$$

The mean and variance of X are

$$\mu = \frac{\alpha + \beta}{2} \quad \text{and} \quad \sigma^2 = \frac{(\beta - \alpha)^2}{12}. \tag{3.14}$$

A special case of $U(\alpha, \beta)$ is the *standard uniform* random variable, $U(0, 1)$ obtained by setting $\alpha = 0$ and $\beta = 1$. The pdf and cdf of $U(0, 1)$ are

$$f(x) = 1, \quad 0 \leq x \leq 1, \tag{3.15}$$

and

$$F(x) = \begin{cases} 0, & \text{if } x < 0 \\ x, & \text{if } 0 \leq x < 1, \\ 1, & \text{if } x \geq 1. \end{cases}$$

Figure 3.2 shows the pdf and cdf of the standard uniform random variable.

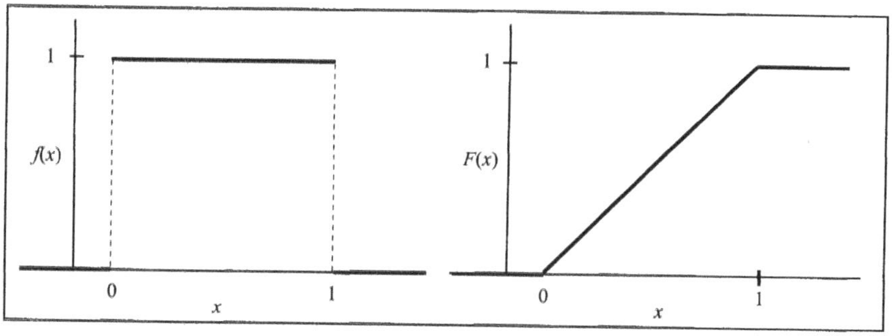

Figure 3.2: The pdf and cdf of the standard uniform random variable.

Example 3.1 (Birth time). If the times of birth are random variables assumed to be uniform on the interval $[0, 24]$, that is, all times in a given 24-hour period are equally possible, then the time of birth X is a uniform random variable, $U(0, 24)$, with pdf

$$f(x) = 1/24, \quad 0 \leq x \leq 24.$$

Note that the uniform model is valid so long as births occur naturally, that is, no induced births, for example. ■

Example 3.2 (Accidents). Let X be the distance in km from a hospital to the location where an accident occurs on a highway of 20 km length. Then, we may assume that X is $U(0, 20)$ random variable. The validity of this assumption requires certain conditions such as the road be straight and homogeneous and the drivers' abilities are constant over the 20-km highway. ■

The family of uniform random variables is stable with respect to changes of location and scale, that is, if X is $U(\alpha, \beta)$, then the variable $Y = cX + d$ is uniform $U(c\alpha + d, c\beta + d)$, see Example 3.22.

Example 3.3 (Temperatures). Suppose that the temperature, in degrees Celsius, at a given time and location is $U(30, 40)$ random variable. Since $F = 1.8C + 32$, where F and C are the temperatures measured in degrees Fahrenheit and Celsius, respectively, then the temperature in degrees Fahrenheit is an $U(86, 104)$ random variable. ■

3.2.2 Exponential Distribution

Let X be the time between two consecutive Poisson events with intensity λ events per unit of time (see Section 2.2.7) such as the time between failures of machines or the time between arrivals at a checkout counter. That is, we start at the time when the first event occurs and measure the time to the next event. In other words, X is the *interarrival* time. Then X is a continuous random variable.

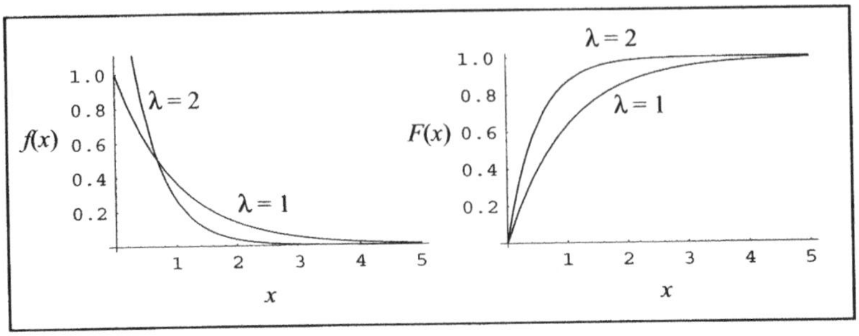

Figure 3.3: An example of the pdf and cdf of two exponential random variables.

What is the pdf and cdf of X? Consider the event that X exceeds x, that is, the second event occurs after time x since the occurrence of the first event. The probability of this event is $\Pr(X > x) = 1 - \Pr(X \leq x) = 1 - F(x)$, where $F(x)$ is the cdf of the random variable X. This event, however, is equivalent to saying that no Poisson events have occurred before time x. Replacing λ by λx in the Poisson pmf in (2.28), the probability of obtaining zero Poisson events is $P(0) = e^{-\lambda x}$. Therefore, we have

$$1 - F(x) = e^{-\lambda x},$$

from which it follows that the cdf of X is

$$F(x) = \begin{cases} 0, & \text{if } \ x < 0, \\ 1 - e^{-\lambda x}, & \text{if } \ x \geq 0, \ \lambda > 0. \end{cases}$$

Taking the derivative of $F(x)$ with respect to x, we obtain the pdf

$$f(x) = \frac{dF(x)}{dx} = \begin{cases} 0, & \text{if } \ x < 0, \\ \lambda e^{-\lambda x}, & \text{if } \ x \geq 0. \end{cases} \tag{3.16}$$

The random variable X whose pdf is given in (3.16) is called an *exponential* random variable with parameter λ and is denoted by $\text{Exp}(\lambda)$.

When X is replaced by $-Y$ in (3.16), we obtain the pdf of the *reversed* exponential random variable,

$$f(y) = \begin{cases} \lambda e^{\lambda y}, & \text{if } \ y < 0, \ \lambda > 0, \\ 0, & \text{if } \ y \geq 0. \end{cases} \tag{3.17}$$

The graphs of the pdf and cdf of two exponential random variables are shown in Figure 3.3. It can be shown that the mean and variance of the exponential random variable are

$$\mu = \frac{1}{\lambda} \quad \text{and} \quad \sigma^2 = \frac{1}{\lambda^2}. \tag{3.18}$$

The pdf of the exponential distribution in (3.16) can also be expressed as

$$f(x) = \begin{cases} 0, & \text{if } \ x < 0, \\ \dfrac{1}{\delta} e^{-x/\delta}, & \text{if } \ x \geq 0, \ \delta > 0. \end{cases} \tag{3.19}$$

This is simply a reparameterization of (3.16), where λ is replaced by $1/\delta$. In this form the cdf is

$$F(x) = \begin{cases} 0, & \text{if} \quad x < 0, \\ 1 - e^{-x/\delta}, & \text{if} \quad x \geq 0,\ \delta > 0, \end{cases}$$

and the mean and variance are simply $\mu = \delta$ and $\sigma^2 = \delta^2$, respectively.

Exponential random variables have the so-called *memoryless* or *no-aging* property, that is,

$$\Pr(X > a + b | X > a) = \Pr(X > b).$$

In words, if X is associated with lifetime the probability of X exceeding a given time b is the same no matter which time origin a is considered, from which the terminology *no-aging* was derived.

Example 3.4 (Waiting time at an intersection). When a car arrives at the intersection of two roads, it stops and then it needs a minimum time of t_0 seconds without passing cars to initiate the movement. If the arrival time, X, is assumed to be exponential with intensity λ cars/second, the probability of the waiting time to be more than t_0 is given by

$$\Pr(X > t_0) = 1 - \Pr(X \leq t_0) = 1 - F(t_0) = e^{-\lambda t_0}.$$

■

Example 3.5 (Time between consecutive storms). Assume that the occurrence of storms is Poissonian with rate λ storms/year. Then, the time until the occurrence of the first storm and the time between consecutive storms are exponential random variables with parameter λ. For example, assume that $\lambda = 5$ storms/year. Then, the probability of the time until the occurrence of the first storm or the time between consecutive storms to be smaller than 1 month is

$$p = \Pr(X < 1/12) = F(1/12) = 1 - e^{-\lambda/12} = 1 - e^{-5/12} = 0.3408.$$

■

For more properties and applications of the exponential distribution, the interested reader may refer to the book by Balakrishnan and Basu (1995).

3.2.3 Gamma Distribution

The Gamma distribution is a generalization of the exponential distribution. Consider a Poisson time process with intensity λ events per unit time. The time it takes for the first event to occur is an exponential random variable with parameter λ. Now, let X be the time up to the occurrence of θ Poisson events. If $\theta = 1$, then X is an exponential random variable, but if $\theta > 1$, then X is a Gamma random variable. What is then the pdf of a Gamma random variable?

To derive the pdf of a Gamma random variable, we first introduce a useful function called the *Gamma function*, which is defined as

$$\Gamma(\theta) = \int_0^{\infty} y^{\theta-1} e^{-y} dy. \tag{3.20}$$

Some important properties of the Gamma function are

$$\Gamma(0.5) = \sqrt{\pi}, \tag{3.21}$$

$$\Gamma(\theta) = (\theta-1)\Gamma(\theta-1), \quad \text{if } \theta > 1, \tag{3.22}$$

$$\Gamma(\theta) = (\theta-1)!, \quad \text{if } \theta \text{ is a positive integer.} \tag{3.23}$$

Now, if X is the time it takes for the θ Poisson events to occur, then the probability that X is in the interval $(x, x+dx]$ is $\Pr(x < X \leq x+dx) = f(x)dx$. But this probability is equal to the probability of having $\theta - 1$ Poisson events occurred in a period of duration x times the probability of the occurrence of one event in a period of duration dx. Thus, we have

$$f(x)dx = \frac{e^{-\lambda x}(\lambda x)^{\theta-1}}{(\theta-1)!} \lambda \, dx,$$

from which we obtain

$$f(x) = \frac{\lambda^{\theta} x^{\theta-1} e^{-\lambda x}}{(\theta-1)!}, \quad 0 \leq x < \infty. \tag{3.24}$$

Using the property of the Gamma function in (3.23), Equation (3.24) can be written as

$$f(x) = \frac{\lambda^{\theta} x^{\theta-1} e^{-\lambda x}}{\Gamma(\theta)}, \quad 0 \leq x < \infty, \tag{3.25}$$

which is valid for any real positive θ. The pdf in (3.25) is known as the Gamma distribution with parameters θ and λ, and is denoted by $G(\theta, \lambda)$. Note that when $\theta = 1$, the pdf in (3.25) becomes (3.16), which is the pdf of the exponential distribution. The pdf of some Gamma distributions are graphed in Figure 3.4.

In general, the cdf of the Gamma distribution

$$F(x) = \int_0^x f(t)dt = \int_0^x \frac{\lambda^{\theta} t^{\theta-1} e^{-\lambda t}}{\Gamma(\theta)} dt, \tag{3.26}$$

which is called the *incomplete Gamma ratio*, does not have a closed form, but can be obtained by numerical integration. For integer θ, (3.26) has a closed-form formula (see Exercise 3.7).

The mean and variance of $G(\theta, \lambda)$ are

$$\mu = \frac{\theta}{\lambda} \quad \text{and} \quad \sigma^2 = \frac{\theta}{\lambda^2}. \tag{3.27}$$

The family of Gamma distributions has the following important properties:

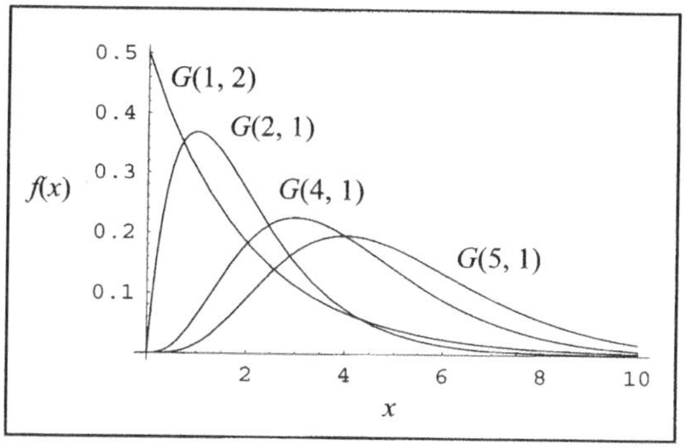

Figure 3.4: Examples of the pdf of some Gamma random variables.

1. It is reproductive with respect to parameter θ, that is, if $X_1 \sim G(\theta_1, \lambda)$ and $X_2 \sim G(\theta_2, \lambda)$ are independent, then

$$X_1 + X_2 \sim G(\theta_1 + \theta_2, \lambda).$$

2. It is stable with respect to scale changes, that is, if X is $G(\theta, \lambda)$, then cX is $G(\theta, \lambda/c)$, see Example 3.21.

3. It is not stable with respect to changes of location. In other words, if X is a Gamma random variable, then $X + a$ is not a Gamma random variable.

Example 3.6 (Parallel system of lamps). A lamp, L, consists of a parallel system of n individual lamps $L_1, \ldots, L_n$ (see Fig. 3.5). A lamp is lit once the previous one fails. Assuming no replacement after failure, the useful life of the lamp is the sum of the lives of the individual lamps. If the life of each of the individual lamps is assumed to be $G(\theta, \lambda)$, then the life of the lamp L is $G(n\theta, \lambda)$ by Property 1. ■

Example 3.7 (Structure). Consider the structure in Figure 3.6, which consists of three bars a, b, and c forming a right-angled triangle with sides $3, 4$, and 5 m, respectively.

The structure is subjected to a vertical pressure or load L. This creates three axial forces (two compression forces F_a and F_b, and one traction force F_c). If the force L is assumed to be $G(2, 1)$, the axial loads F_a, F_b, and F_c are all Gamma random variables. It can be shown that the equilibrium conditions lead to the following system of linear equations:

$$\begin{bmatrix} \cos x & -\cos y & 0 \\ \sin x & \sin y & 0 \\ 0 & \cos y & -1 \end{bmatrix} \begin{bmatrix} F_a \\ F_b \\ F_c \end{bmatrix} = \begin{bmatrix} 0 \\ L \\ 0 \end{bmatrix}. \tag{3.28}$$

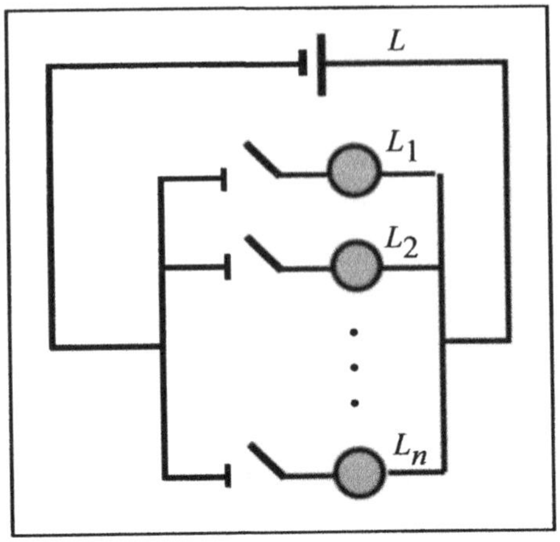

Figure 3.5: A lamp, L consisting of a system of n parallel lamps, $L_1, \ldots, L_n$.

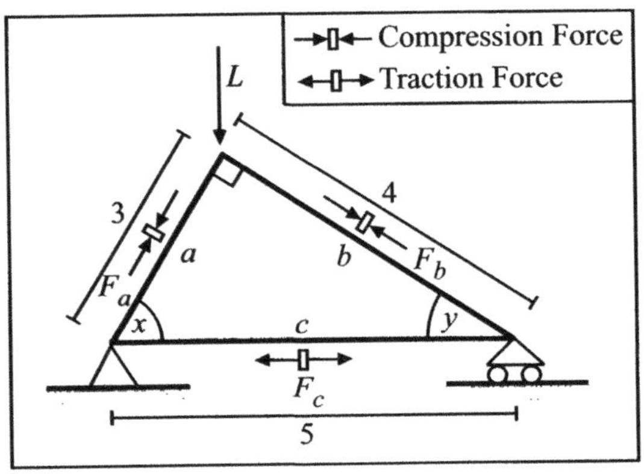

Figure 3.6: A three-bar structure subjected to a vertical force or load L.

Since $\cos x = 3/5$, $\cos y = 4/5$, $\sin x = 4/5$, and $\sin y = 3/5$, the solution of (3.28) gives

$$F_a = \frac{4}{5}\,L, \quad F_b = \frac{3}{5}\,L, \quad \text{and} \quad F_c = \frac{12}{25}\,L.$$

Now, since $G(\theta, \lambda)$ is stable with respect to scale changes, then cL is $G(2, 1/c)$. Therefore, $F_a \sim G(2, 5/4)$, $F_b \sim G(2, 5/3)$ and $F_c \sim G(2, 25/12)$. ■

Example 3.8 (Farmers subsidy). Farmers receive a monetary subsidy after three floods. Assume that the floods are Poisson events with rate 0.5 floods/year. Then, the time to the third flood is a $G(3, 0.5)$. Thus, the mean time until farmers receive subsidy is $\mu = \theta/\lambda = 3/0.5 = 6$ years. Also, the probability that a subsidy will be received after 10 years is

$$\Pr(X > 10) = 1 - F(10) = 1 - \int_0^{10} \frac{0.5(0.5x)^2 e^{-0.5x}}{2!}\, dx = 1 - 0.8753 = 0.1247.$$

■

3.2.4 Log-Gamma Distribution

The standard form of the log-gamma family of distributions has pdf

$$f(x) = \frac{1}{\Gamma(k)}\, e^{kx}\, \exp\left(-e^x\right), \quad -\infty < x < \infty,\ k > 0. \tag{3.29}$$

The pdf in (3.29) can be derived from the logarithmic transformation of gamma distribution, and hence the name *log-gamma distribution.* A *three-parameter* log-gamma family can be obtained from (3.29), by introducing a location parameter μ and a scale parameter σ, with pdf

$$f(x) = \frac{1}{\sigma\Gamma(k)} \exp\left[\frac{k(x-\mu)}{\sigma}\right] \exp\left[-\exp\left(\frac{x-\mu}{\sigma}\right)\right], \quad \begin{array}{r} -\infty < x, \mu < \infty, \\ \sigma, k > 0. \end{array} \tag{3.30}$$

The corresponding cdf is

$$F(x) = I_\gamma(\exp\{(x-\mu)/\sigma\}; k),$$

where $I_\gamma(y;k)$ is the incomplete gamma ratio defined as [see (3.26)]

$$I_\gamma(y;k) = \frac{1}{\Gamma(k)} \int_0^y e^{-t}\, t^{k-1}\, dt, \quad 0 < y < \infty,\ k > 0.$$

The mean and variance of X are

$$E(X) = \mu + \sigma\,\psi(k) \quad \text{and} \quad Var(X) = \sigma^2\,\psi'(k), \tag{3.31}$$

where $\psi(z) = \dfrac{d}{dz} \ln\Gamma(z) = \dfrac{\Gamma'(z)}{\Gamma(z)}$ is the digamma function and $\psi'(z)$ is its derivative (trigamma function). Since, for large k, $\psi(k) \sim \ln k$ and $\psi'(k) \sim 1/k$, Prentice (1974) suggested a reparameterized form of the log-gamma pdf in (3.30) as

$$\begin{aligned} f(x) &= \frac{1}{\sigma\,\Gamma(k)}\, k^{k-\frac{1}{2}}\, e^{\sqrt{k}(x-\mu)/\sigma}\, \exp\{-k e^{(x-\mu)/(\sigma\sqrt{k})}\}, \\ &\qquad -\infty < x < \infty,\ -\infty < \mu < \infty,\ \sigma > 0. \end{aligned} \tag{3.32}$$

It can be shown that as $k \to \infty$, the density in (3.32) tends to the $Normal(\mu, \sigma^2)$ density function.

Lawless (2003) and Nelson (2004) illustrate the usefulness of the log-gamma pdf in (3.32) as a lifetime model. Inferential procedures for this model have been discussed by Balakrishnan and Chan (1994, 1995b,a, 1998), DiCiccio (1987), Lawless (1980), Prentice (1974), and Young and Bakir (1987).

3.2.5 Beta Distribution

The Beta random variable is useful for modeling experimental data with range limited to the interval $[0, 1]$. For example, when X is the proportion of impurities in a chemical product or the proportion of time that a machine is under repair, then, X is such that $0 \le X \le 1$ and a Beta distribution is used to model experimental data collected on such variables. Its name is due to the presence of the *Beta function* in its pdf. The Beta function is defined as

$$\beta(\lambda, \theta) = \int_0^1 x^{\lambda-1}(1-x)^{\theta-1}\,dx, \quad \lambda > 0,\ \theta > 0. \tag{3.33}$$

The Beta function is related to the Gamma function, defined in (3.20), by

$$\beta(\lambda, \theta) = \frac{\Gamma(\lambda)\Gamma(\theta)}{\Gamma(\lambda+\theta)}. \tag{3.34}$$

The pdf of a Beta random variable is given by

$$f(x) = \frac{x^{\lambda-1}(1-x)^{\theta-1}}{\beta(\lambda,\theta)} = \frac{\Gamma(\lambda+\theta)}{\Gamma(\lambda)\Gamma(\theta)} x^{\lambda-1}(1-x)^{\theta-1}, \quad 0 \le x \le 1, \tag{3.35}$$

where $\lambda > 0$ and $\theta > 0$. The Beta random variable is denoted by $Beta(\lambda, \theta)$. The cdf of the $Beta(\lambda, \theta)$ is

$$F(x) = \int_0^x f(t)dt = \int_0^x \frac{t^{\lambda-1}(1-t)^{\theta-1}}{\beta(\lambda,\theta)} dt = I_\beta(x; \lambda, \theta), \tag{3.36}$$

where $I_\beta(x; \lambda, \theta)$ is called the *incomplete Beta ratio*, which can not be given in closed form, but can be obtained by numerical integration.

The mean and variance of the Beta random variable are

$$\mu = \frac{\lambda}{\lambda+\theta} \quad \text{and} \quad \sigma^2 = \frac{\lambda\theta}{(\lambda+\theta+1)(\lambda+\theta)^2},$$

respectively.

The fact that $0 \le X \le 1$ does not restrict the use of the Beta random variable because if Y is a random variable defined on the interval $[a, b]$, then

$$X = \frac{Y-a}{b-a}$$

defines a new variable such that $0 \le X \le 1$. Therefore, the Beta density function can be applied to a random variable defined on the interval $[a, b]$ by translation and a change of scale.

The interest in this variable is also based on its flexibility, because it can take on many different shapes, which can fit different sets of experimental data very well. For example, Figure 3.7 shows different examples of the pdf of the Beta distribution. Two particular cases of the Beta distribution are interesting. Setting $(\lambda = 1, \theta = 1)$ gives the *standard uniform* random variable, $U(0, 1)$, while setting $(\lambda = 2, \theta = 1$ or $\lambda = 1, \theta = 2)$ gives the *triangular* random variable whose cdf is given by $f(x) = 2x$ or $f(x) = 2(1-x)$, $0 \le x \le 1$, respectively.

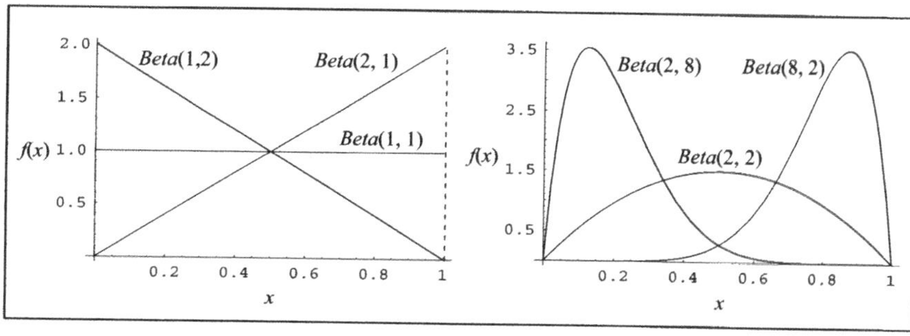

Figure 3.7: Examples showing that the probability density functions of Beta random variables take on wide range of different shapes.

3.2.6 Normal or Gaussian Distribution

One of the most important distributions in probability and statistics is the *normal* distribution (also known as the *Gaussian* distribution), which arises in various applications. For example, consider the random variable, X, which is the sum of n *independently and identically distributed* (iid) random variables $X_1, \ldots, X_n$. Then, by the *central limit theorem*, X is asymptotically (as $n \to \infty$) normal, regardless of the form of the distribution of the random variables $X_1, \ldots, X_n$. In fact, the normal distribution also arises in many cases where the random variables to be summed are dependent.

The normal random variable with mean μ and variance σ^2 is denoted by $X \sim N(\mu, \sigma^2)$ and its pdf is

$$f(x) = \frac{1}{\sigma\sqrt{2\pi}} \exp\left[-\frac{1}{2}\left(\frac{x-\mu}{\sigma}\right)^2\right], \quad -\infty < x < \infty, \tag{3.37}$$

where $-\infty < \mu < \infty$ and $\sigma > 0$. The mean and variance of a normal random variable are μ and σ^2, respectively. Figure 3.8 is a graph of the pdf of a $N(50, 25)$. Note that the pdf is symmetric about the mean $\mu = 50$. Also, the pdf has two inflection points, one on each side of the mean μ and equi-distant from μ. The standard deviation σ is equal to the distance between the mean and the inflection point. Like any other continuous random variable, the area under the curve is 1. The cdf of the normal random variable does not exist in closed form, but can be obtained by numerical integration.

The effect of the parameters μ and σ on the pdf and cdf can be seen in Figure 3.9, which shows the pdf and cdf of two normal random variables with the same mean (zero) but different standard deviations. The higher the standard deviation, the flatter the pdf.

If X is $N(\mu, \sigma^2)$, then the random variable

$$Z = \frac{X - \mu}{\sigma} \tag{3.38}$$

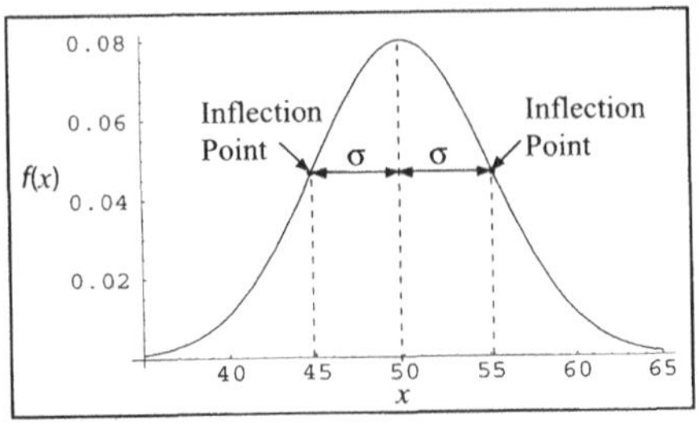

Figure 3.8: The pdf of $N(\mu, \sigma^2)$, where $\mu = 50$ and $\sigma = 5$.

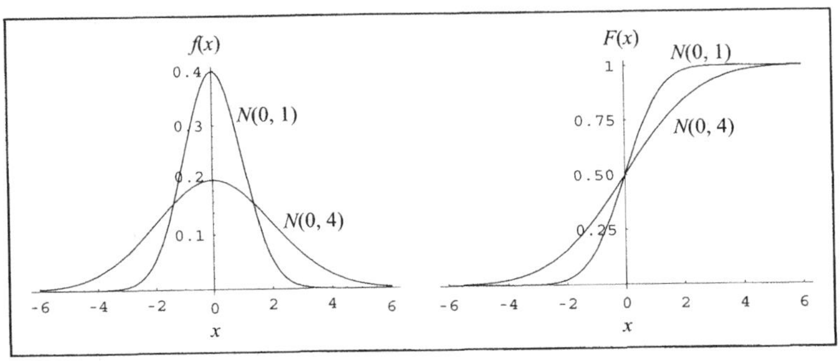

Figure 3.9: Some examples of normal pdfs and cdfs.

is $N(0, 1)$. The normal random variable with mean 0 and standard deviation 1 is called the *standard normal* distribution. From (3.37), the pdf of Z is

$$\phi(z) = \frac{1}{\sqrt{2\pi}} e^{-z^2/2}, \quad -\infty < z < \infty, \tag{3.39}$$

and the corresponding cdf is

$$\Phi(z) = \int_{-\infty}^{z} \frac{1}{\sqrt{2\pi}} e^{-x^2/2} dx, \quad -\infty < z < \infty. \tag{3.40}$$

The pdf and cdf of the standard normal random variable are shown in Figure 3.10. This cdf also does not exist in closed form. However, it has been computed numerically and is given in the Appendix as Table A.1. Note that because of the symmetry of the normal density, we have $\Phi(-z) = 1 - \Phi(z)$.

The main interest of the change of variable in (3.38) is that we can use Table A.1 to calculate probabilities for any other normal distribution. For example, if $X \sim N(\mu, \sigma^2)$, then

$$\Pr(X \le x) = \Pr\left(\frac{X - \mu}{\sigma} \le \frac{x - \mu}{\sigma}\right) = \Pr\left(Z \le \frac{x - \mu}{\sigma}\right) = \Phi\left(\frac{x - \mu}{\sigma}\right),$$

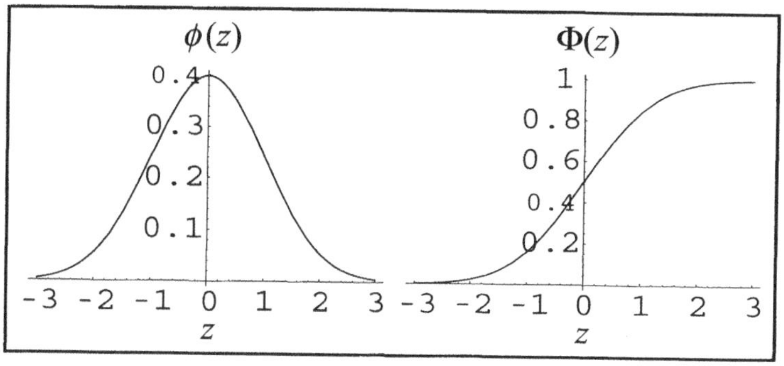

Figure 3.10: The pdf, $\phi(z)$, and the cdf, $\Phi(z)$, of the standard normal random variable, $N(0, 1)$.

where $\Phi(z)$ is the cdf of the standard normal distribution in (3.40), which can be obtained from Table A.1 in the Appendix.

Example 3.9 (Normal distribution). Suppose that a simple compression strength is a normal random variable with mean $\mu = 200$ kg/cm^2 and a standard deviation 40 kg/cm^2. Then, the probability that the compression strength is at most 140 kg/cm^2 is

$$\begin{aligned} \Pr(X \leq 140) &= F(140) = \Phi\left(\frac{140 - 200}{40}\right) \\ &= \Phi(-1.5) = 1 - \Phi(1.5) = 1 - 0.9332 = 0.0668, \end{aligned}$$

where $\Phi(1.5)$ is obtained from Table A.1. Figure 3.11 shows that $\Pr(X \leq 140) = \Pr(Z \leq -1.5)$. This probability is equal to the shaded areas under the two curves. ∎

The family of normal distributions is reproductive with respect to the parameters μ and σ, that is, if $X_1 \sim N(\mu_1, \sigma_1^2)$, $X_2 \sim N(\mu_2, \sigma_2^2)$, and X_1 and X_2 are independent, then

$$X_1 + X_2 \sim N(\mu_1 + \mu_2, \sigma_1^2 + \sigma_2^2).$$

If the random variables X_j, $j = 1, \ldots, n$, are independent and normal $N(\mu_j, \sigma_j^2)$, then the random variable

$$Y = \sum_{j=1}^{n} c_j X_j, \quad c_j \in \mathbb{R}, \quad j = 1, 2, \ldots, n, \tag{3.41}$$

is normal with

$$\mu = \sum_{j=1}^{n} c_j \mu_j \quad \text{and} \quad \sigma^2 = \sum_{j=1}^{n} c_j^2 \sigma_j^2.$$

This shows that the normal family is stable with respect to linear combinations.

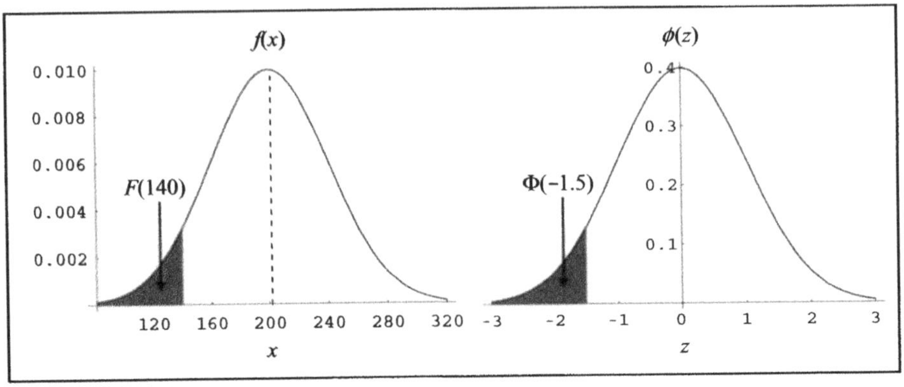

Figure 3.11: The pdf, $f(x)$, where $X \sim N(200, 40^2)$, and the pdf, $\phi(z)$, where $Z \sim N(0,1)$. The shaded area under $f(x)$ to the left of $x = 140$ is equal to the shaded area under $\phi(z)$ to the left of $z = -1.5$.

Normal Approximation to the Binomial Distribution

We know from Section 2.2.3 that the mean and variance of a binomial random variable are $\mu = np$ and $\sigma^2 = np(1-p)$. If the parameter n is large and neither p nor $(1-p)$ are very close to zero, the variable

$$Z = \frac{X - np}{\sqrt{np(1-p)}} \tag{3.42}$$

is approximately $N(0,1)$. This allows approximating the binomial probabilities using the normal probabilities. In practice, good approximations are obtained if $np, n(1-p) > 5$.

Example 3.10 (Normal approximation). Suppose that 30% of patients entering a hospital with myocardial infarction die in the hospital. If 2000 patients enter in one year and X is the number of these patients who will die in the hospital, then X is $B(2000, 0.3)$. Since n is large, $np = 600 > 5$, and $n(1-p) = 1400 > 5$, we can use the normal approximation to the binomial. Since $\mu = 2000 \times 0.3 = 600$ patients and $\sigma^2 = 2000 \times 0.3 \times 0.7 = 420$, then X can be approximated by $N(600, 420)$. Thus, for example, the probability that a maximum of 550 patients will die in the hospital is

$$\begin{aligned} \Pr(X \leq 550) &\approx \Pr\left(Z \leq (550 - 600)/\sqrt{420}\right) \\ &= \Phi(-2.44) = 1 - \Phi(2.44) \\ &= 1 - 0.9927 = 0.0073, \end{aligned}$$

where $\Phi(2.44)$ is obtained from Table A.1 in the Appendix. ■

3.2.7 Log-Normal Distribution

A random variable X is log-normal when its logarithm, $\log(X)$, is normal. The pdf of the log-normal random variable can be expressed as

$$f(x) = \frac{1}{x\sigma\sqrt{2\pi}} \exp\left[-\frac{1}{2}\left(\frac{\log(x)-\mu}{\sigma}\right)^2\right], \quad x \geq 0, \tag{3.43}$$

where the parameters μ and σ are the mean and the standard deviation of the initial normal random variable. The mean and variance of the log-normal random variable are then

$$e^{\mu+\sigma^2/2} \quad \text{and} \quad e^{2\mu}(e^{2\sigma^2} - e^{\sigma^2}). \tag{3.44}$$

In some applications, the random variables of interest are defined to be the products (instead of sums) of iid positive random variables. In these cases, taking the logarithm of the product yields the sum of the logarithms of its components. Thus, by the central limit theorem, the logarithm of the product of n iid random variables is asymptotically normal.

The log-normal random variable is not reproductive with respect to its parameters μ and σ^2, but stable with respect to products of independent variables, that is, if $X_1 \sim LN(\mu_1, \sigma_1^2)$ and $X_2 \sim LN(\mu_2, \sigma_2^2)$, then

$$X_1 \times X_2 \sim LN(\mu_1 + \mu_2, \sigma_1^2 + \sigma_2^2).$$

3.2.8 Logistic Distribution

A random variable X is said to have a logistic distribution if its cdf is given by

$$F(x) = \frac{1}{1 + \exp\{-(x-\alpha)/\beta\}}, \quad \beta > 0, \quad -\infty < x, \alpha < \infty, \tag{3.45}$$

where α and β are location and scale parameters, respectively. Note that the logistic distribution in (3.45) is symmetric about $x = \alpha$ and has a shape similar to that of the normal distribution.

The use of logistic function as a growth curve can be justified as follows. Consider the differential equation:

$$\frac{dF(x)}{dx} = k\left[F(x) - a\right]\left[b - F(x)\right], \tag{3.46}$$

where k, a, and b are constants with $k > 0$ and $b > a$. In other words, the rate of growth is equal to the excess over the initial asymptotic value a times the deficiency compared with final asymptotic value b. The solution of the differential equation (3.46) with $a = 0$ and $b = 1$ (the asymptotic limits of the cdf) is

$$F(x) = \left(1 + ce^{-x/k}\right)^{-1},$$

where c is a constant. This is the same as the logistic distribution in (3.45) with $k = \beta$ and $c = e^{\alpha/\beta}$. Equation (3.46) is used as a model of autocatalysis (see Johnson, Kotz, and Balakrishnan (1995)).

From (3.45), the pdf of the logistic random variable is

$$f(x) = \frac{\exp\{-(x-\alpha)/\beta\}}{\beta\left[1+\exp\{-(x-\alpha)/\beta\}\right]^2}, \quad \beta > 0, \quad -\infty < x, \alpha < \infty. \tag{3.47}$$

The mean and variance of the logistic random variable are

$$\mu = \alpha \quad \text{and} \quad \sigma^2 = \frac{\pi^2\beta^2}{3}. \tag{3.48}$$

A simple relationship between the cdf (3.45) and the pdf (3.47) is

$$f(x) = \frac{1}{\beta} F(x)[1 - F(x)].$$

This relation is useful to establish several properties of the logistic distribution; see, for example, Balakrishnan (1992).

3.2.9 Chi-Square and Chi Distributions

Let $Y_1, \ldots, Y_n$ be independent random variables, where Y_i is distributed as $N(\mu_i, 1)$. Then, the variable

$$X = \sum_{i=1}^{n} Y_i^2$$

is called a *noncentral* χ^2 random variable with n degrees of freedom and *noncentrality parameter* $\lambda = \sum_{i=1}^{n} \mu_i^2$. It is denoted by $\chi_n^2(\lambda)$. When $\mu_i = 0$ for all i, then $\lambda = 0$ and we obtain the *central* χ^2 random variable, which is denoted by χ_n^2. The pdf of the central χ^2 random variable with n degrees of freedom is

$$f(x) = \frac{x^{(n/2)-1}\, e^{-x/2}}{2^{n/2}\, \Gamma(n/2)}, \quad x \geq 0, \tag{3.49}$$

where $\Gamma(.)$ is the Gamma function defined in (3.20). The cdf $F(x)$ can not be given in closed form in general. However, it is available numerically and is given in the Appendix as Table A.3. The mean and variance of a $\chi_n^2(\lambda)$ random variable are

$$\mu = n + \lambda \quad \text{and} \quad \sigma^2 = 2(n + 2\lambda). \tag{3.50}$$

The $\chi_n^2(\lambda)$ variable is reproductive with respect to n and λ, that is, if $X_1 \sim \chi_{n_1}^2(\lambda_1)$ and $X_2 \sim \chi_{n_2}^2(\lambda_2)$, then

$$X_1 + X_2 \sim \chi_{n_1+n_2}^2(\lambda_1 + \lambda_2).$$

The positive square root of a $\chi_n^2(\lambda)$ random variable is called a χ random variable and is denoted by $\chi_n(\lambda)$.

3.2.10 Rayleigh Distribution

An interesting particular case of the χ_n random variable is the *Rayleigh* random variable, which is obtained when $n = 2$. The pdf of the Rayleigh random variable is given by

$$f(x) = \frac{1}{\delta^2} x \exp\left[-\frac{1}{2}\left(\frac{x}{\delta}\right)^2\right], \quad \delta > 0, x \geq 0. \tag{3.51}$$

The corresponding cdf is

$$F(x) = 1 - \exp\left[-\frac{1}{2}\left(\frac{x}{\delta}\right)^2\right]. \tag{3.52}$$

The mean and variance are

$$\mu = \delta\sqrt{\pi/2} \quad \text{and} \quad \sigma^2 = \delta^2(4-\pi)/2. \tag{3.53}$$

The Rayleigh distribution is used, for example, to model wave heights; see, for example, Longuet-Higgins (1975).

3.2.11 Student's t Distribution

Let Z and Y be $N(\lambda, 1)$ and χ_n^2 random variables, respectively. If Z and Y are independent, then the random variable

$$X = \frac{Z}{\sqrt{Y/n}}$$

is called the *noncentral Student's t* random variable with n degrees of freedom and noncentrality parameter λ and is denoted by $t_n(\lambda)$. When $\lambda = 0$, we obtain the *central Student's t* random variable, which is denoted by t_n, and its pdf is

$$f(x) = \frac{\Gamma(n+1)/2}{\Gamma(n/2)\sqrt{n\pi}}\left(1 + \frac{x^2}{n}\right)^{-(n+1)/2}, \quad -\infty < x < \infty, \tag{3.54}$$

where $\Gamma(.)$ is the Gamma function defined in (3.20). The cdf $F(x)$ is not simple. However, it is available numerically and is given in the Appendix as Table A.2. The mean and variance of the central t random variable are $\mu = 0$, for $n > 1$, and $\sigma^2 = n/(n-2)$, for $n > 2$, respectively.

3.2.12 F Distribution

Let Y_1 and Y_2 be χ_m^2 and χ_n^2 random variables, respectively, where m and n are positive integers. If Y_1 and Y_2 are independent, then the random variable

$$X = \frac{Y_1/m}{Y_2/n}$$

has an F distribution with m and n degrees of freedom, and is denoted by $F_{(m,n)}$. The corresponding pdf is

$$f(x) = \frac{\Gamma((m+n)/2)}{\Gamma(m/2)\Gamma(n/2)} \left(\frac{m}{n}\right)^{m/2} \frac{x^{(m/2)-1}}{[1+(m/n)x]^{(m+n)/2}}, \quad x > 0, \tag{3.55}$$

where $\Gamma(.)$ is the Gamma function defined in (3.20). The cdf $F(x)$ is available numerically and three quantiles of which are given in Tables A.4–A.6 in the Appendix. The mean and variance of F random variable are

$$\mu = \frac{n}{n-2}, \quad \text{for } n > 2,$$

and

$$\sigma^2 = \frac{2n^2(n+m-2)}{m(n-2)^2(n-4)}, \quad \text{for } n > 4,$$

respectively.

3.2.13 Weibull Distribution

The Weibull distribution appears very frequently in practical problems when we observe data that represent minima values. The reason for this is given in Chapter 9, where it is shown that, for many parent populations with limited left tail, the limit of the minima of independent samples converges to a Weibull distribution. The pdf of the Weibull random variable is given by

$$f(x) = \frac{\beta}{\delta} \exp\left[-\left(\frac{x-\lambda}{\delta}\right)^{\beta}\right] \left(\frac{x-\lambda}{\delta}\right)^{\beta-1}, \quad x > \lambda, \tag{3.56}$$

and the cdf by

$$F(x) = 1 - \exp\left[-\left(\frac{x-\lambda}{\delta}\right)^{\beta}\right], \quad x \geq \lambda, \tag{3.57}$$

with mean and variance

$$\mu = \lambda + \delta\Gamma\left(1+\frac{1}{\beta}\right) \quad \text{and} \quad \sigma^2 = \delta^2\left[\Gamma\left(1+\frac{2}{\beta}\right) - \Gamma^2\left(1+\frac{1}{\beta}\right)\right]. \tag{3.58}$$

Also of interest is the *reversed* Weibull distribution with pdf

$$f(x) = \frac{\beta}{\delta} \exp\left[-\left(\frac{\lambda-x}{\delta}\right)^{\beta}\right] \left(\frac{\lambda-x}{\delta}\right)^{\beta-1}, \quad x < \lambda, \tag{3.59}$$

and cdf

$$F(x) = \exp\left[-\left(\frac{\lambda-x}{\delta}\right)^{\beta}\right], \quad x \leq \lambda, \tag{3.60}$$

with mean and variance

$$\mu = \lambda - \delta\Gamma\left(1+\frac{1}{\beta}\right) \quad \text{and} \quad \sigma^2 = \delta^2\left[\Gamma\left(1+\frac{2}{\beta}\right) - \Gamma^2\left(1+\frac{1}{\beta}\right)\right]. \tag{3.61}$$

3.2.14 Gumbel Distribution

The Gumbel distribution appears very frequently in practical problems when we observe data that represent maxima values. The reason for this is presented in Chapter 9, where it is shown that for many parent populations with limited or unlimited left tail, the limit of the maxima of independent samples converges to a Gumbel distribution. The pdf of the Gumbel random variable is given by

$$f(x) = \frac{1}{\delta} \exp\left[\frac{\lambda - x}{\delta} - \exp\left(\frac{\lambda - x}{\delta}\right)\right], \quad -\infty < x < \infty, \tag{3.62}$$

and the cdf by

$$F(x) = \exp\left[-\exp\left(\frac{\lambda - x}{\delta}\right)\right], \quad -\infty < x < \infty, \tag{3.63}$$

with mean and variance

$$\mu = \lambda + 0.57772\delta \quad \text{and} \quad \sigma^2 = \frac{\pi^2\delta^2}{6}. \tag{3.64}$$

Also of interest is the *reversed* Gumbel distribution with pdf

$$f(x) = \frac{1}{\delta} \exp\left[\frac{x - \lambda}{\delta} - \exp\left(\frac{x - \lambda}{\delta}\right)\right], \quad -\infty < x < \infty, \tag{3.65}$$

and cdf

$$F(x) = 1 - \exp\left[-\exp\left(\frac{x - \lambda}{\delta}\right)\right], \quad -\infty < x < \infty, \tag{3.66}$$

with mean and variance

$$\mu = \lambda - 0.57772\delta \quad \text{and} \quad \sigma^2 = \pi^2\delta^2/6. \tag{3.67}$$

3.2.15 Fréchet Distribution

The Fréchet distribution appears very frequently in practical problems when we observe data that represent maxima values. The reason for this is provided in Chapter 9, where it is shown that for many parent populations with unlimited left tail, the limit of the maxima of independent samples converges to a Fréchet distribution.

The pdf of the Fréchet random variable is given by

$$f(x) = \frac{\beta\delta}{(x-\lambda)^2} \exp\left[-\left(\frac{\delta}{x-\lambda}\right)^{\beta}\right]\left(\frac{\delta}{x-\lambda}\right)^{\beta-1}, \quad x > \lambda, \tag{3.68}$$

and the cdf by

$$F(x) = \exp\left[-\left(\frac{\delta}{x-\lambda}\right)^{\beta}\right], \quad x > \lambda, \tag{3.69}$$

with mean and variance

$$\mu = \lambda + \delta\Gamma\left(1 - \frac{1}{\beta}\right), \quad \text{for } \beta > 1, \tag{3.70}$$

and

$$\sigma^2 = \delta^2\left[\Gamma\left(1 - \frac{2}{\beta}\right) - \Gamma^2\left(1 - \frac{1}{\beta}\right)\right], \quad \text{for } \beta > 2. \tag{3.71}$$

Also of interest is the *reversed* Fréchet distribution with pdf

$$f(x) = \frac{\beta\delta}{(\lambda - x)^2}\exp\left[-\left(\frac{\delta}{\lambda - x}\right)^{\beta}\right]\left(\frac{\delta}{\lambda - x}\right)^{\beta - 1}, \quad x < \lambda, \tag{3.72}$$

and cdf

$$F(x) = 1 - \exp\left[-\left(\frac{\delta}{\lambda - x}\right)^{\beta}\right], \quad x < \lambda, \tag{3.73}$$

with mean and variance

$$\mu = \lambda - \delta\Gamma\left(1 - \frac{1}{\beta}\right), \quad \text{for } \beta > 1, \tag{3.74}$$

and

$$\sigma^2 = \delta^2\left[\Gamma\left(1 - \frac{2}{\beta}\right) - \Gamma^2\left(1 - \frac{1}{\beta}\right)\right], \quad \text{for } \beta > 2. \tag{3.75}$$

3.2.16 Generalized Extreme Value Distributions

The generalized extreme value distributions include all distributions that can be obtained as the limit of sequences of maxima and minima values (see Chapter 9). The pdf of the *maximal generalized extreme value distribution* (GEVD) for $\kappa \neq 0$ is given by

$$f(x; \lambda, \delta, \kappa) = \exp\left[-\left[1 - \kappa\left(\frac{x - \lambda}{\delta}\right)\right]^{1/\kappa}\right]\left[1 - \kappa\left(\frac{x - \lambda}{\delta}\right)\right]^{1/\kappa - 1}\frac{1}{\delta}, \tag{3.76}$$

where the support is $x \leq \lambda + \delta/\kappa$, if $\kappa > 0$, or $x \geq \lambda + \delta/\kappa$, if $\kappa < 0$. For $\kappa = 0$, the pdf is

$$f(x; \lambda, \delta) = \exp\left[-\exp\left(\frac{\lambda - x}{\delta}\right)\right]\exp\left(\frac{\lambda - x}{\delta}\right)\frac{1}{\delta}. \tag{3.77}$$

The cdf of the maximal GEVD is

$$F(x; \lambda, \delta, \kappa) = \begin{cases} \exp\left\{-\left[1 - \kappa\left(\frac{x - \lambda}{\delta}\right)\right]^{1/\kappa}\right\}, & 1 - \kappa\left(\frac{x - \lambda}{\delta}\right) \geq 0, \kappa \neq 0, \\ \exp\left\{-\exp\left(\frac{\lambda - x}{\delta}\right)\right\}, & -\infty < x < \infty, \kappa = 0. \end{cases} \tag{3.78}$$

The corresponding p-quantile is

$$x_p = \begin{cases} \lambda + \delta\left[1 - (-\log p)^{\kappa}\right]/\kappa, & \text{if } \kappa \neq 0, \\ \lambda - \delta \log(-\log p), & \text{if } \kappa = 0. \end{cases} \tag{3.79}$$

The Gumbel, reversed Weibull and Fréchet distributions are particular cases of the maximal GEVD.

Also of interest is the *minimal* GEVD. Its pdf for $\kappa \neq 0$ is given by

$$f(x;\lambda,\delta,\kappa) = \exp\left[-\left[1+\kappa\left(\frac{x-\lambda}{\delta}\right)\right]^{1/\kappa}\right]\left[1+\kappa\left(\frac{x-\lambda}{\delta}\right)\right]^{1/\kappa-1}\frac{1}{\delta}, \tag{3.80}$$

where the support is $x \geq \lambda - \delta/\kappa$ if $\kappa > 0$, or $x \leq \lambda - \delta/\kappa$ if $\kappa < 0$. For $\kappa = 0$, the pdf is

$$f(x;\lambda,\delta) = \exp\left[-\exp\left(\frac{x-\lambda}{\delta}\right)\right]\exp\left(\frac{x-\lambda}{\delta}\right)\frac{1}{\delta}. \tag{3.81}$$

The cdf of the minimal GEVD, $F(x;\lambda,\delta,\kappa)$, is

$$\begin{cases} 1 - \exp\left\{-\left[1+\kappa\left(\frac{x-\lambda}{\delta}\right)\right]^{1/\kappa}\right\}, & 1+\kappa\left(\frac{x-\lambda}{\delta}\right) \geq 0, \kappa \neq 0, \\ 1 - \exp\left[-\exp\left(\frac{x-\lambda}{\delta}\right)\right], & -\infty < x < \infty. \end{cases} \tag{3.82}$$

The corresponding p-quantile is

$$x_p = \begin{cases} \lambda - \delta\left[1 - (-\log(1-p))^{\kappa}\right]/\kappa, & \text{if } \kappa \neq 0, \\ \lambda + \delta \log(-\log(1-p)), & \text{if } \kappa = 0. \end{cases} \tag{3.83}$$

The reversed Gumbel, Weibull, and reversed Fréchet distributions are particular cases of the minimal GEVD.

3.2.17 Generalized Pareto Distributions

As we shall see in (see Chapter 10), the generalized Pareto distribution (GPD) arises when you consider excesses of a random variable above or below given thresholds. There are two versions of the GPD, maximal and minimal. The pdf of the *maximal* GPD is

$$f(x;\lambda,\kappa) = \begin{cases} \frac{1}{\lambda}\left(1-\frac{\kappa x}{\lambda}\right)^{1/\kappa-1}, & \left(1-\frac{\kappa x}{\lambda}\right) \geq 0, \\ & \kappa \neq 0, \lambda > 0, \\ \frac{1}{\lambda}e^{-x/\lambda}, & x \geq 0, \kappa = 0, \lambda > 0, \end{cases} \tag{3.84}$$

where λ and κ are scale and shape parameters, respectively. For $\kappa \neq 0$, the range of x is $0 \leq x \leq \lambda/\kappa$ if $\kappa > 0$, and $x \geq 0$ if $\kappa \leq 0$. The cdf of the maximal GPD is

$$F(x;\lambda,\kappa) = \begin{cases} 1-\left(1-\dfrac{\kappa x}{\lambda}\right)^{1/\kappa}, & \left(1-\dfrac{\kappa x}{\lambda}\right) \geq 0, \\ & \kappa \neq 0, \lambda > 0, \\ 1-e^{-x/\lambda}, & x \geq 0, \kappa = 0, \lambda > 0. \end{cases} \tag{3.85}$$

The p-quantile of the GPD is

$$x_p = \begin{cases} \lambda\left[1-(1-p)^{\kappa}\right]/\kappa, & \text{if} \quad \kappa \neq 0, \\ -\lambda \log(1-p), & \text{if} \quad \kappa = 0. \end{cases} \tag{3.86}$$

Also of interest is the *minimal* or *reversed* generalized Pareto distribution with cdf

$$f(x;\lambda,\kappa) = \begin{cases} \dfrac{1}{\lambda}\left(1+\dfrac{\kappa x}{\lambda}\right)^{(1/\kappa)-1}, & \left(1+\dfrac{\kappa x}{\lambda}\right) \geq 0, \\ & \kappa \neq 0, \lambda > 0, \\ \dfrac{1}{\lambda}e^{x/\lambda}, & x \leq 0, \kappa = 0, \lambda > 0, \end{cases} \tag{3.87}$$

where λ and κ are scale and shape parameters, respectively. For $\kappa \neq 0$, the range of x is $-\lambda/\kappa \leq x \leq 0$ if $\kappa > 0$, and $x < 0$ if $\kappa \leq 0$. The cdf of the reversed GPD is

$$F(x;\lambda,\kappa) = \begin{cases} \left(1+\dfrac{\kappa x}{\lambda}\right)^{1/\kappa}, & \left(1+\dfrac{\kappa x}{\lambda}\right) \geq 0, \\ & \kappa \neq 0, \lambda > 0, \\ e^{x/\lambda}, & x \leq 0, \kappa = 0, \lambda > 0, \end{cases} \tag{3.88}$$

The p-quantile of the reversed GPD is

$$x_p = \begin{cases} \lambda\left(p^{\kappa}-1\right)/\kappa, & \text{if} \quad \kappa \neq 0, \\ \lambda \log(p), & \text{if} \quad \kappa = 0. \end{cases} \tag{3.89}$$

Finally, we conclude this section with a summary of all the univariate continuous distributions so far discussed in Tables 3.1, 3.2, and 3.3.

3.3 Truncated Distributions

In this section, we introduce truncated distributions that are very useful when dealing with extremes wherein only values above or below certain threshold values are of interest.

Table 3.1: Probability Density Functions of Some Continuous Random Variables that Frequently Arise in Engineering Applications.

Distribution	$f(x)$
Uniform	$(\beta-\alpha)^{-1}$
Exponential	$\lambda e^{-\lambda x}$
Gamma	$\dfrac{\lambda^\theta x^{\theta-1} e^{-\lambda x}}{\Gamma(\theta)}$
Log-Gamma	$\frac{1}{\sigma\Gamma(k)} \exp\left[\dfrac{k(x-\mu)}{\sigma}\right] \exp\left[-\exp\left(\dfrac{x-\mu}{\sigma}\right)\right]$
Beta	$\dfrac{x^{\lambda-1}(1-x)^{\theta-1}}{\beta(\lambda,\theta)}$
Normal	$\dfrac{1}{\sigma\sqrt{2\pi}} \exp\left[-\dfrac{(x-\mu)^2}{2\sigma^2}\right]$
Log-normal	$\dfrac{1}{x\sigma\sqrt{2\pi}} \exp\left[-\dfrac{(\log(x)-\mu)^2}{2\sigma^2}\right]$
Logistic	$\dfrac{\exp\{-(x-\alpha)/\beta\}}{\beta\left[1+\exp\{-(x-\alpha)/\beta\}\right]^2}$
Central χ^2	$\dfrac{x^{(n/2)-1}\, e^{-x/2}}{2^{n/2}\,\Gamma(n/2)}$
Rayleigh	$\dfrac{x}{\delta^2} \exp\left[-\dfrac{1}{2}\left(\dfrac{x}{\delta}\right)^2\right]$
Student's t	$\dfrac{\Gamma(n+1)/2}{\Gamma(n/2)\sqrt{n\pi}} \left(1+\dfrac{x^2}{n}\right)^{-(n+1)/2}$

Definition 3.1 (Truncated distributions). *Let X be a random variable. We call the random variables $X|X \le x_0$, $X|X > x_0$, and $X|x_0 < X \le x_1$ truncated at x_0 from the right, from the left, or truncated at x_0 from the left and at x_1 from the right, respectively.*

The following theorem gives the corresponding cdfs as a function of the cdf $F_X(x)$ of X.

Table 3.2: Probability Density Functions of Some Continuous Random Variables that Frequently Arise in Engineering Applications.

Distribution	$f(x)$
Weibull	$\frac{\beta}{\delta} \exp\left[-\left(\frac{x-\lambda}{\delta}\right)^{\beta}\right]\left(\frac{x-\lambda}{\delta}\right)^{\beta-1}$
Reversed Weibull	$\frac{\beta}{\delta} \exp\left[-\left(\frac{\lambda-x}{\delta}\right)^{\beta}\right]\left(\frac{\lambda-x}{\delta}\right)^{\beta-1}$
Gumbel	$\frac{1}{\delta} \exp\left[\frac{\lambda-x}{\delta} - \exp\left(\frac{\lambda-x}{\delta}\right)\right]$
Reversed Gumbel	$\frac{1}{\delta} \exp\left[\frac{x-\lambda}{\delta} - \exp\left(\frac{x-\lambda}{\delta}\right)\right]$
Fréchet	$\frac{\beta\delta}{(x-\lambda)^2} \exp\left[-\left(\frac{\delta}{x-\lambda}\right)^{\beta}\right]\left(\frac{\delta}{x-\lambda}\right)^{\beta-1}$
Reversed Fréchet	$\frac{\beta\delta}{(\lambda-x)^2} \exp\left[-\left(\frac{\delta}{\lambda-x}\right)^{\beta}\right]\left(\frac{\delta}{\lambda-x}\right)^{\beta-1}$
Maximal GEVD	Equations (3.76)–(3.77)
Minimal GEVD	Equations (3.80)–(3.81)
Maximal GPD	$\frac{1}{\lambda}\left(1-\frac{\kappa x}{\lambda}\right)^{(1/\kappa)-1}$
Minimal GPD	$\frac{1}{\lambda}\left(1+\frac{\kappa x}{\lambda}\right)^{(1/\kappa)-1}$

Theorem 3.1 (Cdf of a truncated distribution). *The cdf of the truncated random variable $X|X \le x_0$ is*

$$F_{X|X\le x_0}(x) = \begin{cases} F_X(x)/F_X(x_0), & \textit{if } x < x_0, \\ 1, & \textit{if } x \ge x_0. \end{cases} \tag{3.90}$$

The cdf of the truncated random variable $X|X > x_0$ is

$$F_{X|X> x_0}(x) = \begin{cases} 0, & \textit{if } x \le x_0, \\ \dfrac{F_X(x) - F_X(x_0)}{1 - F_X(x_0)}, & \textit{if } x > x_0. \end{cases} \tag{3.91}$$

Table 3.3: The Means and Variances of Some Continuous Random Variables that Frequently Arise in Engineering Applications.

Distribution	Mean	Variance
Uniform	$(\alpha+\beta)/2$	$(\beta-\alpha)^2/12$
Exponential	$1/\lambda$	$1/\lambda^2$
Gamma	θ/λ	θ/λ^2
Beta	$\dfrac{\lambda}{\lambda+\theta}$	$\dfrac{\lambda\theta}{(\lambda+\theta+1)(\lambda+\theta)^2}$
Normal	μ	σ^2
Log-normal	$e^{\mu+\sigma^2/2}$	$e^{2\mu}(e^{2\sigma^2}-e^{\sigma^2})$
Central χ^2	n	$2n$
Rayleigh	$\delta\sqrt{\pi/2}$	$\delta^2(4-\pi)/2$
Student's t	0	$n/(n-2)$
Weibull	$\lambda+\delta\Gamma\left(1+\dfrac{1}{\beta}\right)$	$\delta^2\left[\Gamma\left(1+\dfrac{2}{\beta}\right)-\Gamma^2\left(1+\dfrac{1}{\beta}\right)\right]$
Reversed Weibull	$\lambda-\delta\Gamma\left(1+\dfrac{1}{\beta}\right)$	$\delta^2\left[\Gamma\left(1+\dfrac{2}{\beta}\right)-\Gamma^2\left(1+\dfrac{1}{\beta}\right)\right]$
Gumbel	$\lambda+0.57772\delta$	$\pi^2\delta^2/6$
Reversed Gumbel	$\lambda-0.57772\delta$	$\pi^2\delta^2/6$
Fréchet	$\lambda+\delta\Gamma\left(1-\dfrac{1}{\beta}\right)$	$\delta^2\left[\Gamma\left(1-\dfrac{2}{\beta}\right)-\Gamma^2\left(1-\dfrac{1}{\beta}\right)\right]$
Reversed Fréchet	$\lambda-\delta\Gamma\left(1-\dfrac{1}{\beta}\right)$	$\delta^2\left[\Gamma\left(1-\dfrac{2}{\beta}\right)-\Gamma^2\left(1-\dfrac{1}{\beta}\right)\right]$
Logistic	α	$\pi^2\beta^2/3$

Finally, the cdf of the truncated random variable $X|x_0 < X \le x_1$ *is*

$$F_{X|x_0<X\le x_1}(x) = \begin{cases} 0, & \textit{if } x \le x_0, \\ \dfrac{F_X(x) - F_X(x_0)}{F_X(x_1) - F_X(x_0)}, & \textit{if } x_0 < x \le x_1, \\ 1, & \textit{if } x > x_1. \end{cases}$$

Proof. For right truncation, we have

$$\begin{aligned} F_{X|X\le x_0}(x) &= \Pr(X \le x | X \le x_0) \\ &= \frac{\Pr\left((X \le x) \cap (X \le x_0)\right)}{\Pr(X \le x_0)} \\ &= \frac{\Pr\left((X \le x) \cap (X \le x_0)\right)}{F_X(x_0)} \\ &= \begin{cases} \dfrac{\Pr(X \le x)}{F_X(x_0)} = \dfrac{F_X(x)}{F_X(x_0)}, & \text{if } x < x_0, \\ \dfrac{\Pr(X \le x_0)}{F_X(x_0)} = 1, & \text{if } x \ge x_0. \end{cases} \end{aligned}$$

For the left truncation, we have

$$\begin{aligned} F_{X|X> x_0}(x) &= \Pr(X \le x | X > x_0) \\ &= \frac{\Pr\left((X \le x) \cap (X > x_0)\right)}{\Pr(X > x_0)} \\ &= \frac{\Pr(x_0 < X \le x)}{1 - F_X(x_0)} \\ &= \begin{cases} 0, & \text{if } x \le x_0, \\ \dfrac{F_X(x) - F_X(x_0)}{1 - F_X(x_0)}, & \text{if } x > x_0. \end{cases} \end{aligned}$$

Finally, for truncation on both sides, we have

$$\begin{aligned} F_{X|x_0<X\le x_1}(x) &= \Pr(X \le x | x_0 < X \le x_1) \\ &= \frac{\Pr\left((X \le x) \cap (x_0 < X \le x_1)\right)}{\Pr(x_0 < X \le x_1)} \\ &= \frac{\Pr(x_0 < X \le \min(x, x_1))}{F_X(x_1) - F_X(x_0)} \\ &= \begin{cases} 0, & \text{if } x \le x_0, \\ \dfrac{F_X(x) - F_X(x_0)}{F_X(x_1) - F_X(x_0)}, & \text{if } x_0 < x \le x_1, \\ 1, & \text{if } x > x_1. \end{cases} \end{aligned}$$

■

Example 3.11 (Lifetime). The remaining lifetime, X, in years of a patient after suffering a heart attack has a cdf

$$F_X(x) = \begin{cases} 0, & \text{if } x < 0, \\ \dfrac{1 - e^{-\lambda x}}{1 - e^{-50\lambda}}, & \text{if } 0 \le x < 50, \\ 1, & \text{if } x \ge 50. \end{cases}$$

If a given patient suffered a heart attack 30 years ago, determine the cdf of the remaining lifetime.

Before solving this problem, it is worthwhile mentioning that the given variable has a right-truncated exponential distribution, which implies that no patient can survive above 50 years after suffering a heart attack.

Since the patient suffered a heart attack 30 years ago, we must determine the cdf of the random variable $(X - 30)$ conditioned on $X \ge 30$. Then, we ask for the truncated distribution on the left at 30 and translated by 30 units, that is,

$$F_X(x) = \begin{cases} 0, & \text{if } x < 0, \\ \dfrac{e^{-30\lambda} - e^{-\lambda(x+30)}}{e^{-30\lambda} - e^{-50\lambda}}, & \text{if } 0 \le x < 20, \\ 1, & \text{if } x \ge 20. \end{cases}$$

■

Example 3.12 (Hospital). Suppose that the age, X (in years) of patients entering a hospital has the following pdf:

$$f_X(x) = \begin{cases} \dfrac{\pi}{200} \sin\left(\dfrac{\pi x}{100}\right), & \text{if } 0 \le x < 100, \\ 0, & \text{otherwise.} \end{cases}$$

Then, the pdf for children younger than 5 years old that enter the hospital is the same density but truncated on the right at $X = 5$. Thus, we have

$$f_X(x) = \begin{cases} \dfrac{\pi \sin(\pi x/100)}{100\,[1 - \cos(5\pi/100)]}, & \text{if } x < 5, \\ 0, & \text{otherwise.} \end{cases}$$

Similarly, the pdf for patients above 60 years of age is the same density but truncated on the left at $X = 60$. Thus, we have

$$f_Z(z) = \begin{cases} \dfrac{\pi \sin(\pi z/100)}{100\,[1 + \cos(60\pi/100)]}, & \text{if } 60 \le z < 100, \\ 0, & \text{otherwise.} \end{cases}$$

■

Example 3.13 (Screw strength). A factory producing screws states that the strength of screws, R^*, in kg/cm^2 has an exponential distribution $E(\lambda)$. If all the screws are subject to a quality test consisting of applying a test stress of 10 kg/cm^2 and those failing are discarded, determine the pdf of the strength, R, of the accepted screws.

Since after the test the screws with a strength less than 10 kg/cm^2 are discarded, the resulting strength is truncated on the left at 10 kg/cm^2, so that we have

$$f_R(x) = \begin{cases} \dfrac{\lambda e^{-\lambda x}}{e^{-10\lambda}} = \lambda e^{-\lambda(x-10)}, & \text{if } x > 10 \text{ kg/cm}^2, \\ 0, & \text{otherwise.} \end{cases}$$

Note that this is just an exponential distribution $E(\lambda)$ with a location shift of $10 kg/cm^2$. ■

3.4 Some Other Important Functions

In this section, we present four important functions associated with random variables: the survival, hazard, moment generating, and characteristic functions.

3.4.1 Survival and Hazard Functions

Let X be a nonnegative random variable with pdf $f(x)$ and cdf $F(x)$. This happens, for example, when the random variable X is the lifetime of an object (e.g., a person or a machine). We have the following definitions.

Definition 3.2 (Survival function). *The function*

$$S(x) = \Pr(X > x) = 1 - F(x) \tag{3.92}$$

is called the survival function.

The function $S(x)$ is called a survival function because it gives the probability that the object will survive beyond time x.

Definition 3.3 (Hazard function). *The hazard function (hazard rate or mortality function) is defined as*

$$H(x) = \frac{f(x)}{\Pr(X > x)} = \frac{f(x)}{1 - F(x)}. \tag{3.93}$$

Assume that X is the lifetime of an element. Then, the hazard function can be interpreted as the probability, per unit time, that the item will fail just after time x given that the item has survived up to time x:

$$\begin{aligned} H(x) &= \lim_{\varepsilon \to 0} \frac{\Pr(x < X \le x + \varepsilon | X > x)}{\varepsilon} \\ &= \frac{\lim_{\varepsilon \to 0} \dfrac{F(x+\varepsilon) - F(x)}{\varepsilon}}{1 - F(x)} \end{aligned}$$

$$= \frac{\frac{d}{dx}F(x)}{1-F(x)} = \frac{f(x)}{1-F(x)}.$$

In other words, the hazard function can be interpreted as the probability of instantaneous failure given that the item has survived up to time x.

From (3.92) and (3.93), it can be seen that

$$f(x) = H(x)S(x), \tag{3.94}$$

that is, the pdf is the product of the hazard and survival functions. There is also a one-to-one correspondence between the cdf, $F(x)$, and the hazard function, $H(x)$. To see this, note that

$$\frac{d}{dx}\log(1-F(x)) = -H(x),$$

and integrating from 0 to x we have,

$$\log(1-F(x)) = -\int_0^x H(t)dt,$$

which yields

$$F(x) = 1 - \exp\left\{-\int_0^x H(t)dt\right\}. \tag{3.95}$$

From the relationship between $F(x)$ and $H(x)$ in (3.95), one can be obtained from the other. Note also that a comparison between (3.92) and (3.95) suggests the following relationship between the survival and hazard functions:

$$S(x) = \exp\left\{-\int_0^x H(t)dt\right\}. \tag{3.96}$$

Example 3.14 (Weibull). Consider a Weibull random variable with cdf

$$F(x) = 1 - \exp(-x^\beta), \quad \beta > 0, x > 0.$$

The corresponding pdf is given by

$$f(x) = \beta x^{\beta-1}\exp(-x^\beta), \quad \beta > 0, x > 0.$$

Then, the hazard function is

$$H(x) = \frac{f(x)}{1-F(x)} = \frac{\beta x^{\beta-1}\exp(-x^\beta)}{\exp(-x^\beta)} = \beta x^{\beta-1}.$$

Note that $H(x)$ is increasing, constant, or decreasing according to $\beta > 1$, $\beta = 1$, or $\beta < 1$, respectively. ■

3.4.2 Moment Generating Function

As we have already seen in this chapter and in Chapter 2, every random variable (discrete or continuous) has a cumulative distribution function $F(x)$ and an associated pmf ($P(x)$ in the discrete case) or pdf ($f(x)$ in the continuous case). In addition to these functions, random variable have two other important functions; the *moment generating function* (MGF) and the *characteristic function* (CF). These are discussed below.

Definition 3.4 (Moment generating function). *Let X be a univariate random variable with distribution function $F(x)$. The MGF of X, denoted by $M_X(t)$, is defined for the discrete case as*

$$M_X(t) = E\left(e^{tX}\right) = \sum_{x \in S(X)} e^{tx} P(x), \tag{3.97}$$

where $P(x)$ is the pmf of X and $S(X)$ is the support of $P(x)$. For the continuous case, the MGF is defined as

$$M_X(t) = E\left(e^{tX}\right) = \int_{-\infty}^{\infty} e^{tx} f(x) dx, \tag{3.98}$$

where $f(x)$ is the pdf of X and $M_X(t)$ is a function of t.

For simplicity of notation we shall use $M(t)$ instead of $M_X(t)$, unless otherwise needed. The MGF for some other random variables are given in Table 3.4.

The function $M(t)$ is called the moment generating function because it generates all the moments of the random variable X. Namely, the kth order moment of the random variable X is given by

$$E(X^k) = \left.\frac{d^{(k)}}{dt} M(t)\right|_{t=0}. \tag{3.99}$$

In other words, the kth order moment of the random variable X is the kth derivative of the MGF with respect to t, evaluated at $t = 0$. Note that (3.99) is valid for both discrete and continuous random variables. This is illustrated by two examples.

Example 3.15 (The MGF of a discrete random variable). From the pdf of the binomial random variable in (2.15), we obtain the MGF

$$M(t) = E(e^{tx}) = \sum_{x=0}^{n} e^{tx} \binom{n}{x} p^x q^{n-x} = \sum_{x=0}^{n} \binom{n}{x} \left(p\, e^t\right)^x q^{n-x} = \left(p\, e^t + q\right)^n.$$

The first two derivatives of $M(t)$ are

$$\frac{d}{dt} M(t) = n\left(pe^t + (1-p)\right)^{n-1} pe^t$$

Table 3.4: Moment Generating Functions of Some Common Random Variables.

Discrete Distribution		Continuous Distribution	
Name	$M(t)$	Name	$M(t)$
Uniform	$\dfrac{e^t(1-e^{nt})}{n(1-e^t)}$	Uniform	$\dfrac{e^{\beta t}-e^{\alpha t}}{t(\beta-\alpha)}$
Dirac	e^t	Exponential	$\dfrac{\lambda}{\lambda-t}$
Bernoulli	$pe^t+(1-p)$	Gamma	$\left(\dfrac{\lambda}{\lambda-t}\right)^\theta$
Binomial	$\left[pe^t+(1-p)\right]^n$	Log-gamma	$e^{t\mu}\Gamma(k+t\sigma)/\Gamma(k)$
Geometric	$\dfrac{pe^t}{1-(1-p)e^t}$	Normal	$\exp\left\{t\mu+\dfrac{\sigma^2t^2}{2}\right\}$
Negative Binomial	$\left[\dfrac{pe^t}{1-(1-p)e^t}\right]^r$	χ^2_n	$(1-2t)^{-n/2}$
Poisson	$\exp\left\{\lambda(e^t-1)\right\}$	Logistic	$e^{t\alpha}\Gamma(1+\beta t)\Gamma(1-\beta t)$

and

$$\frac{d^{(2)}}{dt}M(t)=n\left(pe^t+(1-p)\right)^{n-1}pe^t+n(n-1)\left(pe^t+(1-p)\right)^{n-2}p^2e^{2t},$$

respectively. Then, substituting zero for t in the above two equations, we obtain the first two moments, that is,

$$E(X)=np \tag{3.100}$$

and

$$E(X^2)=np+n(n-1)p^2, \tag{3.101}$$

from which the variance is obtained as

$$\sigma^2=E(X^2)-[E(X)]^2=np(1-p). \tag{3.102}$$

■

Example 3.16 (The MGF of a continuous random variable). From the pdf of the Gamma random variable in (3.25), we derive the MGF as follows:

$$M(t)=\int_0^\infty e^{tx}\frac{\lambda^\theta x^{\theta-1}e^{-\lambda x}}{\Gamma(\theta)}\,dx=\frac{\lambda^\theta}{\Gamma(\theta)}\int_0^\infty x^{\theta-1}e^{-(\lambda-t)x}dx. \tag{3.103}$$

To calculate the integral in (3.103), we use the change of variable technique and let $y = (\lambda - t)x$, from which we obtain

$$x = \frac{y}{\lambda - t} \quad \text{and} \quad dx = \frac{dy}{\lambda - t}.$$

Substituting these in (3.103), we obtain

$$\begin{aligned} M(t) &= \frac{\lambda^\theta}{\Gamma(\theta)} \int_0^\infty \left(\frac{y}{\lambda - t}\right)^{\theta - 1} e^{-y} \frac{1}{\lambda - t} dy \\ &= \frac{\lambda^\theta}{\Gamma(\theta)(\lambda - t)^\theta} \int_0^\infty y^{\theta - 1} e^{-y} dy \\ &= \left(\frac{\lambda}{\lambda - t}\right)^\theta, \end{aligned} \tag{3.104}$$

where the last identity is obtained because the integral on the right-hand side of (3.104) is equal to $\Gamma(\theta)$ [see (3.20)].

The first two derivatives of $M(t)$ are

$$\frac{d}{dt} M(t) = \theta \lambda^\theta (\lambda - t)^{-\theta - 1}$$

and

$$\frac{d^{(2)}}{dt} M(t) = \theta(\theta + 1) \lambda^\theta (\lambda - t)^{-\theta - 2},$$

respectively. Then, substituting zero for t in the above two equations, we obtain the first two moments, that is,

$$E(X) = \theta / \lambda \tag{3.105}$$

and

$$E(X^2) = \theta(\theta + 1)/\lambda^2, \tag{3.106}$$

from which the variance is obtained as

$$\sigma^2 = E(X^2) - [E(X)]^2 = \theta / \lambda^2, \tag{3.107}$$

as shown in Table 3.3. ■

3.4.3 Characteristic Function

The MGF does not always exist, which means that not every random variable has a MGF. A function that always exists is the characteristic function, which is defined next.

Definition 3.5 (Characteristic function). *Let X be a univariate random variable with distribution function $F(x)$; its characteristic function, denoted by $\psi_X(t)$, is defined for the discrete case as*

$$\psi_X(t) = \sum_{x \in S(X)} e^{itx} P(x), \tag{3.108}$$

where $P(x)$ is the pmf of X and $S(X)$ is the support of $P(x)$. For the continuous case, it is defined as

$$\psi_X(t) = \int_{-\infty}^{\infty} e^{itx} f(x) dx, \tag{3.109}$$

where $f(x)$ is the pdf of X. Note that, in both cases, $\psi_X(t)$ is a complex function.

For simplicity of notation, we shall use $\psi(t)$ instead of $\psi_X(t)$, unless otherwise needed. Before we discuss the importance of the characteristic function, let us derive it for some random variables.

Example 3.17 (Characteristic function of a discrete uniform random variable). The characteristic function of the discrete uniform random variable (see Section 2.2.1), which has the pmf

$$P(x) = \frac{1}{n}, \quad x = 1, 2, \ldots, n,$$

is

$$\begin{aligned} \psi(t) &= \sum_x e^{itx} P(x) = \sum_x e^{itx} \frac{1}{n} \\ &= \frac{e^{it} + e^{2it} + \ldots + e^{nit}}{n} = \frac{e^{it}(1 - e^{nit})}{n(1 - e^{it})}. \end{aligned}$$

■

Example 3.18 (Characteristic function of a continuous uniform random variable). The characteristic function of the continuous uniform random variable, $U(0, \beta)$, with pdf

$$f(x) = \frac{1}{\beta}, \quad 0 \le x \le \beta,$$

is

$$\psi(t) = \int_0^{\beta} e^{itx} \frac{1}{\beta} dx = \frac{1}{\beta} \frac{e^{itx}}{it}\Big|_0^{\beta} = \frac{e^{it\beta} - 1}{it\beta},$$

which is a special case of the characteristic function of a continuous uniform family, which is shown in Table 3.5. ■

Example 3.19 (Characteristic function of a binomial random variable). The characteristic function of the binomial random variable $B(n, p)$ with pmf

$$P(x) = \binom{n}{x} p^x q^{n-x}, \quad x = 0, 1, \ldots, n,$$

is

$$\psi(t) = \sum_{x=0}^{n} e^{itx} \binom{n}{x} p^x q^{n-x} = \sum_{x=0}^{n} \binom{n}{x} (p e^{it})^x q^{n-x} = (p e^{it} + q)^n.$$

■

Table 3.5: Characteristic Functions of Some Common Random Variables.

Discrete Distribution		Continuous Distribution	
Name	$\psi(t)$	Name	$\psi(t)$
Uniform	$\dfrac{e^{it}(1-e^{nit})}{n(1-e^{it})}$	Uniform	$\dfrac{e^{it\beta}-e^{it\alpha}}{it(\beta-\alpha)}$
Dirac	e^{it}	Exponential	$\left(\dfrac{\lambda}{\lambda-it}\right)^{-1}$
Bernoulli	$pe^{it}+(1-p)$	Gamma	$\left(\dfrac{\lambda}{\lambda-it}\right)^{\theta}$
Binomial	$\left(pe^{it}+(1-p)\right)^n$	Log-gamma	$e^{it\mu}\Gamma(k+it\sigma)/\Gamma(k)$
Geometric	$\dfrac{pe^{it}}{1-(1-p)e^{it}}$	Normal	$\exp\left\{it\mu-\dfrac{\sigma^2t^2}{2}\right\}$
Negative binomial	$\left[\dfrac{pe^{it}}{1-(1-p)e^{it}}\right]^r$	χ^2_n	$(1-2it)^{-n/2}$
Poisson	$\exp\left\{\lambda(e^{it}-1)\right\}$	Logistic	$e^{it\alpha}\Gamma(1+i\beta t)\Gamma(1-i\beta t)$

Table 3.5 gives the characteristic functions of the most common distributions. The characteristic function makes the calculations of moments easy. It also helps sometimes in the identification of distributions of sums of independent random variables.

The most important properties (applications) of the characteristic function are:

1. The characteristic function always exists.

2. $\psi(0)=1$.

3. $-1 \le \psi(t) \le 1$.

4. If $Z=aX+b$, where X is a random variable and a and b are real constants, we have
$$\psi_Z(t)=e^{itb}\psi_X(at), \tag{3.110}$$
where $\psi_Z(t)$ and $\psi_X(t)$ are the characteristic functions of Z and X, respectively.

5. The characteristic function of the sum of two independent random variables is the product of their characteristic functions:
$$\psi_{X+Y}(t)=\psi_X(t)\psi_Y(t). \tag{3.111}$$

6. Suppose that $X_1, \ldots, X_n$ is a set of n independent random variables with characteristic functions $\psi_{X_1}(t), \ldots, \psi_{X_n}(t)$, respectively. Let $C = \sum_{i=1}^{n} a_i X_i$ be a linear combination of the random variables. Then, the characteristic function of C is given by

$$\psi_C(t) = \prod_{i=1}^{n} \psi_{X_i}(a_i t). \tag{3.112}$$

7. The characteristic function of the random variable R, which is the sum of a random number N of identically and independently distributed random variables $X_1, \ldots, X_N$ is given by

$$\psi_R(t) = \psi_N\left(\frac{\log \psi_X(t)}{i}\right), \tag{3.113}$$

where $\psi_X(t)$, $\psi_R(t)$ and $\psi_N(t)$ are the characteristic functions of X_i, R, and N, respectively.

Example 3.20 (Sum of normal random variables). Let $Z_1, \ldots, Z_n$ be independent standard normal random variables with characteristic function

$$\psi_Z(t) = e^{-t^2/2}.$$

Also, let $S = Z_1 + \ldots + Z_n$ and $W = S/\sqrt{n}$. Then, according to Property 5, the characteristic function of S is

$$\psi_S(t) = [\psi_Z(t)]^n,$$

and, according to Property 4, the characteristic function of W is

$$\begin{aligned} \psi_W(t) &= \psi_S\left(t/\sqrt{n}\right) = \left[\psi_Z\left(t/\sqrt{n}\right)\right]^n \\ &= \exp\left(-\frac{1}{2}\left(\frac{t}{\sqrt{n}}\right)^2 n\right) = e^{-t^2/2}, \end{aligned} \tag{3.114}$$

which shows that $W = (Z_1 + \ldots + Z_n)/\sqrt{n}$ has the same characteristic function as Z_i. Hence, W has a standard normal distribution. ■

Example 3.21 (Stability of the Gamma family with respect to scale changes). Let X be a Gamma $G(\theta, \lambda)$ random variable with characteristic function

$$\psi_X(t) = \left(\frac{\lambda}{\lambda - it}\right)^\theta$$

and $Y = cX$. Then, by Property 4 we have

$$\begin{aligned} \psi_Y(t) &= \psi_X(ct) = \psi_{G(\theta,\lambda)}(ct) = \left(\frac{\lambda}{\lambda - ict}\right)^\theta \\ &= \left(\frac{\lambda/c}{\lambda/c - it}\right)^\theta = \psi_{G(\theta,\lambda/c)}(t), \end{aligned}$$

which shows that the random variable Y is $G(\theta, \lambda/c)$, that is, the Gamma family is stable with respect to scale changes. ■

Example 3.22 (Stability of the uniform family). Let X be uniform $U(\alpha, \beta)$, with characteristic function

$$\psi_X(t) = \frac{e^{it\beta} - e^{it\alpha}}{it(\beta - \alpha)}.$$

Then, by Property 4 the characteristic function of the random variable $Y = cX + d$ is

$$\begin{aligned}\psi_Y(t) &= e^{itd}\,\psi_X(ct) = e^{itd}\,\frac{e^{itc\beta} - e^{itc\alpha}}{ict(\beta - \alpha)}\\ &= \frac{e^{it(c\beta+d)} - e^{it(c\alpha+d)}}{it(c\beta - c\alpha)} = \frac{e^{it(c\beta+d)} - e^{it(c\alpha+d)}}{it[(c\beta + d) - (c\alpha + d)]},\end{aligned}$$

which shows that $Y \sim U(c\alpha + d, c\beta + d)$. ■

As indicated, the kth order moment $E(X^k)$ can be easily calculated from the characteristic function as

$$E(X^k) = \frac{\psi^{(k)}(0)}{i^k}. \tag{3.115}$$

Example 3.23 (Moments of the Bernoulli random variable). Since the characteristic function of the Bernoulli random variable is

$$\psi(t) = pe^{it} + q,$$

using (3.115) we have

$$E(X^k) = \frac{\psi^{(k)}(0)}{i^k} = \frac{pi^k}{i^k} = p.$$

This shows that all moments are equal to p. ■

Example 3.24 (Moments of the Gamma random variable). Since the characteristic function of the Gamma $G(\theta, \lambda)$ random variable is

$$\psi(t) = \left(\frac{\lambda}{\lambda - it}\right)^{\theta},$$

its moments with respect to the origin are

$$E(X^k) = \frac{\psi^{(k)}(0)}{i^k} = \frac{\theta\,(\theta + 1)\cdots(\theta + k - 1)}{\lambda^k}.$$

■

Exercises

3.1 Show that:

(a) The mean and variance of the uniform $U(\alpha, \beta)$ random variable are

$$\mu = \frac{\alpha + \beta}{2} \quad \text{and} \quad \sigma^2 = \frac{(\beta - \alpha)^2}{12}. \tag{3.116}$$

(b) The mean and variance of an exponential random variable with parameter λ are

$$\mu = \frac{1}{\lambda} \quad \text{and} \quad \sigma^2 = \frac{1}{\lambda^2}. \tag{3.117}$$

3.2 The simple compression strength (measured in kg/cm^2) of a given concrete is a normal random variable:

(a) If the mean is 300 kg/cm^2 and the standard deviation is 40 kg/cm^2, determine the 15th percentile.

(b) If the mean is 200 kg/cm^2 and the standard deviation is 30 kg/cm^2, give the probability associated with a strength of at most 250 kg/cm^2.

(c) If the mean is 300 kg/cm^2, obtain the standard deviation if the 80th percentile is 400 kg/cm^2.

(d) If an engineer states that 400 kg/cm^2 is the 20th percentile in the previous case, is he right?

3.3 The occurrence of earthquakes of intensity above five in a given region is Poissonian with mean rate 0.5 earthquakes/year.

(a) Determine the pdf of the time between consecutive earthquakes.

(b) If an engineering work fails after five earthquakes of such an intensity, obtain the pdf of the lifetime of such a work in years.

(c) Obtain the pmf of the number of earthquakes (of intensity five or larger) that occur in that region during a period of 10 years.

3.4 The arrivals of cars to a gas station follow a Poisson law of mean rate five cars per hour. Determine:

(a) The probability of five arrivals between 17.00 and 17.30.

(b) The pdf of the time up to the first arrival.

(c) The pdf of the time until the arrival of the fifth car.

3.5 If the height, T, of an asphalt layer is normal with mean 6 cm and standard deviation 0.5 cm, determine:

(a) The pdf value $f_T(5)$.

(b) The probability $\Pr(T \leq 5)$.

(c) The conditional probability $\Pr(|T-6|<2.5)$.

(d) The conditional probability $\Pr(|T-6|<2.5|T\leq 5)$.

3.6 Show that, as $k\to\infty$, the log-gamma pdf in (3.32) tends to the $N(\mu,\sigma^2)$ density function.

3.7 Show that the cdf of Gamma in (3.26) has the following closed form for integer θ:

$$F(x)=1-\sum_{y=0}^{\theta-1}\frac{e^{-x}x^{y}}{y!}, \tag{3.118}$$

which shows that $F(x)$ is related to the Poisson probabilities.

3.8 Starting with the gamma pdf

$$f(y)=\frac{1}{\Gamma(k)}\,e^{-y}\,y^{k-1},\quad y>0,\ k>0,$$

show that the pdf in (3.29) is obtained by a logarithmic transformation.

3.9 A random variable X has the density in (3.29). Show that $Y=-X$ has the pdf

$$f(y)=\frac{1}{\Gamma(k)}\,e^{-ky}\,\exp\{-e^{-y}\},\quad -\infty<y<\infty,\ k>0,$$

which includes the Gumbel extreme value distribution in (3.65) as a special case. [Hint: The shape parameter $k=1$.]

3.10 A random variable X has the pdf in (3.30). Show that the MGF of X is

$$E\left(e^{tX}\right)=e^{t\mu}\,\Gamma(k+t\sigma)/\Gamma(k). \tag{3.119}$$

3.11 Show that a generalized Pareto distribution truncated from the left is also a generalized Pareto distribution.

3.12 The grades obtained by students in a statistics course follow a distribution with cdf

$$F(x)=\begin{cases}0, & \text{if } x<0,\\ \dfrac{x}{10}, & \text{if } 0\leq x<10,\\ 1, & \text{if } x\geq 10.\end{cases}$$

(a) Obtain the cdf of the students with grade below 5.

(b) If the students receive at least one point just for participating in the evaluation, obtain the new cdf for this case.

3.13 Obtain the hazard function of the exponential distribution. Discuss the result.

3.14 A cumulative distribution function $F(x)$ is said to be an increasing (IHR) or decreasing (DHR) hazard ratio, if its hazard function is nondecreasing or nonincreasing in x, respectively. Show that the following properties hold:

(a) If X_i, $i = 1, 2$, are IHR random variables with hazard functions given by $H_i(x)$, $i = 1, 2$, then the random variable $X = X_1 + X_2$ is also IHR with hazard function $H_X(x) \leq \min\{H_1(x), H_2(x)\}$.

(b) A mixture of DHR distributions is also DHR. This property is not necessarily true for IHR distributions.

(c) Parallel and series systems of identical IHR units are IHR. For the series systems, the units do not have to have identical distributions.

3.15 Let X be a random variable with survival function defined by (see Glen and Leemis (1997))

$$S(x) = \Pr(X > x) = \frac{\arctan[\alpha(\phi - x)] + (\pi/2)}{\arctan(\alpha\phi) + (\pi/2)}, \quad x \geq 0,$$

where $\alpha > 0$ and $-\infty < \phi < \infty$.

(a) Show that $S(x)$ is a genuine survival function, that is, it satisfies the conditions $S(0) = 1$, $\lim_{x\to\infty} S(x) = 0$ and $S(x)$ is nonincreasing.

(b) Show that the hazard function

$$H(x) = \frac{\alpha}{\{\arctan[\alpha(\phi - x)] + (\pi/2)\}\{1 + \alpha^2(x - \phi)^2\}}, \quad x \geq 0,$$

has an upside-down bathtub form.

3.16 Use MGFs in Table 3.4 to derive the mean and variance of the corresponding random variables in Table 3.3. [Hint: Find the first two derivatives of the MGF.]

3.17 Use CFs in Table 3.5 to derive the mean and variance of the corresponding random variables in Table 3.3. [Hint: Find the first two derivatives of the CF.]

3.18 Let X and Y be independent random variables.

(a) Show that the characteristic function of the random variable $Z = aX + bY$ is $\psi_Z(t) = \psi_X(at)\psi_Y(bt)$, where $\psi_Z(t)$ is the characteristic function of the random variable Z.

(b) Use this property to establish that a linear combination of normal random variables is normal.

3.19 Use the properties of the characteristic function to show that a linear combination of independent normal random variables is another normal random variable.

3.20 Let X_1 and X_2 be iid random variables with Gumbel cdf, which is given by

$$F(x_i) = \exp[-\exp(-x_i)], \quad -\infty < x_i < \infty, \quad i = 1, 2.$$

Show that the random variable $X = X_1 - X_2$ has a logistic distribution with cdf

$$F_X(x) = \frac{1}{1 + e^{-x}}, \quad -\infty < x < \infty.$$

[Hint: Use the characteristic function.]

Chapter 4

Multivariate Probabilistic Models

When we deal with a single random quantity, we have a *univariate* random variable. When we deal with two or more random quantities simultaneously, we have a *multivariate* random variable. In Chapters 2 and 3 we introduced both discrete and continuous univariate probabilistic models. In this chapter, we extend the discussions to the more general case of multivariate random variables.

Section 4.1 presents some ways to deal with multivariate discrete random variables. Some frequently encountered multivariate discrete random models are discussed briefly in Section 4.2. Section 4.3 presents some ways to deal with multivariate continuous random variables, while Section 4.4 discusses some commonly used multivariate continuous models.

4.1 Multivariate Discrete Random Variables

In Section 2.2, we have dealt with discrete random variables individually, that is, one random quantity at a time. In some practical situations, we may need to deal with several random quantities simultaneously. In this section, we describe some ways to deal with multivariate random variables. For a detailed discussion on various multivariate discrete models, see the book by Johnson, Kotz, and Balakrishnan (1997).

4.1.1 Joint Probability Mass Function

Let $X = \{X_1, X_2, \ldots, X_n\}$ be a multivariate discrete random variable of dimension n, taking on values $x_i \in S(X_i), i = 1, 2, \ldots, n$. The pmf of this multivariate random variable is denoted by $P(x_1, x_2, \ldots, x_n)$, which means $\Pr(X_1 = x_1, X_2 = x_2, \ldots, X_n = x_n)$. This is called the *joint* probability mass function. The joint pmf has n arguments, $x_1, x_2, \ldots, x_n$, one for each variable. When $n = 2$, we have a *bivariate* random variable.

Table 4.1: The Joint Probability Mass Function and the Marginal Probability Mass Functions of (X_1, X_2) in Example 4.1.

	X_2			
X_1	1	2	3	$P_1(x_1)$
0	0.1	0.3	0.2	0.6
1	0.2	0.1	0.1	0.4
$P_2(x_2)$	0.3	0.4	0.3	1.0

Example 4.1 (Bivariate pmf). Suppose that $n = 2$ and the supports of X_1 and X_2 are $S(X_1) = \{0, 1\}$ and $S(X_2) = \{1, 2, 3\}$, respectively. The joint pmf can be displayed in a table such as the one given in Table 4.1. It has two arguments, $x_1 = 0, 1$ and $x_2 = 1, 2, 3$. From Table 4.1 we see, for example, that $P(0, 1) = 0.1$ and $P(0, 3) = 0.2$. ■

4.1.2 Marginal Probability Mass Function

From the joint pmf, we can obtain *marginal* probability mass functions, one marginal for each variable. The marginal pmf of X_1, $P_1(x_1)$, is shown in the last column in Table 4.1, which is obtained by adding across the rows. Similarly, the marginal pmf of X_2, $P_2(x_2)$, is shown in the last row in Table 4.1. It is obtained by adding across the columns. More generally, the marginal of the jth variable, X_j, is obtained by summing the joint pmf over all possible values of all other variables. For example, the marginal pmf of X_1 is

$$P_1(x_1) = \sum_{x_2 \in S(X_2)} \cdots \sum_{x_n \in S(X_n)} P(x_1, x_2, \ldots, x_n), \quad x_1 \in S(X_1) \tag{4.1}$$

and the marginal pmf of (X_1, X_2) is

$$P_{12}(x_1, x_2) = \sum_{x_3 \in S(X_3)} \cdots \sum_{x_n \in S(X_n)} P(x_1, x_2, \ldots, x_n), \; x_1 \in S(X_1), \; x_2 \in S(X_2). \tag{4.2}$$

4.1.3 Conditional Probability Mass Function

In some situations, we wish to compute the pmf of some random variables given that some other variables are known to have certain values. For example, in Example 4.1, we may wish to find the pmf of X_2 given that $X_1 = 0$. This is known as the *conditional* pmf and is denoted by $P(x_2|x_1)$, which means $\Pr(X_2 = x_2|X_1 = x_1)$. The conditional pmf is the ratio of the joint pmf to the marginal pmf, that is,

$$P(x_2|x_1) = \frac{P(x_1, x_2)}{P(x_1)}, \quad x_2 \in S(X_2), \tag{4.3}$$

where $P(x_1, x_2)$ is the joint density of X_1 and X_2 and x_1 is assumed to be given.

Thus, for example, $P(1|1) = 0.2/0.4 = 0.5$, $P(2|1) = 0.1/0.4 = 0.25$, and $P(3|1) = 0.1/0.4 = 0.25$. Note that

$$\sum_{x_2=1}^{3} P(x_2|x_1) = 1 \quad \forall x_1 \in S(X_1)$$

because every conditional pmf is a pmf, that is, $P(x_2|x_1)$ must satisfy (2.1).

4.1.4 Covariance and Correlation

We have seen that from the joint pmf one can obtain the marginal pmf for each of the variables, $P_1(x_1), P_2(x_2), \ldots, P_n(x_n)$. From these marginals, one can compute the means, $\mu_1, \mu_2, \ldots, \mu_n$, and variances, $\sigma_1^2, \sigma_2^2, \ldots, \sigma_n^2$, using (2.4) and (2.7), respectively. In addition to the means and variances, one can also compute the *covariance* between every pair of variables. The covariance between X_i and X_j, denoted by σ_{ij}, is defined as

$$\begin{aligned} \sigma_{ij} &= E(X_i - \mu_i)(X_j - \mu_j) && (4.4) \\ &= \sum_{x_i \in S(X_i)} \sum_{x_j \in S(X_j)} (x_i - \mu_i)(x_j - \mu_j) P(x_i, x_j), && (4.5) \end{aligned}$$

where $P(x_i, x_j)$ is the joint pmf of X_i and X_j, which is obtained by summing the joint pmf over all possible values of all variables other than X_i and X_j. Note that

$$\sigma_{ii} = E(X_i - \mu_i)(X_i - \mu_i) = E(X_i - \mu_i)^2 = \sigma_i^2,$$

which shows that the covariance of a variable with itself is simply the variance of that variable.

Example 4.2 (Means, variances, and covariances). Consider the joint pmf in Table 4.1. The computations for the means, variances, and covariance are shown in Tables 4.2 and 4.3, from which we can see that

$$\mu_1 = 0.4, \quad \mu_2 = 2, \quad \sigma_1^2 = 0.240, \quad \sigma_2^2 = 0.6, \quad \text{and} \quad \sigma_{12} = -0.1. \qquad (4.6)$$

■

The covariance between two variables gives information about the *direction* of the relationship between the two variables. If it is positive, the two variables are said to be *positively correlated* and, if it is negative, they are said to be *negatively correlated*. Because σ_{12} in the above example is negative, X_1 and X_2 are negatively correlated.

A graphical interpretation of the covariance between two variables X and Y is as follows. Let us draw all points with positive probabilities in a Cartesian plane. A typical point (x, y) is shown in Figure 4.1. A vertical line at $x = \mu_X$ and a horizontal line at $y = \mu_Y$ divide the plane into four quadrants. Note that

Table 4.2: Computations of the Means and Variances.

Variable X_1					
x_1	$P_1(x_1)$	$x_1P_1(x_1)$	$x_1-\mu_1$	$(x_1-\mu_1)^2$	$(x_1-\mu_1)^2P_1(x_1)$
0	0.6	0.0	−0.4	0.16	0.096
1	0.4	0.4	0.6	0.36	0.144
Total	1.0	0.4			0.240
Variable X_2					
x_2	$P_2(x_2)$	$x_2P_2(x_2)$	$x_2-\mu_2$	$(x_2-\mu_2)^2$	$(x_2-\mu_2)^2P_2(x_2)$
1	0.3	0.3	−1	1	0.3
2	0.4	0.8	0	0	0.0
3	0.3	0.9	1	1	0.3
Total	1.0	2.0			0.6

Table 4.3: Computations of the Covariance Between X_1 and X_2.

x_1	x_2	$P(x_1,x_2)$	$x_1-\mu_1$	$x_2-\mu_2$	$(x_1-\mu_1)(x_2-\mu_2)P(x_1,x_2)$
0	1	0.1	−0.4	−1	0.04
0	2	0.3	−0.4	0	0.00
0	3	0.2	−0.4	1	−0.08
1	1	0.2	0.6	−1	−0.12
1	2	0.1	0.6	0	0.00
1	3	0.1	0.6	1	0.06
		1			−0.10

the absolute value of the product $(x-\mu_X)(y-\mu_Y)$ is equal to the area of the shaded rectangle shown in Figure 4.1. Note that this area is zero when $x=\mu_X$ or $y=\mu_Y$. The area gets larger as the point (x,y) gets farther away from the point (μ_X,μ_Y). Note also that the product $(x-\mu_X)(y-\mu_Y)$ is positive in the first and third quadrants and negative in the second and fourth quadrants. This is indicated by the + and − signs in Figure 4.1. The covariance is the weighted sum of these products with weights equal to $\Pr(X=x,Y=y)$. If the sum of the weighted positive terms (those in the first and third quadrants) is equal to the sum of the weighted negative terms (those in the second and fourth quadrants), then the covariance is zero (the negative terms annihilate the positive ones). On the other hand, if sum of the weighted positive terms exceeds that of the sum of the weighted negative terms, then the covariance is positive; otherwise, it is negative.

Although the covariance between two variables gives information about the direction of the relationship between the two variables, it does not tell us much

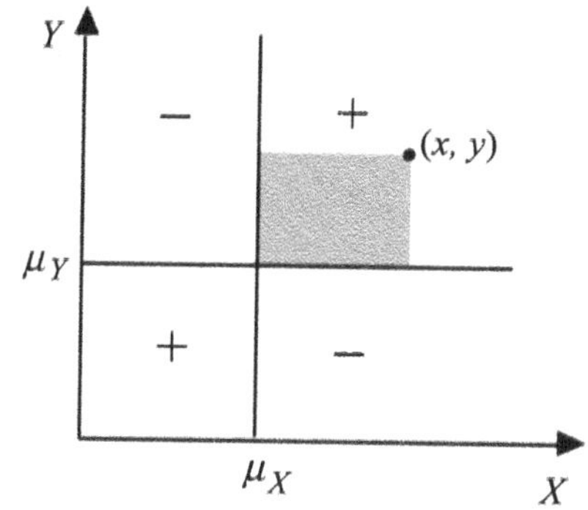

Figure 4.1: A graphical illustration of the covariance between X and Y.

about the strength of the relationship between the two variables because it is affected by the unit of measurements. That is, if we change the unit of measurement (e.g., from dollars to thousands of dollars), the covariance will change accordingly.

A measure of association that is not affected by changes in unit of measurement is the *correlation coefficient.* The correlation coefficient between two variables X_i and X_j, denoted by ρ_{ij}, is defined as

$$\rho_{ij} = \frac{\sigma_{ij}}{\sigma_i \sigma_j}, \tag{4.7}$$

that is, it is the covariance divided by the product of the two standard deviations.

It can be shown that $-1 \leq \rho_{ij} \leq 1$. The correlation ρ_{ij} measures linear association between the two variables. That is, if $\rho_{ij} = \pm 1$, then one variable is a linear function of the other. If $\rho_{ij} = 0$, it means only that the two variables are not linearly related (they may be nonlinearly related, however). In the above example, $\rho_{12} = -0.1/(\sqrt{0.240 \times 0.6}) = -0.264$, hence X_1 and X_2 are negatively, but mildly correlated.

All considerations made for the graphical interpretation of the covariance are also valid for the correlation coefficient because of its definition. Figure 4.2 is an illustration showing the correspondence between the scatter diagram and the values of σ_X, σ_Y, and ρ_{XY}.

When we deal with a multivariate random variable $X = \{X_1, X_2, \ldots, X_k\}$, it is convenient to summarize their means, variances, covariances, and correlations as follows. The means are displayed in a $k \times 1$ vector, and the variances-covariances and correlation coefficients are displayed in $k \times k$ matrices as follows:

$$\boldsymbol{\mu} = \begin{bmatrix} \mu_1 \\ \mu_2 \\ \vdots \\ \mu_k \end{bmatrix}, \quad \boldsymbol{\Sigma} = \begin{bmatrix} \sigma_1^2 & \sigma_{12} & \cdots & \sigma_{1k} \\ \sigma_{21} & \sigma_2^2 & \cdots & \sigma_{2k} \\ \vdots & \vdots & \ddots & \vdots \\ \sigma_{k1} & \sigma_{k2} & \cdots & \sigma_k^2 \end{bmatrix}, \quad \boldsymbol{\rho} = \begin{bmatrix} 1 & \rho_{12} & \cdots & \rho_{1k} \\ \rho_{21} & 1 & \cdots & \rho_{2k} \\ \vdots & \vdots & \ddots & \vdots \\ \rho_{k1} & \rho_{k2} & \cdots & 1 \end{bmatrix}. \tag{4.8}$$

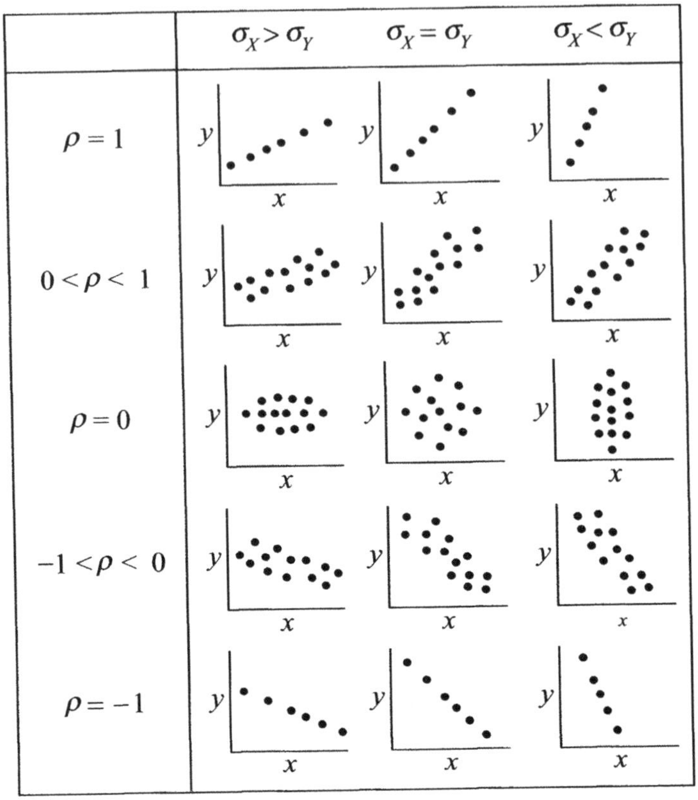

Figure 4.2: Graphical illustrations of the correlation coefficient.

The vector $\boldsymbol{\mu}$ is called the *mean vector*, the matrix $\boldsymbol{\Sigma}$ is called the *variance-covariance* matrix or just the *covariance matrix*, and $\boldsymbol{\rho}$ is called the *correlation* matrix. The covariance matrix contains the variances on the diagonal and the covariances on the off-diagonal. Note that the covariance and correlation matrices are both symmetric, that is, $\sigma_{ij} = \sigma_{ji}$ and $\rho_{ij} = \rho_{ji}$. Note also that the diagonal elements of $\boldsymbol{\rho}$ are all ones because X_i has a perfect correlation with itself.

For example, the means, variances, covariances, and correlation coefficients in Example 4.2, can be summarized as

$$\boldsymbol{\mu} = \begin{bmatrix} 0.4 \\ 2.0 \end{bmatrix}, \quad \boldsymbol{\Sigma} = \begin{bmatrix} 0.240 & -0.1 \\ -0.1 & 0.6 \end{bmatrix}, \text{and} \quad \boldsymbol{\rho} = \begin{bmatrix} 1 & -0.264 \\ -0.264 & 1 \end{bmatrix}.$$

4.2 Common Multivariate Discrete Models

In this section, we describe with some multivariate discrete models of interest.

4.2.1 Multinomial Distribution

We have seen in Section 2.2.3 that the binomial random variable results from random experiments that each has two possible outcomes. If a random experiment has more than two outcomes, the resultant random variable is called a *multinomial* random variable. Suppose that we perform an experiment with k possible outcomes $r_1, \ldots, r_k$, with probabilities $p_1, \ldots, p_k$, respectively. Since the outcomes are mutually exclusive and collectively exhaustive, these probabilities must satisfy $\sum_{i=1}^{k} p_i = 1$. If we repeat this experiment n times and X_i denotes the numbers of times we obtain outcomes r_i, for $i = 1, \ldots, k$, then $X = \{X_1, \ldots, X_k\}$ is a multinomial random variable, which is denoted by $M(n; p_1, \ldots, p_k)$. The pmf of $M(n; p_1, \ldots, p_k)$ is

$$P(x_1, x_2, \ldots, x_k) = \frac{n!}{x_1! x_2! \ldots x_k!} p_1^{x_1}\, p_2^{x_2} \ldots p_k^{x_k}, \tag{4.9}$$

where $n!/(x_1!x_2! \ldots x_k!)$ is the number of possible combinations that lead to the desired outcome. Note that $P(x_1, x_2, \ldots, x_k)$ means $\Pr(X_1 = x_1, X_2 = x_2, \ldots, X_k = x_k)$. Note also that the support of the multinomial random variable X, whose pmf is given by (4.9), is

$$S(X) = \{x_1, x_2, \ldots, x_k : x_i = 0, 1, \ldots, n, \sum_{i=1}^{k} x_i = n\}. \tag{4.10}$$

Example 4.3 (Different failure types). Suppose that we wish to determine the strengths of six plates of fiberglass. Suppose also that there are three possible types of failures. The probabilities of these types of failures are $0.2, 0.3$, and 0.5, respectively. Let X_i be the number of plates with failure of type i. Then, $X = \{X_1, X_2, X_3\}$ is $M(6; 0.2, 0.3, 0.5)$. Thus, for example, the probability of having $2, 1$, and 3 failures of the three types is given by

$$P(2, 1, 3) = \frac{6!}{2!1!3!}\, 0.2^2\, 0.3^1\, 0.5^3 = 0.09.$$

■

Example 4.4 (Traffic in a square). A car when arriving at a square can choose among four different streets S_1, S_2, S_3, and S_4 with probabilities p_1, p_2, p_3, and p_4, respectively. Find the probability that of 10 cars arriving at the square, 3 will take street S_1, 4 will take street S_2, 2 will take street S_3, and 1 will take street S_4.

Since we have four possible outcomes per car, the random variable of the number of cars taking each of the four streets is $M(10; p_1, p_2, p_3, p_4)$. Then, the required probability is

$$P(3, 4, 2, 1) = \frac{10!}{3!4!2!1!}\, p_1^3\, p_2^4\, p_3^2\, p_4^1.$$

■

In a multinomial random variable, the mean and variance of X_i, and the covariance of X_i and X_j are

$$\mu_i = np_i, \quad \sigma_{ii}^2 = np_i(1-p_i), \quad \text{and} \quad \sigma_{ij} = -np_ip_j,$$

respectively. Thus, all pairs of variables are negatively correlated. This is expected because they are nonnegative integers that sum to n; hence, if one has a large value, the others must necessarily have small values.

The multinomial family of random variables is reproductive with respect to parameter n, that is, if $X_1 \sim M(n_1; p_1, \ldots, p_k)$ and $X_2 \sim M(n_2; p_1, \ldots, p_k)$, then

$$X_1 + X_2 \sim M(n_1 + n_2; p_1, \ldots, p_k).$$

4.2.2 Multivariate Hypergeometric Distribution

We have seen in Section 2.2.6 that the hypergeometric distribution arises from sampling without replacement from a finite population with two groups (defective and nondefective). If the finite population consists of k groups, the resulting distribution is a *multivariate hypergeometric* distribution. Suppose the population consists of N products of which $D_1, D_2, \ldots, D_k$ are of the k types, with $\sum_{i=1}^k D_i = N$. Suppose we wish to draw a random sample of size $n < N$ from this population without replacement. The random variable $X = \{X_1, \ldots, X_k\}$, where X_i is the number of products of type i in the sample, is the multivariate hypergeometric variable and is denoted by $MHG(N; p_1, \ldots, p_k, n)$, where $p_i = D_i/N$ is the proportion of products of type i in the population. The pmf of $MHG(N; p_1, \ldots, p_k, n)$ is

$$P(x_1, \ldots, x_k) = \frac{\binom{D_1}{x_1} \cdots \binom{D_k}{x_k}}{\binom{N}{n}}, \quad \sum_{i=1}^k x_i = n, \quad \sum_{i=1}^k D_i = N. \tag{4.11}$$

Example 4.5 (Urn problem). Suppose an urn contains 20 balls of which 5 are white, 10 are black, and 5 are red. We draw a sample of size 10 without replacement. What is the probability that the drawn sample contains exactly 2 white, 5 black and 3 red balls? Here, $N = 20$, $p_1 = 5/20 = 1/4$, $p_2 = 10/20 = 1/2$, $p_3 = 5/20 = 1/4$, and $n = 10$. Letting $X = \{X_1, X_2, X_3\}$ be the number of white, black and red balls, then X is $MHG(20; 1/4, 1/2, 1/4, 10)$. From (4.11), we then have

$$P(2,5,3) = \frac{\binom{5}{2}\binom{10}{5}\binom{5}{3}}{\binom{20}{10}} = \frac{6300}{46189} = 0.136.$$

■

4.3 Multivariate Continuous Random Variables

In Section 3.2, we dealt with continuous random variables, one at a time. In some practical situations, we may need to deal with several random variables simultaneously. In this section, we describe some ways to deal with multivariate continuous random variables.

4.3.1 Joint Probability Density Function

Let $X = \{X_1, \ldots, X_n\}$ be multivariate continuous random variable of dimension n, with support $x_i \in S(X_i), i = 1, 2, \ldots, n$. The pdf of this multivariate continuous random variable is given by $f(x_1, \ldots, x_n)$. This is called the *joint* pdf. The joint pdf has n arguments, $x_1, \ldots, x_n$, one for each of the variables.

The pdf satisfies the following properties:

$$f(x_1, \ldots, x_n) \geq 0,$$

and

$$\int_{-\infty}^{\infty} \cdots \int_{-\infty}^{\infty} f(x_1, \ldots, x_n) dx_1 \ldots dx_n = 1.$$

Note that for any multivariate continuous random variable, we have $\Pr(X_1 = x_1, \ldots, X_n = x_n) = 0$.

4.3.2 Joint Cumulative Distribution Function

The joint cdf is defined in an analogous manner to that used for the univariate case:

$$F(\mathbf{x}) = \Pr(X_1 \leq x_1, \ldots, X_n \leq x_n) = \int_{-\infty}^{x_n} \cdots \int_{-\infty}^{x_1} f(t_1, \ldots, t_n) dt_1 \ldots dt_n.$$

When $n = 2$, we have a *bivariate* continuous random variable.

Example 4.6 (Bivariate cumulative distribution function). The cdf of a bivariate random variable (X_1, X_2) is

$$F(x_1, x_2) = \int_{-\infty}^{x_2} \int_{-\infty}^{x_1} f(t_1, t_2) dt_1 dt_2.$$

The relationship between the pdf and cdf is

$$f(x_1, x_2) = \frac{\partial^2 F(x_1, x_2)}{\partial x_1 \, \partial x_2}.$$

Among other properties of a bivariate cdf, we mention the following:

1. $F(\infty, \infty) = 1$.

2. $F(-\infty, x_2) = F(x_1, -\infty) = 0$.

3. $F(x_1 + a_1, x_2 + a_2) \geq F(x_1, x_2)$, where $a_1, a_2 \geq 0$.

4. $P(a_1 < X_1 \leq b_1, a_2 < X_2 \leq b_2) = F(b_1, b_2) - F(a_1, b_2) - F(b_1, a_2) + F(a_1, a_2)$. This formula is illustrated in Figure 4.3.

∎

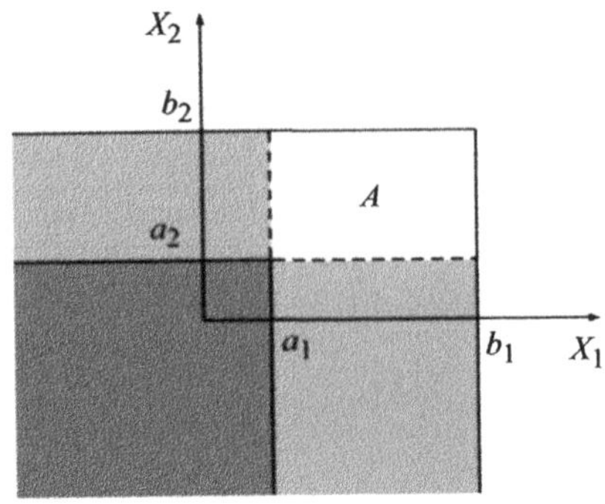

Figure 4.3: An illustration of computing the probability of a rectangle in terms of the joint cdf associated with its vertices.

4.3.3 Marginal Probability Density Functions

From the joint pdf we can obtain *marginal* probability density functions, one marginal for each variable. The marginal pdf of one variable is obtained by integrating the joint pdf over all other variables. Thus, for example, the marginal pdf of X_1, is

$$f(x_1) = \int_{-\infty}^{\infty} \cdots \int_{-\infty}^{\infty} f(x_1, \ldots, x_n) dx_2 \ldots dx_n.$$

A multivariate random variable is said to be continuous if its marginals are all continuous.

The probabilities of any set B of possible values of a multivariate continuous variable can be calculated if the pdf, written as $f(\mathbf{x})$ or $f(x_1, \ldots, x_n)$, is known just by integrating it in the set B. For example, the probability that X_i belongs to a given region, say, $a_i < X_i \le b_i$ for all i, is the integral

$$\Pr(a_1 < X_1 \le b_1, \ldots, a_n < X_n \le b_n) = \int_{a_n}^{b_n} \cdots \int_{a_1}^{b_1} f(x_1, \ldots, x_n) dx_1 \ldots dx_n.$$

4.3.4 Conditional Probability Density Functions

We define the conditional pdf for the case of bivariate random variables. The extension to the multivariate case is straightforward. For simplicity of notation we use (X, Y) instead of (X_1, X_2). Let then (X, Y) be a bivariate random variable. The random variable Y given $X = x$ is denoted by $(Y|X = x)$. The corresponding probability density and distribution functions are called the *conditional* pdf and cdf, respectively. The following expressions give the conditional pdf for the random variables $(Y|X = x)$ and $(X|Y = y)$:

$$f_{(Y|X=x)}(y|x) = \frac{f_{(X,Y)}(x, y)}{f_X(x)}, \quad f_X(x) > 0,$$

and

$$f_{(X|Y=y)}(x|y) = \frac{f_{(X,Y)}(x,y)}{f_Y(y)}, \quad f_Y(y) > 0.$$

It may also be of interest to compute the pdf conditioned on events different from $Y = y$. For example, for the event $Y \le y$, we get the conditional cdf:

$$F_{(X|Y\le y)}(x,y) = \Pr(X \le x|Y \le y) = \frac{\Pr(X \le x, Y \le y)}{\Pr(Y \le y)} = \frac{F_{(X,Y)}(x,y)}{F_Y(y)}.$$

The corresponding pdf is given by

$$f_{(X|Y\le y)}(x,y) = \frac{\partial}{\partial x} F_{(X|Y\le y)}(x,y) = \frac{\partial}{\partial x} \frac{F_{(X,Y)}(x,y)}{F_Y(y)}.$$

Two random variables X and X are said to be *independent* if

$$f_{(Y|X=x)}(y|x) = f_Y(y), \tag{4.12}$$

or

$$f_{(X|Y=y)}(x) = f_X(x), \tag{4.13}$$

otherwise, they are said to be *dependent*. This means that X and Y are independent if the conditional pdf is equal to the marginal pdf. Note that (4.12) and (4.13) are equivalent to

$$f_{(X,Y)}(x,y) = f_X(x) f_Y(y), \tag{4.14}$$

that is, if two variables are independent, then their joint pdf is equal to the product of their marginals. This is also true for $n > 2$ random variables. That is, if $X_1, \ldots, X_n$ are independent random variables, then

$$f(x_1, \ldots, x_n) = f_{X_1}(x_1) \ldots f_{X_n}(x_n). \tag{4.15}$$

4.3.5 Covariance and Correlation

Using the marginal pdfs, $f_{X_1}(x_1), \ldots, f_{X_n}(x_n)$, we can compute the means, $\mu_1, \ldots, \mu_n$, and variances, $\sigma_1^2, \ldots, \sigma_n^2$ using (3.9) and (3.11), respectively. We can also compute the *covariance* between every pair of variables. The covariance between X_i and X_j, denoted by σ_{ij}, is defined as

$$\sigma_{ij} = \int_{x_i \in S(X_i)} \int_{x_j \in S(X_j)} (x_i - \mu_i)(x_j - \mu_j) f(x_i, x_j) dx_j dx_i, \tag{4.16}$$

where $f(x_i, x_j)$ is the joint pdf of X_i and X_j, which is obtained by integrating the joint pdf over all variables other than X_i and X_j. As in the discrete case, the *correlation coefficient* is

$$\rho_{ij} = \frac{\sigma_{ij}}{\sigma_i \sigma_j}. \tag{4.17}$$

For convenience, we usually arrange the means, variances, covariances, and correlations in matrices. The means are arranged in a column vector, $\boldsymbol{\mu}$, and the variances and covariances are arranged in a matrix, $\boldsymbol{\Sigma}$, as follows:

$$\boldsymbol{\mu} = \begin{pmatrix} \mu_1 \\ \mu_2 \\ \vdots \\ \mu_n \end{pmatrix} \quad \text{and} \quad \boldsymbol{\Sigma} = \begin{pmatrix} \sigma_{11} & \sigma_{12} & \cdots & \sigma_{1n} \\ \sigma_{21} & \sigma_{22} & \cdots & \sigma_{2n} \\ \vdots & \vdots & \ddots & \vdots \\ \sigma_{n1} & \sigma_{n2} & \cdots & \sigma_{nn} \end{pmatrix}, \tag{4.18}$$

where we use σ_{ii} instead of σ_i^2, for convenience. The vector $\boldsymbol{\mu}$ is known as the mean vector and the matrix $\boldsymbol{\Sigma}$ is known as the variance-covariance matrix. Similarly, the correlation coefficients can be arranged in a matrix

$$\boldsymbol{\rho} = \begin{pmatrix} 1 & \rho_{12} & \cdots & \rho_{1n} \\ \rho_{21} & 1 & \cdots & \rho_{2n} \\ \vdots & \vdots & \ddots & \vdots \\ \rho_{n1} & \rho_{n2} & \cdots & 1 \end{pmatrix}, \tag{4.19}$$

which is known as the *correlation* matrix. Note that both $\boldsymbol{\Sigma}$ and $\boldsymbol{\rho}$ are symmetric matrices. The relationship between them is

$$\boldsymbol{\rho} = \mathbf{D}\boldsymbol{\Sigma}\,\mathbf{D}, \tag{4.20}$$

where $\mathbf{D}$ is a diagonal matrix whose ith diagonal element is $1/\sqrt{\sigma_{ii}}$.

4.3.6 The Autocorrelation Function

In this section, we introduce the concept of autocorrelation function, that will be used in dependent models to be described later in Chapter 9.

Definition 4.1 (Autocorrelation function). *Let $X_1, \ldots$ be a sequence of random variables with the same mean and variance, and given by $E(X_i) = \mu$ and $Var(X_i) = \sigma_X^2$. The covariance between the random variables X_i and X_{i+k} separated by k intervals (of time), which under the stationarity assumption must be the same for all i, is called the autocovariance at lag k and is defined by*

$$\gamma_k = cov(X_i, X_{i+k}) = E[(X_i - \mu)(X_{i+k} - \mu)]. \tag{4.21}$$

The autocorrelation function at lag k is

$$\rho_k = \frac{E[(X_i - \mu)(X_{i+k} - \mu)]}{\sqrt{E[(X_i - \mu)^2]E[(X_{i+k} - \mu)^2]}} = \frac{\gamma_k}{\sigma_X^2}. \tag{4.22}$$

4.3.7 Bivariate Survival and Hazard Functions

Let (X, Y) be a bivariate random variable, where X and Y are nonnegative lifetime random variables, and let $F(x, y)$ be an absolutely continuous bivariate distribution function with density function $f(x, y)$.

Definition 4.2 (Bivariate survival function). *The bivariate survival function is given by*

$$S(x, y) = \Pr(X > x, Y > y). \tag{4.23}$$

Thus, the bivariate survival function gives the probability that the object X will survive beyond time x and the object Y will survive beyond time y.

Definition 4.3 (Bivariate hazard function). *The bivariate hazard function or bivariate failure rate is given by*

$$H(x, y) = \frac{f(x, y)}{\Pr(X > x, Y > y)}. \tag{4.24}$$

The above definition is due to Basu (1971). From (4.23) and (4.24), we see that

$$H(x, y) = \frac{f(x, y)}{S(x, y)}. \tag{4.25}$$

Note that if X and Y are independent random variables, then we have

$$H(x, y) = H_X(x) H_Y(y),$$

where $H_X(x)$ and $H_Y(y)$ are the corresponding univariate hazard functions. Similar to the univariate case, $H(x, y)$ can be interpreted as the probability of failure of both items in the intervals of time $[x, x + \varepsilon_1)$ and $[y, y + \varepsilon_2)$, on the condition that they did not fail before time x and time y, respectively:

$$\begin{aligned} H(x, y) &= \lim_{x \to \varepsilon_1, y \to \varepsilon_2} \frac{\Pr(x < X \le x + \varepsilon_1, y < Y \le y + \varepsilon_2 | X > x, Y > y)}{\varepsilon_1 \varepsilon_2} \\ &= \frac{f(x, y)}{\Pr(X > x, Y > y)}. \end{aligned}$$

Unlike in the univariate case, the bivariate hazard function does not define $F(x, y)$, and so some other types of hazard functions may be taken into consideration.

Example 4.7 (Bivariate survival function). Consider a bivariate random variable with bivariate survival function

$$S(x, y) = \Pr(X > x, Y > y) = (1 + x + y)^{-\alpha}, \quad x, y > 0, \ \alpha > 0.$$

The joint pdf is

$$f(x, y) = \frac{\partial^2 S(x, y)}{\partial x \partial y} = \alpha(\alpha + 1)(1 + x + y)^{-(\alpha+2)}, \quad x, y > 0, \ \alpha > 0,$$

and the bivariate hazard function is

$$H(x, y) = \frac{f(x, y)}{\Pr(X > x, Y > y)} = \alpha(\alpha + 1)(1 + x + y)^{-2}, \quad x, y > 0, \ \alpha > 0.$$

■

4.3.8 Bivariate CDF and Survival Function

Let (X, Y) be a bivariate random variable with joint cdf

$$F(x,y) = \Pr(X \le x, Y \le y),$$

and joint survival function

$$S(x,y) = \Pr(X > x, Y > y).$$

The relationship between $S(x,y)$ and $F(x,y)$ is given by (see Fig. 4.3)

$$S(x,y) = 1 + F(x,y) - F_X(x) - F_Y(y), \tag{4.26}$$

where $F_X(x)$ and $F_Y(y)$ are the cdf of the marginals.

4.3.9 Joint Characteristic Function

The characteristic function can be generalized to n dimensions as follows.

Definition 4.4 (Multivariate characteristic function). *Let $X = (X_1, \ldots, X_n)$ be a multivariate random variable of dimension n. Its joint characteristic function is defined as*

$$\psi_{\mathbf{X}}(\mathbf{t}) = \int_{-\infty}^{\infty}\int_{-\infty}^{\infty}\cdots\int_{-\infty}^{\infty} e^{i(t_1 x_1 + \ldots + t_n x_n)} dF_{\mathbf{X}}(x_1, \ldots, x_n),$$

where $F_{\mathbf{X}}(x_1, \ldots, x_n)$ is the cdf of $\mathbf{X}$ and $\mathbf{t} = (t_1, \ldots, t_n)$.

4.4 Common Multivariate Continuous Models

4.4.1 Bivariate Logistic Distribution

The joint cdf of the bivariate logistic distribution is

$$F(x,y) = \frac{1}{1 + e^{-(x-\lambda)/\sigma} + e^{-(y-\delta)/\tau}}, \quad -\infty < x, y < \infty, \tag{4.27}$$

where $-\infty < \lambda, \delta < \infty$ and $\sigma, \tau > 0$, was introduced by Gumbel (1961). The corresponding joint density function is

$$f(x,y) = \frac{2e^{-(x-\lambda)/\sigma-(y-\delta)/\tau}}{\sigma\tau\left[1 + e^{-(x-\lambda)/\sigma} + e^{-(y-\delta)/\tau}\right]^3}, \quad -\infty < x, y < \infty. \tag{4.28}$$

From Equation (4.27), by letting x or y go to ∞, we obtain the marginal cumulative distribution functions of X and Y as

$$F_X(x) = \frac{1}{1 + e^{-(x-\lambda)/\sigma}} \quad \text{and} \quad F_Y(y) = \frac{1}{1 + e^{-(y-\delta)/\tau}}, \tag{4.29}$$

which are univariate logistic distributions. The conditional density function of $X|Y$ is

$$f_{X|Y}(x|y) = \frac{2e^{-(x-\lambda)/\sigma}(1+e^{-(y-\delta)/\tau})^2}{\sigma\left[1+e^{-(x-\lambda)/\sigma}+e^{-(y-\delta)/\tau}\right]^3}, \tag{4.30}$$

and the conditional mean of X given $Y = y$ is

$$E(X|Y=y) = \lambda + \sigma - \sigma\log(1+e^{-(y-\delta)/\tau}). \tag{4.31}$$

4.4.2 Multinormal Distribution

dimension n, which is denoted by $N(\boldsymbol{\mu}, \boldsymbol{\Sigma})$, where $\boldsymbol{\mu}$ and $\boldsymbol{\Sigma}$ are the mean vector and covariance matrix, respectively. The pdf of $\mathbf{X}$ is given by

$$f(\mathbf{x}) = \frac{1}{(2\pi)^{n/2}\sqrt{\det(\boldsymbol{\Sigma})}}\exp\left[-(\mathbf{x}-\boldsymbol{\mu})^T\boldsymbol{\Sigma}^{-1}(\mathbf{x}-\boldsymbol{\mu})/2\right],$$

where $\mathbf{x} = (x_1, \ldots, x_n)$ and $\det(\boldsymbol{\Sigma})$ is the determinant of $\boldsymbol{\Sigma}$. The following theorem gives the conditional mean and variance-covariance matrix of any conditional variable, which is normal.

Theorem 4.1 (Conditional mean and covariance matrix). *Let Y and Z be two sets of random variables jointly having a multivariate normal distribution with mean vector and covariance matrix given in partitioned forms by*

$$\boldsymbol{\mu} = \begin{pmatrix} \boldsymbol{\mu}_Y \\ \boldsymbol{\mu}_Z \end{pmatrix} \quad \text{and} \quad \boldsymbol{\Sigma} = \begin{pmatrix} \boldsymbol{\Sigma}_{YY} & \boldsymbol{\Sigma}_{YZ} \\ \boldsymbol{\Sigma}_{ZY} & \boldsymbol{\Sigma}_{ZZ} \end{pmatrix},$$

where $\boldsymbol{\mu}_Y$ and $\boldsymbol{\Sigma}_{YY}$ are the mean vector and covariance matrix of Y, $\boldsymbol{\mu}_Z$ and $\boldsymbol{\Sigma}_{ZZ}$ are the mean vector and covariance matrix of Z, and $\boldsymbol{\Sigma}_{YZ}$ is the covariance of Y and Z. Then the conditional pdf of Y given $Z = z$ is multivariate normal with mean vector $\boldsymbol{\mu}_{Y|Z=z}$ and covariance matrix $\boldsymbol{\Sigma}_{Y|Z=z}$, where

$$\begin{aligned} \boldsymbol{\mu}_{Y|Z=z} &= \boldsymbol{\mu}_Y + \boldsymbol{\Sigma}_{YZ}\,\boldsymbol{\Sigma}_{ZZ}^{-1}\,(z-\boldsymbol{\mu}_Z), \\ \boldsymbol{\Sigma}_{Y|Z=z} &= \boldsymbol{\Sigma}_{YY} - \boldsymbol{\Sigma}_{YZ}\,\boldsymbol{\Sigma}_{ZZ}^{-1}\,\boldsymbol{\Sigma}_{ZY}. \end{aligned} \tag{4.32}$$

For other properties of multivariate normal random variables, one may refer to any multivariate analysis book such as Rencher (2002) or the multivariate distribution theory book by Kotz, Balakrishnan, and Johnson (2000).

4.4.3 Marshall-Olkin Distribution

Due to its importance, we include here the Marshall-Olkin distribution (see Marshall and Olkin (1967)), which has several interesting physical interpretations. One such interpretation is as follows. Suppose we have a system with two components in series. Both components are subject to a Poissonian processes of fatal shocks, such that if a component is affected by a shock it fails.

Component 1 is subject to a Poissonian process with intensity λ_1, Component 2 is subject to a Poissonian process with intensity λ_2, and both components are subject to a Poissonian process with intensity λ_{12}. Let $N_1(x;\lambda_1)$, $N_2(y;\lambda_2)$, and $N_{12}(\max(x,y);\lambda_{12})$ be the number of shocks associated with first, second, and third Poissonian processes during a period of duration x, y, and $\max(x,y)$, respectively. Then, $N_1(x;\lambda_1)$, $N_2(y;\lambda_2)$, and $N_{12}(\max(x,y);\lambda_{12})$ are Poisson random variables with means $x\lambda_1$, $y\lambda_2$, and $\max(x,y)\lambda_{12}$, respectively. Thus, it follows from the pmf of the Poisson random variable in (2.28) that the bivariate survival function in (4.23) in this case becomes

$$
\begin{aligned}
S(x,y) &= \Pr(X > x, Y > y) \\
&= \Pr\left[N_1(x;\lambda_1) = 0, N_2(y;\lambda_2) = 0, N_{12}(\max(x,y);\lambda_{12}) = 0\right] \\
&= \exp\left[-\lambda_1 x - \lambda_2 y - \lambda_{12}\max(x,y)\right]. \qquad (4.33)
\end{aligned}
$$

This model has another interpretation using nonfatal shocks. Consider the same model as before, but now the shocks are not fatal. Once a shock coming from the Poisson process with intensity λ_1 has occurred, there is a probability p_1 of failure of Component 1; once a shock coming from the Poisson process with intensity λ_2 has occurred, there is a probability p_2 of failure of Component 2; and finally, once a shock coming from the Poisson process with intensity λ_{12} has occurred, there are probabilities p_{00}, p_{01}, p_{10}, and p_{11} of failure of both components, only Component 1, only Component 2, and no failure, respectively. In this case, we have

$$
\begin{aligned}
S(x,y) &= \Pr(X > x, Y > y) \\
&= \exp\left[-\delta_1 x - \delta_2 y - \delta_{12}\max(x,y)\right], \qquad (4.34)
\end{aligned}
$$

where

$$
\delta_1 = \lambda_1 p_1 + \lambda_{12} p_{01}, \quad \delta_2 = \lambda_2 p_2 + \lambda_{12} p_{10}, \quad \text{and} \quad \delta_{12} = \lambda_{12} p_{00}. \qquad (4.35)
$$

The following is a straightforward generalization of this model to n dimensions:

$$
\begin{aligned}
S(x_1,\ldots,x_n) = \exp\Bigg[&-\sum_{i=1}^{n}\lambda_i x_i - \sum_{i<j}\lambda_{ij}\max(x_i,x_j) \\
&- \sum_{i<j<k}\lambda_{ijk}\max(x_i,x_j,x_k) - \ldots - \lambda_{12\ldots n}\max(x_1,\ldots,x_n)\Bigg].
\end{aligned}
$$

4.4.4 Freund's Bivariate Exponential Distribution

Freund (1961) constructed an alternate bivariate exponential model in the following manner. Suppose a system has two components (C_1 and C_2) with their lifetimes X_1 and X_2 having exponential densities

$$
f_{X_i}(x) = \theta_i \exp\{-\theta_i x\}, \quad x \geq 0, \; \theta_i > 0 \; (i = 1, 2). \qquad (4.36)
$$

The dependence between X_1 and X_2 is introduced by the assumption that when Component C_i (with lifetime X_i) fails, the parameter for X_{3-i} changes from θ_{3-i}

to θ'_{3-i} (for $i = 1, 2$). In this set-up, the joint density function of X_1 and X_2 is

$$f_{X_1,X_2}(x_1,x_2) = \begin{cases} \theta_1\ \theta'_2\ \exp\left(-\theta'_2\ x_2 - \gamma_2\ x_1\right), & 0 \le x_1 < x_2, \\ \theta'_1\ \theta_2\ \exp\left(-\theta'_1\ x_1 - \gamma_1\ x_2\right), & 0 \le x_2 < x_1, \end{cases} \tag{4.37}$$

where $\gamma_i = \theta_1 + \theta_2 - \theta'_i$ ($i = 1, 2$). The corresponding joint survival function is

$$S_{X_1,X_2}(x_1,x_2) = \begin{cases} \frac{1}{\gamma_2}\left(\theta_1\ e^{-\theta'_2 x_2 - \gamma_2 x_1} + (\theta_2 - \theta'_2)\ e^{-(\theta_1+\theta_2)x_2}\right), & 0 \le x_1 < x_2, \\ \frac{1}{\gamma_1}\left(\theta_2\ e^{-\theta'_1 x_1 - \gamma_1 x_2} + (\theta_1 - \theta'_1)\ e^{-(\theta_1+\theta_2)x_1}\right), & 0 \le x_2 < x_1. \end{cases} \tag{4.38}$$

It should be noted that, under this model, the probability that Component C_i is the first to fail is $\theta_i/(\theta_1 + \theta_2), i = 1, 2$, and that the time to first failure is distributed as Exp$(\theta_1 + \theta_2)$. Further, the distribution of the time from first failure to failure of the other component is a mixture of Exp(θ'_1) and Exp(θ'_2) with proportions $\theta_2/(\theta_1 + \theta_2)$ and $\theta_1/(\theta_1 + \theta_2)$, respectively.

Block and Basu (1974) constructed a system of absolutely continuous bivariate exponential distributions by modifying the above presented Marshall-Olkin's bivariate exponential distributions (which do have singular part). This system is a reparameterization of Freund's bivariate exponential distribution in (4.38) with

$$\lambda_i = \theta_1 + \theta_2 - \theta'_{3-i}\ (i = 1, 2) \quad \text{and} \quad \lambda_{12} = \theta'_1 + \theta'_2 - \theta_1 - \theta_2.$$

For an elaborate discussion on various forms of bivariate and multivariate exponential distributions and their properties, one may refer to Chapter 47 of Kotz, Balakrishnan, and Johnson (2000).

Exercises

4.1 Compute the mean vector, covariance matrix, and correlation matrix for each of the following multinomial random variables:

   (a) The three variables in Example 4.3.

   (b) The four variables in Example 4.4.

4.2 For the multinomial distribution, $M(n; p_i, \ldots, p_k)$, show that the correlation coefficient between X_i and X_j is

$$\rho_{ij} = -\sqrt{p_i\ p_j/[(1 - p_i)(1 - p_j)]}.$$

4.3 Derive the mean vector, covariance matrix, and correlation matrix for the multivariate hypergeometric distribution in (4.11).

4.4 Show that the multivariate hypergeometric distribution in (4.11) tends to the multinomial distribution in (4.9), as $N, D_1, ..., D_k \to \infty$ and $D_i/N \to p_i$ $(i = 1, ..., k)$.

4.5 For the Marshall-Olkin distribution in (4.33), derive:

(a) The marginal distributions of X and Y;

(b) The means and variances of X and Y, and the correlation coefficient between X and Y.

4.6 The bivariate survival function of the Gumbel bivariate exponential distribution given by

$$S(x, y) = \exp(-\theta_1 x - \theta_2 y - \rho xy), \quad x, y > 0, \tag{4.39}$$

where $\theta_1, \theta_2 > 0$ and $0 \le \rho \le \theta_1\theta_2$.

(a) Obtain the marginal pdfs and show that X and Y are independent if and only if $\rho = 0$.

(b) Show that the bivariate hazard function is given by

$$H(x, y) = (\theta_1 + \rho x)(\theta_2 + \rho y) - \rho.$$

4.7 As special case of the bivariate survival function in (4.39) is

$$S(x, y) = \exp(-x - y + \rho xy).$$

Obtain:

(a) The cdf $F_{X,Y}(x, y)$.

(b) The marginal cdfs $F_X(x)$ and $F_Y(y)$.

(c) The conditional cdfs $F_{X|Y=y}(x|y)$ and $F_{Y|X=x}(y|x)$.

4.8 The bivariate survival function of the Morgenstern distribution is

$$S(x, y) = \exp(-x - y)\left\{1 + \alpha\left[1 - \exp(-x)\right]\left[1 - \exp(-y)\right]\right\}.$$

Obtain:

(a) The cdf $F_{X,Y}(x, y)$.

(b) The marginal cdfs $F_X(x)$ and $F_Y(y)$.

(c) The conditional cdfs $F_{X|Y=y}(x|y)$ and $F_{Y|X=x}(y|x)$.

4.9 Show that the joint survival function corresponding to (4.37) is as given in (4.38).

4.10 For the jointly distributed random variables with pdf in (4.37), show that, when $\theta_1 + \theta_2 \neq \theta_i'$ $(i = 1, 2)$, the marginal density function of X_i $(i = 1, 2)$ is

$$f_{X_i}(x) = \frac{1}{\theta_1 + \theta_2 - \theta_i'} \left\{ (\theta_i - \theta_i')(\theta_1 + \theta_2)\ e^{-(\theta_1+\theta_2)x} + \theta_i'\theta_{3-i} e^{-\theta_i' x} \right\},\ x \geq 0,$$

which is indeed a mixture of two exponential distributions if $\theta_i > \theta_i'$.

4.11 For the bivariate exponential distribution in (4.37), show that the joint MGF of (X_1, X_2) is

$$E\left(e^{s_1 X_1 + s_2 X_2}\right) = \frac{1}{\theta_1 + \theta_2 - s_1 - s_2} \left\{ \frac{\theta_2}{1 - s_1/\theta_1'} + \frac{\theta_1}{1 - s_2/\theta_2'} \right\}.$$

Find $E(X_i)$, $\text{Var}(X_i)$, $\text{Cov}(X_1,\ X_2)$, and $\text{Corr}(X_1,\ X_2)$.

4.12 Consider the general logistic model

$$F(x, y) = \exp\left\{ -\left[\left(1 - \kappa_1 \frac{x - \lambda_1}{\delta_1}\right)^{1/\alpha\kappa_1} + \left(1 - \kappa_2 \frac{y - \lambda_2}{\delta_2}\right)^{1/\alpha\kappa_2} \right]^{\alpha} \right\}$$

(a) Discuss the valid ranges of all the parameters $\lambda_1, \lambda_2, \delta_1, \delta_2, \kappa_1, \kappa_2, \alpha$.

(b) Determine the domain of definition of the random variable (X, Y).

(c) Obtain the two marginals and identify them.

(d) Obtain the conditionals $X|Y = y$ and $Y|X = x$ and identify them.

(e) Obtain the bivariate and the marginal survival functions.

4.13 Suggest different methods to simulate a sample drawn from the following multivariate models

(a) A normal $N(\boldsymbol{\mu}, \boldsymbol{\Sigma})$.

(b) A bivariate logistic distribution.

(c) A Marshall-Olkin distribution.

4.14 Find changes of variables to transform

(a) A Freund's bivariate distribution into an independent bivariate Unit-Fréchet marginals distribution.

(b) A bivariate normal distribution into an independent bivariate Unit-Fréchet marginals distribution.

Part III

Model Estimation, Selection, and Validation

Chapter 5

Model Estimation

The previous chapters dealt with the specification of some important families of discrete and continuous probabilistic models. These families depend on parameters which are usually unknown in practice. In this chapter, we discuss the following methods for the point and interval estimation of the model parameters from data: Maximum Likelihood Method (MLM), the Method of Moments (MOM), the Probability-Weighted Moments (PWM) method, the Elemental Percentile Method (EPM), and the Quantile Least Squares Method (QLSM). For recent discussions on parameter estimation related to extremes and related problems see, for example, Lu and Peng (2002), Matthys and Beirlant (2003), Wu and Li (2003), and Yun (2002). Since quantiles are essential in extreme value theory and practice, methods for estimating quantiles are also given. All these methods are illustrated with examples. They are also applied to extreme value models in later chapters.

To set the stage up for this chapter, we assume that $X = \{X_1, X_2, \ldots, X_n\}$ are independently and identically distributed (iid) random variables having a common parametric family of pdfs $f(x; \boldsymbol{\theta})$ and cdfs $F(x; \boldsymbol{\theta})$, where

$$\boldsymbol{\theta} = \{\theta_1, \theta_2, \ldots, \theta_k\} \in \boldsymbol{\Theta}$$

is a vector-valued parameter of dimension k in the parameter space $\boldsymbol{\Theta}$. These variables represent a random sample of size n drawn from $f(x; \boldsymbol{\theta})$. Arrange $(X_1, \ldots, X_n)$ in an increasing order of magnitude and let $X_{1:n} \leq \ldots \leq X_{n:n}$ be the ordered values. The rth element of this sequence, $X_{r:n}$, is called the rth *order statistic* in the sample. Now, let $x_{1:n} \leq x_{2:n} \leq \ldots \leq x_{n:n}$ be the observed order statistics in a random sample drawn from $F(x; \boldsymbol{\theta})$. Define the so-called *plotting positions* by

$$p_{i:n} = \frac{i - \alpha}{n + \beta}, \quad i = 1, \ldots, n, \tag{5.1}$$

for appropriate choices of $\alpha \geq 0$ and $\beta \geq 0$. Plotting positions (α and β values) can be chosen empirically (depending on the data, the type of distribution, the estimation method to be used, etc.). Here we use $\alpha = 0$ and $\beta = 1$, that is,

$$p_{i:n} = i/(n+1), \quad i = 1, \ldots, n. \tag{5.2}$$

Other alternative plotting positions include

$$p_{i:n} = \frac{i - 0.375}{n + 0.25}, \quad p_{i:n} = \frac{i - 0.5}{n}, \quad \text{and} \quad p_{i:n} = \frac{i - 0.44}{n + 0.12}. \tag{5.3}$$

For justifications of these formulas see, for example, Castillo (1988), pp. 161–166. Other references for plotting positions include Barnett (1975), Blom (1958), Cunnane (1978), Evans, Hastings, and Peacock (2000), Gringorten (1963), and Harter (1984).

5.1 The Maximum Likelihood Method

The maximum likelihood method is based on maximizing the likelihood of the observed sample. It can be used to derive point and interval estimates, as described below.

5.1.1 Point Estimation

Since the variables in X are independent, their joint probability density function is

$$f(x|\boldsymbol{\theta}) = \prod_{i=1}^{n} f(x_i; \boldsymbol{\theta}). \tag{5.4}$$

After the sample has been collected, the values of $x = \{x_1, x_2, \ldots, x_n\}$ become known and the above function can be viewed as a function of $\boldsymbol{\theta}$ given x. This function is called the *likelihood function* and is written as

$$L(\boldsymbol{\theta}|x) = \prod_{i=1}^{n} f(x_i; \boldsymbol{\theta}). \tag{5.5}$$

It is equivalent but often mathematically easier to deal with the *loglikelihood function* instead of the likelihood function itself. The loglikelihood function is given by

$$\ell(\boldsymbol{\theta}|x) = \log L(\boldsymbol{\theta}|x) = \sum_{i=1}^{n} \log f(x_i; \boldsymbol{\theta}). \tag{5.6}$$

The *maximum likelihood estimate* (MLE) of $\boldsymbol{\theta}$ is obtained by maximizing the likelihood function in (5.5), or equivalently, the loglikelihood function in (5.6), with respect to $\boldsymbol{\theta}$. The MLE is denoted by $\hat{\boldsymbol{\theta}}_{\text{MLE}}$, but for simplicity of notation in this section, we use $\hat{\boldsymbol{\theta}}$ instead of $\hat{\boldsymbol{\theta}}_{\text{MLE}}$. Thus,

$$\max_{\boldsymbol{\theta}} \ell(\boldsymbol{\theta}|x) = \ell(\hat{\boldsymbol{\theta}}|x). \tag{5.7}$$

If there exists a regular relative maximum $\hat{\boldsymbol{\theta}}$, the maximum likelihood estimator is obtained by solving the system of equations

$$\frac{\partial \ell(\boldsymbol{\theta}|x)}{\partial \theta_j} = 0, \quad j = 1, 2, \ldots, k, \tag{5.8}$$

where k is the dimension of the vector $\boldsymbol{\theta}$.

Example 5.1 (MLE for the exponential distribution). Let $x = \{x_1, x_2, \ldots, x_n\}$ be a sample from the exponential distribution with pdf

$$f(x;\theta) = (1/\theta)\, e^{-x/\theta}, \quad x \geq 0,\ \theta > 0, \tag{5.9}$$

with *quantile* function

$$x_p = F^{-1}(p;\theta) = -\theta \log(1-p), \tag{5.10}$$

where $F^{-1}(p;\theta)$ is the inverse of the cdf $F(x;\theta)$. Then, the likelihood function is

$$L(\theta|x) = \prod_{i=1}^{n} f(x_i;\theta) = \frac{1}{\theta^n} \exp\left(-\frac{1}{\theta}\sum_{i=1}^{n} x_i\right). \tag{5.11}$$

Taking the log of $L(\theta|x)$, we obtain the loglikelihood

$$\ell(\theta|x) = \log L(\theta|x) = -n\log\theta - \frac{1}{\theta}\sum_{i=1}^{n} x_i. \tag{5.12}$$

The first derivative of $\ell(\theta|x)$ is

$$\frac{\partial \ell(\theta|x)}{\partial \theta} = -\frac{n}{\theta} + \frac{1}{\theta^2}\sum_{i=1}^{n} x_i. \tag{5.13}$$

Setting the first derivative equal to zero, and solving for θ, we obtain the MLE of θ as

$$\hat{\theta} = \frac{1}{n}\sum_{i=1}^{n} x_i = \bar{x}, \tag{5.14}$$

where $\bar{x}$ is the sample mean. Since

$$\frac{\partial^2 \ell(\theta|x)}{\partial \theta^2} = \left(\frac{n}{\theta^2} - \frac{2}{\theta^3}\sum_{i=1}^{n} x_i\right)\Bigg|_{\theta=\bar{x}} = -\frac{n}{\bar{x}^2} < 0,$$

then $\hat{\theta}$ in (5.14) is a relative maximum.

For example, assuming that the times between 48 consecutive telephone calls, given in Table 1.8, are exponentially distributed, then the MLE of θ is found to be $\hat{\theta} = 0.934$ minutes. ∎

5.1.2 Some Properties of the MLE

Under some regularity conditions, the maximum likelihood estimates have several desirable properties. These include:

P1: The maximum likelihood equations in (5.8) have a consistent root for $\boldsymbol{\theta}$ with a probability that tends to 1 as $n \to \infty$. Thus, $\hat{\boldsymbol{\theta}} \to \boldsymbol{\theta}$ as $n \to \infty$, that is, $\hat{\boldsymbol{\theta}}$ is a consistent (its variance goes to zero as $n \to \infty$) estimator of $\boldsymbol{\theta}$.

P2: The MLEs are asymptotically efficient (they are often efficient, that is, they have minimum variance).

P3: If $h(\boldsymbol{\theta})$ is a function of $\boldsymbol{\theta}$, then the MLE of $h(\boldsymbol{\theta})$ is $h(\hat{\boldsymbol{\theta}})$, that is, the function h evaluated at $\boldsymbol{\theta} = \hat{\boldsymbol{\theta}}$. For example, an estimate of the pth quantile can be obtained as

$$\hat{x}_p = F^{-1}(p; \hat{\boldsymbol{\theta}}) = h(p; \hat{\boldsymbol{\theta}}), \quad 0 < p < 1. \tag{5.15}$$

P4: All consistent solutions of the maximum likelihood equations are asymptotically normally distributed, that is,

$$\hat{\boldsymbol{\theta}} \to N_k\left(\boldsymbol{\theta}, \boldsymbol{\Sigma}_{\hat{\boldsymbol{\theta}}}\right), \tag{5.16}$$

where $N_k(\boldsymbol{\theta}, \boldsymbol{\Sigma}_{\hat{\boldsymbol{\theta}}})$ denotes the k-dimensional normal distribution with mean vector $\boldsymbol{\theta}$ and covariance matrix $\text{Cov}(\hat{\boldsymbol{\theta}}) = \boldsymbol{\Sigma}_{\hat{\boldsymbol{\theta}}}$, and $\to$ means that the cdf of $\hat{\boldsymbol{\theta}}$ converges to the cdf of $N_k\left(\boldsymbol{\theta}, \boldsymbol{\Sigma}_{\hat{\boldsymbol{\theta}}}\right)$ when $n \to \infty$. The covariance matrix $\boldsymbol{\Sigma}_{\hat{\boldsymbol{\theta}}}$ is the inverse of the Fisher *information matrix*, $\mathbf{I}_{\boldsymbol{\theta}}$, whose (r, j)th element is given by

$$i_{rj} = E\left(\frac{\partial \ell(\boldsymbol{\theta}|x)}{\partial \theta_r} \frac{\partial \ell(\boldsymbol{\theta}|x)}{\partial \theta_j}\right). \tag{5.17}$$

Thus, the inverse of the Fisher information matrix gives the $\text{Cov}(\hat{\boldsymbol{\theta}})$, the matrix of variances and covariances of the parameter estimators, which is given by

$$\text{Cov}(\hat{\boldsymbol{\theta}}) = \boldsymbol{\Sigma}_{\hat{\boldsymbol{\theta}}} = \mathbf{I}_{\boldsymbol{\theta}}^{-1}. \tag{5.18}$$

An estimate of $\text{Cov}(\hat{\boldsymbol{\theta}})$ can be obtained by replacing $\boldsymbol{\theta}$ by its MLE, $\hat{\boldsymbol{\theta}}$, that is,

$$\widehat{\text{Cov}}(\hat{\boldsymbol{\theta}}) = \hat{\boldsymbol{\Sigma}}_{\hat{\boldsymbol{\theta}}}. \tag{5.19}$$

Under certain regularity conditions that are often satisfied, i_{rj} can be written as

$$i_{rj} = -E\left(\frac{\partial^2 \ell(\boldsymbol{\theta}|x)}{\partial \theta_r \partial \theta_j}\right). \tag{5.20}$$

Furthermore, if the loglikelihood $\ell(\boldsymbol{\theta}|x)$, is approximately quadratic in a neighborhood of the maximum, then i_{rj} is given by

$$i_{rj} = -\left.\frac{\partial^2 \ell(\boldsymbol{\theta}|x)}{\partial\theta_r \partial\theta_j}\right|_{\boldsymbol{\theta}=\hat{\boldsymbol{\theta}}}. \tag{5.21}$$

It follows from (5.16) that each component $\hat{\theta}_j$ of $\hat{\boldsymbol{\theta}}$ is asymptotically normal, that is,

$$\hat{\theta}_j \to N(\theta_j, \sigma^2_{\hat{\theta}_j}), \tag{5.22}$$

where $\sigma^2_{\hat{\theta}_j}$ is the variance of $\hat{\theta}_j$, which is the jth diagonal element of $\boldsymbol{\Sigma}_{\hat{\boldsymbol{\theta}}}$.

When $\boldsymbol{\theta}$ is a scalar, (5.16) reduces to

$$\hat{\theta} \to N\left(\theta, \sigma^2_{\hat{\theta}}\right), \tag{5.23}$$

where

$$\sigma^2_{\hat{\theta}} = -\left[E\left(\frac{\partial^2 \ell(\theta|x)}{\partial\theta^2}\right)\right]^{-1}. \tag{5.24}$$

An estimate of the variance of $\hat{\theta}$ is obtained by evaluating the second derivative in (5.24) at $\theta = \hat{\theta}$:

$$\hat{\sigma}^2_{\hat{\theta}} = -\left[E\left(\frac{\partial^2 \ell(\theta|x)}{\partial\theta^2}\right)\bigg|_{\theta=\hat{\theta}}\right]^{-1}. \tag{5.25}$$

Example 5.2 (Distribution of $\hat{\theta}$ in the exponential distribution). Consider the case of exponential distribution in Example 5.1. Taking the derivative of (5.13) with respect to θ, we obtain the second derivative

$$\frac{\partial^2 \ell(\theta|x)}{\partial\theta^2} = \frac{n}{\theta^2} - \frac{2}{\theta^3}\sum_{i=1}^{n} x_i, \tag{5.26}$$

and the Fisher information matrix (a scalar) then becomes

$$I_\theta = -E\left(\frac{\partial^2 \ell(\theta|x)}{\partial\theta^2}\right) = -\frac{n}{\theta^2} + \frac{2}{\theta^3}E\left(\sum_{i=1}^{n} X_i\right) = -\frac{n}{\theta^2} + \frac{2}{\theta^3}n\theta = \frac{n}{\theta^2}. \tag{5.27}$$

From (5.24), the variance of $\hat{\theta}$ is

$$\sigma^2_{\hat{\theta}} = \theta^2/n. \tag{5.28}$$

Consequently, using (5.23), as $n \to \infty$, we have

$$\hat{\theta} \sim N\left(\theta, \theta^2/n\right). \tag{5.29}$$

An estimate of $\sigma^2_{\hat{\theta}}$ is

$$\hat{\sigma}^2_{\hat{\theta}} = \hat{\theta}^2/n. \tag{5.30}$$

■

Remark. Since $\hat{\theta} = \bar{x}$, the variance of MLE $\hat{\theta}$ can also be obtained exactly in this case:

$$\text{Var}(\hat{\theta}) = \text{Var}(\bar{X}) = \frac{1}{n^2}\sum_{i=1}^{n}\text{Var}(X_i) = \frac{n\theta^2}{n^2} = \frac{\theta^2}{n},$$

which is exactly the same as obtained in (5.28) from the Fisher information. Furthermore, since $\sum_{i=1}^{n} X_i$ is distributed as Gamma $G(n, 1/\theta)$, it can also be readily seen here that $2n\hat{\theta}/\theta$ has a central χ^2 distribution with $2n$ degrees of freedom which indeed tends to a normal distribution as $n \to \infty$ [see (5.29)].

The above discussion shows how one can measure uncertainty about the MLE $\hat{\boldsymbol{\theta}}$. In practice, one is also interested in estimating a given function, $\boldsymbol{\psi} = h(\boldsymbol{\theta})$, of the parameters. For example, the pth quantile is related to $\boldsymbol{\theta}$ by

$$x_p = F^{-1}(p; \boldsymbol{\theta}) = h(\boldsymbol{\theta}; p), \quad 0 < p < 1. \tag{5.31}$$

By Property P3, the MLE of $\boldsymbol{\psi}$ is $\hat{\boldsymbol{\psi}} = h(\hat{\boldsymbol{\theta}}; p)$. The following method can then be used to obtain the variance of $\hat{\boldsymbol{\psi}}$.

5.1.3 The Delta Method

The *delta* method can be used to obtain the variance-covariance matrix of $h(\hat{\boldsymbol{\theta}})$ as given in the following theorem.

Theorem 5.1 (The Delta method). *Let $\hat{\boldsymbol{\theta}}$ be the maximum likelihood estimate of $\boldsymbol{\theta}$ with variance-covariance matrix $\boldsymbol{\Sigma}_{\hat{\boldsymbol{\theta}}}$. Let $\boldsymbol{\psi} = h(\boldsymbol{\theta})$ be the new parameter vector of dimension r defined as a function of $\boldsymbol{\theta}$. Then, for large sample size n, $\hat{\boldsymbol{\psi}} = h(\hat{\boldsymbol{\theta}})$ is asymptotically normal, that is,*

$$\hat{\boldsymbol{\psi}} \sim N_r\left(\boldsymbol{\psi}; \nabla_{\boldsymbol{\theta}}^T\boldsymbol{\psi}\boldsymbol{\Sigma}_{\hat{\boldsymbol{\theta}}}\nabla_{\boldsymbol{\theta}}\boldsymbol{\psi}\right), \tag{5.32}$$

where $\nabla_{\boldsymbol{\theta}}\boldsymbol{\psi}$ is the $k \times r$ matrix of partial derivatives of $\boldsymbol{\psi}$ with respect to $\boldsymbol{\theta}$, which is given by

$$\nabla_{\boldsymbol{\theta}}\boldsymbol{\psi} = \begin{bmatrix} \frac{\partial\psi_1}{\partial\theta_1} & \frac{\partial\psi_2}{\partial\theta_1} & \cdots & \frac{\partial\psi_r}{\partial\theta_1} \\ \frac{\partial\psi_1}{\partial\theta_2} & \frac{\partial\psi_2}{\partial\theta_2} & \cdots & \frac{\partial\psi_r}{\partial\theta_2} \\ \vdots & \vdots & \ddots & \vdots \\ \frac{\partial\psi_1}{\partial\theta_k} & \frac{\partial\psi_2}{\partial\theta_k} & \cdots & \frac{\partial\psi_r}{\partial\theta_k} \end{bmatrix}. \tag{5.33}$$

Thus, the covariance matrix of $\hat{\boldsymbol{\psi}}$ is

$$\text{Cov}(\hat{\boldsymbol{\psi}}) = \nabla_{\boldsymbol{\theta}}^T\boldsymbol{\psi}\boldsymbol{\Sigma}_{\hat{\boldsymbol{\theta}}}\nabla_{\boldsymbol{\theta}}\boldsymbol{\psi}. \tag{5.34}$$

An estimate of $\text{Cov}(\hat{\boldsymbol{\psi}})$ can be obtained by replacing $\boldsymbol{\theta}$ in (5.34) by its MLE, $\hat{\boldsymbol{\theta}}$, that is,

$$\widehat{\text{Cov}}(\hat{\boldsymbol{\psi}}) = \nabla_{\hat{\boldsymbol{\theta}}}^T\boldsymbol{\psi}\hat{\boldsymbol{\Sigma}}_{\hat{\boldsymbol{\theta}}}\nabla_{\hat{\boldsymbol{\theta}}}\boldsymbol{\psi}. \tag{5.35}$$

It follows from (5.32) that each component $\hat{\psi}_i$ of $\hat{\boldsymbol{\psi}}$ is asymptotically normal, that is,

$$\hat{\psi}_i \sim N(\psi_i, \sigma^2_{\hat{\psi}_i}), \tag{5.36}$$

where $\sigma^2_{\hat{\psi}}$ is the ith diagonal element of $\text{Cov}(\hat{\boldsymbol{\psi}})$. Further, if $\boldsymbol{\psi}$ is a scalar (e.g., when $\psi = x_p$, the pth quantile of X), then $\nabla_{\boldsymbol{\theta}}\psi$ in (5.33) becomes

$$\nabla_{\boldsymbol{\theta}}\psi = \begin{bmatrix} \dfrac{\partial \psi}{\partial \theta_1} \\ \dfrac{\partial \psi}{\partial \theta_2} \\ \vdots \\ \dfrac{\partial \psi}{\partial \theta_k} \end{bmatrix}. \tag{5.37}$$

If $\nabla_{\boldsymbol{\theta}}\psi$ exists and is not zero, then when $n \to \infty$, we have

$$\hat{\psi} \sim N\left(\psi; \nabla^T_{\boldsymbol{\theta}}\psi \boldsymbol{\Sigma}_{\hat{\boldsymbol{\theta}}} \nabla_{\boldsymbol{\theta}}\psi\right). \tag{5.38}$$

A particular case of (5.38) is obtained when $\boldsymbol{\theta}$ is also a scalar. In this case, we have

$$\hat{\psi} \sim N\left(\psi, [h'(\theta)]^2 \sigma^2_{\hat{\theta}}\right), \tag{5.39}$$

where $\sigma^2_{\hat{\theta}}$ is as given in (5.24).

Example 5.3 (Distribution of the estimate of the quantile of the exponential distribution). Consider estimating the pth quantile of the exponential distribution with mean θ. The pth quantile is

$$x_p = F^{-1}(p; \theta) = -\theta \log(1-p) = h(\theta) = \psi, \quad 0 < p < 1. \tag{5.40}$$

The MLE of the pth quantile is then

$$\hat{x}_p = h(\hat{\theta}) = -\hat{\theta} \log(1-p). \tag{5.41}$$

Since $h'(\theta) = -\log(1-p)$, (5.28) and (5.39) lead to

$$\hat{x}_p \sim N\left(-\theta \log(1-p); [\theta \log(1-p)]^2/n\right). \tag{5.42}$$

In this case, the variance is exact. ■

5.1.4 Interval Estimation

Now that we have the asymptotic distributions of the MLE of parameters and quantiles, we can compute confidence regions for the population parameters and quantiles.

Since, by (5.22), $\hat{\theta}_j$ is asymptotically normal, an approximate $(1-\alpha)100\%$ confidence interval for θ_j is given by

$$\theta_j \in \left(\hat{\theta}_j \pm z_{\alpha/2}\, \hat{\sigma}_{\hat{\theta}_j}\right), \quad j = 1, 2, \ldots, k, \tag{5.43}$$

where $\hat{\sigma}_{\hat{\theta}_j}$ is the jth diagonal element of $\hat{\boldsymbol{\Sigma}}_{\hat{\boldsymbol{\theta}}}$ in (5.19) and $z_{\alpha/2}$ is the $(1-\alpha/2)$ quantile of the standard normal distribution, which can be obtained from Table A.1 in the Appendix.

When $\boldsymbol{\theta}$ is a scalar, (5.43) reduces to

$$\theta \in \left(\hat{\theta} \pm z_{\alpha/2}\, \hat{\sigma}_{\hat{\theta}}\right), \tag{5.44}$$

where $\hat{\sigma}^2_{\hat{\theta}}$ is as defined in (5.25). For example, for the exponential distribution with mean θ, we have $\hat{\theta} = \overline{x}$ and from (5.30), $\hat{\sigma}^2_{\hat{\theta}} = \hat{\theta}^2/n = \overline{x}^2/n$, and hence a $(1-\alpha)100\%$ confidence interval for θ is

$$\theta \in \left(\overline{x} \pm z_{\alpha/2}\, \overline{x}/\sqrt{n}\right).$$

To obtain an approximate $(1-\alpha)100\%$ confidence interval for the pth quantile of X, we use (5.18) and (5.38) and get

$$x_p \in \left(\hat{x}_p \pm z_{\alpha/2}\sqrt{\nabla^T_{\hat{\boldsymbol{\theta}}} x_p \hat{\boldsymbol{\Sigma}}_{\hat{\boldsymbol{\theta}}} \nabla_{\hat{\boldsymbol{\theta}}} x_p}\right), \tag{5.45}$$

where

$$\nabla_{\boldsymbol{\theta}} x_p = \left(\frac{\partial x_p}{\partial \theta_1} \quad \frac{\partial x_p}{\partial \theta_2} \quad \cdots \quad \frac{\partial x_p}{\partial \theta_k}\right)^T, \tag{5.46}$$

which is obtained from (5.37) by replacing ψ by x_p in (5.31).

Example 5.4 (Confidence interval for the quantile of the exponential distribution). For the exponential distribution with mean θ, using (5.42), we obtain a $(1-\alpha)100\%$ confidence interval for x_p:

$$x_p \in \left(-\overline{x}\, \log(1-p) \pm z_{\alpha/2}\, \overline{x}\, \log(1-p)/\sqrt{n}\right). \tag{5.47}$$

■

5.1.5 The Deviance Function

An alternative method for measuring the uncertainty of the MLE $\hat{\boldsymbol{\theta}}$ is the *deviance function*, which is defined below.

Definition 5.1 (Deviance function). *Let $\hat{\boldsymbol{\theta}}$ be the maximum likelihood estimate of $\boldsymbol{\theta}$. Then, for any $\boldsymbol{\theta} \in \boldsymbol{\Theta}$, the deviance function is defined as*

$$D(\boldsymbol{\theta}) = 2\left[\ell(\hat{\boldsymbol{\theta}}|x) - \ell(\boldsymbol{\theta}|x)\right], \tag{5.48}$$

where $\ell(\boldsymbol{\theta}|x)$ is the loglikelihood function defined in (5.6).

Note that the deviance function is a nonnegative function because $\ell(\hat{\boldsymbol{\theta}}|x) \geq \ell(\boldsymbol{\theta}|x)$, for all $\boldsymbol{\theta} \in \boldsymbol{\Theta}$. Thus, $D(\boldsymbol{\theta})$ measures departures from maximum likelihood. To find the asymptotic distribution of the deviance, we need the following definition.

Definition 5.2 (Profile loglikelihood function). *Let $f(x;\boldsymbol{\theta})$ be a parametric pdf family, where $\boldsymbol{\theta} \in \boldsymbol{\Theta}$. Partition the parameter $\boldsymbol{\theta}$ as $\boldsymbol{\theta} = (\boldsymbol{\theta}_1, \boldsymbol{\theta}_2)$, where $\boldsymbol{\theta}_1$ and $\boldsymbol{\theta}_2$ are of dimensions k_1 and k_2, respectively. The profile likelihood function for $\boldsymbol{\theta}_1$ is defined as*

$$\ell_p(\boldsymbol{\theta}_1|x) = \max_{\boldsymbol{\theta}_2} \ell(\boldsymbol{\theta}_1, \boldsymbol{\theta}_2|x), \tag{5.49}$$

where the subscript p refers to profile.

Thus, the profile likelihood function is a function of $\boldsymbol{\theta}_1$ and obtained by maximizing the loglikelihood function only with respect to the k_2 parameters in $\boldsymbol{\theta}_2$. Note that the loglikelihood function in (5.6) is a particular case of $\ell_p(\boldsymbol{\theta}_1|x)$, for $k_2 = 0$, that is, when $\boldsymbol{\theta}_2$ is empty.

The following theorem facilitates the task of obtaining critical values for the rejection regions and confidence intervals.

Theorem 5.2 (Limit distribution of the deviance function). *If the parametric family of pdf $f(x;\boldsymbol{\theta})$, where $\boldsymbol{\theta} \in \boldsymbol{\Theta}$, is partitioned as in Definition 5.2, then under some regularity conditions and for large sample size n,*

$$D_p(\boldsymbol{\theta}_1) = 2\left[\ell(\hat{\boldsymbol{\theta}}|x) - \ell_p(\boldsymbol{\theta}_1|x)\right] \sim \chi^2_{k_1}. \tag{5.50}$$

The following conclusions can be made from Theorem 5.2:

1. For $k_1 = k$ ($\boldsymbol{\theta}_2$ is empty): One can obtain the following approximate $(1-\alpha)100\%$ confidence region for $\boldsymbol{\theta}$:

$$\{\boldsymbol{\theta} : D(\boldsymbol{\theta}) \leq \chi^2_k(1-\alpha)\}, \tag{5.51}$$

   where $\chi^2_k(1-\alpha)$ is the $(1-\alpha)100\%$ quantile of the χ^2 distribution with k degrees of freedom.

2. For $k_1 = k = 1$ ($\boldsymbol{\theta}_1$ is a scalar and $\boldsymbol{\theta}_2$ is empty): One can obtain the following approximate $(1-\alpha)100\%$ confidence region for θ_1:

$$\{\theta_1 : D_p(\theta_1) \leq \chi^2_1(1-\alpha)\}. \tag{5.52}$$

   This is a better region than the confidence interval in (5.44) (see Coles (2001)).

For example, for the exponential distribution with mean θ, we have $\hat{\theta} = \overline{x}$,

$$\ell(\theta|x) = -n\log\theta - \frac{1}{\theta}\sum_{i=1}^{n} x_i, \tag{5.53}$$

and

$$D(\theta) = 2n\left[-1 + \frac{\bar{x}}{\theta} - \log\left(\frac{\bar{x}}{\theta}\right)\right].$$

Accordingly, a $(1-\alpha)100\%$ confidence interval for θ obtained using (5.51) is

$$\{\theta : D(\theta) \le \chi_1^2(1-\alpha)\}.$$

In some cases, the loglikelihood function may not be differentiable with respect to the parameters, in which case the MLEs and their variances may have to be obtained by different arguments (as the Fisher information does not exist in such cases). The following example illustrates this point.

Example 5.5 (MLEs for the two-parameter exponential distribution). Let $x = \{x_1, x_2, \ldots, x_n\}$ be a sample from the two-parameter exponential distribution with pdf

$$f(x; \mu, \theta) = \frac{1}{\theta} e^{-(x-\mu)/\theta}, \quad x \ge \mu, \theta > 0,$$

with quantile function

$$x_p = F^{-1}(p; \mu, \theta) = \mu - \theta \log(1-p),$$

where $F^{-1}(p; \mu, \theta)$ is the inverse of the cdf $F(x; \mu, \theta)$. Then, the likelihood function is

$$\begin{aligned} L(\mu, \theta|x) &= \prod_{i=1}^{n} f(x_i; \mu, \theta) = \frac{1}{\theta^n} \exp\left\{-\frac{1}{\theta}\sum_{i=1}^{n}(x_i - \mu)\right\} \\ &= \frac{1}{\theta^n} \exp\left\{\frac{n\mu}{\theta} - \frac{1}{\theta}\sum_{i=1}^{n} x_i\right\}, \quad \mu \le \min(x_1, \ldots, x_n),\ \theta > 0. \end{aligned}$$

It can be seen from the above expression that $L(\mu, \theta|x)$ is a monotone increasing function of μ and, therefore, the MLE of μ is $\hat{\mu} = \min(x_1, \ldots, x_n)$, which is the maximum possible value of μ given the sample. Upon using this MLE of μ, we have:

$$\log L(\hat{\mu}, \theta|x) = -n \log \theta - \frac{1}{\theta}\sum_{i=1}^{n}\{x_i - \min(x_1, \ldots, x_n)\}, \quad \theta > 0,$$

which readily yields the MLE of θ to be

$$\hat{\theta} = \frac{1}{n}\sum_{i=1}^{n}\{x_i - \min(x_1, \ldots, x_n)\}.$$

■

Even though the Fisher information does not exist in this case and so the variances of the MLEs of μ and θ cannot be derived by using Property P4 mentioned earlier, the variances and even the distributions of the MLEs, $\hat{\mu}$ and $\hat{\theta}$, can be explicitly derived by using the properties of exponential order statistics. For example, it can be shown that

$$P_1 = 2n\left(\frac{\hat{\mu}-\mu}{\theta}\right) \qquad \text{and} \qquad P_2 = 2n\left(\frac{\hat{\theta}}{\theta}\right)$$

are independently distributed as central χ^2 with 2 and $2n-2$ degrees of freedom, respectively.

5.2 The Method of Moments

The Method of Moments (MOM) estimators are obtained by first setting the first k moments[1] of the random variable equal to the corresponding sample moments, then solving the resultant system of equations. That is, the MOM estimators are the solution of the system of equations

$$E(X^r) = \frac{1}{n}\sum_{i=1}^{n} x_i^r, \quad r = 1, 2, \ldots, k, \tag{5.54}$$

where k is the dimension of $\boldsymbol{\theta}$. The MOM estimators are denoted by $\hat{\boldsymbol{\theta}}_{\text{MOM}}$.

Example 5.6 (MOM estimates for the exponential distribution). Let $x = \{x_1, \ldots, x_n\}$ be a sample from the exponential distribution with pdf

$$f(x;\theta) = (1/\theta)\, e^{-x/\theta}, \quad x \geq 0,\ \theta > 0. \tag{5.55}$$

Since the pdf of this random variable depends on only one parameter, that is, $k = 1$, and $E(X) = \theta$, the system in (5.54) becomes

$$E(X) = \frac{1}{n}\sum_{i=1}^{n} x_i,$$

from which we have $\theta = \bar{x}$, where $\bar{x}$ is the sample mean. Thus, the MOM estimator of θ is $\hat{\theta}_{\text{MOM}} = \bar{x}$. Thus, for the exponential distribution, the MLE and MOM are the same. ∎

5.3 The Probability-Weighted Moments Method

The probability-weighted moments (PWM) method is a variation of the method of moments. For a continuous random variable X with a pdf $f(x;\boldsymbol{\theta})$ and a cumulative distribution function $F(x;\boldsymbol{\theta})$, the PWM estimators are obtained

[1]See (3.8) for definition.

by setting the first k weighted-moments of the random variable equal to the corresponding weighted-sample moments, then solving the resultant system of equations. More precisely, let

$$M(r,s,t) = E\left\{X^r[F(X;\boldsymbol{\theta})]^s\ [1-F(X;\boldsymbol{\theta})]^t\right\}, \tag{5.56}$$

where r, s, and t are real numbers, be the *probability-weighted moments* of order r, s, and t of the random variable X (Greenwood et al. (1979)). Probability-weighted moments are most useful when the inverse distribution function $X_p = F^{-1}(p;\boldsymbol{\theta})$ can be written in closed form, for then we may use (3.7) and write

$$\begin{aligned} M(r,s,t) &= \int_{-\infty}^{\infty} x^r [F(x;\boldsymbol{\theta})]^s\ [1-F(x;\boldsymbol{\theta})]^t f(x;\boldsymbol{\theta})dx \\ &= \int_0^1 X_p^r p^s (1-p)^t dp, \end{aligned} \tag{5.57}$$

where we have made the change of variable $p = F(x_p;\boldsymbol{\theta})$. The corresponding weighted-sample moments are

$$m(r,s,t) = \frac{1}{n}\sum_{i=1}^{n} x_{i:n}^r p_{i:n}^s (1-p_{i:n})^t, \tag{5.58}$$

where $x_{i:n}$ is the ith sample order statistic and $p_{i:n}$ is a corresponding plotting position, such as the one defined in (5.2). The PWM estimators are then found by solving the system of equations

$$M(r,s,t) = m(r,s,t), \tag{5.59}$$

and the resultant estimators are denoted by $\hat{\boldsymbol{\theta}}_{\text{PWM}}$. Note that $M(r,0,0)$, $r = 1,2,\ldots,k$, are the usual moments[2] of the random variable X.

Greenwood et al. (1979) consider several distributions for which the relationship between the parameters of the distribution and the PWMs, $M(1,s,t)$, is simpler than that between the parameters and the conventional moments $M(r,0,0)$. Three particular cases of (5.56) are obtained in the case where $\boldsymbol{\theta}$ is a scalar:

$$\begin{aligned} M(1,s,0) &= E[X\{F(X;\theta)\}^s], \\ M(1,0,t) &= E[X\{1-F(X;\theta)\}^t], \\ M(1,s,t) &= E[X\{F(X;\theta)\}^s\{1-F(X;\theta)\}^t]. \end{aligned} \tag{5.60}$$

The corresponding sample versions are

$$\begin{aligned} m(1,s,0) &= \frac{1}{n}\sum_{i=1}^{n} x_{i:n}p_{i:n}^s, \\ m(1,0,t) &= \frac{1}{n}\sum_{i=1}^{n} x_{i:n}(1-p_{i:n})^t, \\ m(1,s,t) &= \frac{1}{n}\sum_{i=1}^{n} x_{i:n}p_{i:n}^s(1-p_{i:n})^t. \end{aligned} \tag{5.61}$$

[2]See (3.8) for definition.

These lead to three equations

$$\begin{aligned} M(1,s,0) &= m(1,s,0), \\ M(1,0,t) &= m(1,0,t), \\ M(1,s,t) &= m(1,s,t), \end{aligned} \tag{5.62}$$

the solution of each of which gives a PWM estimator of the scalar parameter θ.

Example 5.7 (PWM estimates for the exponential distribution). Continuing with the exponential distribution with mean θ, the cdf is

$$F(x;\theta) = 1 - e^{-x/\theta}, \quad x \geq 0, \ \theta > 0. \tag{5.63}$$

For $s = t = 1$, we have

$$M(1,1,0) = 3\theta/4, \quad M(1,0,1) = \theta/4, \quad M(1,1,1) = 5\theta/36, \tag{5.64}$$

and (5.61) becomes

$$\begin{aligned} S_{10} &= m(1,1,0) = \frac{1}{n}\sum_{i=1}^{n} x_{i:n}p_{i:n}, \\ S_{01} &= m(1,0,1) = \frac{1}{n}\sum_{i=1}^{n} x_{i:n}(1-p_{i:n}), \\ S_{11} &= m(1,1,1) = \frac{1}{n}\sum_{i=1}^{n} x_{i:n}p_{i:n}(1-p_{i:n}). \end{aligned} \tag{5.65}$$

Setting the population weighted moments in (5.64) to the corresponding sample weighted moments in (5.65), we obtain three PWM estimates of θ:

$$\begin{aligned} \hat{\theta}_{10} &= 4\ S_{10}/3, \\ \hat{\theta}_{01} &= 4\ S_{01}, \\ \hat{\theta}_{11} &= 36\ S_{11}/5. \end{aligned} \tag{5.66}$$

Assuming that the times between 48 consecutive telephone calls, given in Table 1.8, are exponentially distributed, then the PWM estimates of θ in (5.66) are

$$\hat{\theta}_{10} = 0.904, \quad \hat{\theta}_{01} = 1.026, \quad \text{and} \quad \hat{\theta}_{11} = 0.996$$

minutes. Recall that both MLE and MOM estimates of θ are 0.934 minutes. Now, knowing that these data have been generated from an exponential distribution with $\theta = 1$, we see that all methods provide reasonable estimates of θ.

■

5.4 The Elemental Percentile Method

The classical MLE and moments-based estimation methods may have problems, when it comes to applications in extremes, either because the ranges of the distributions depend on the parameters (see Hall and Wang (1999)) or because the moments do not exist in certain regions of the parameter space. For example, for the GEVD in (3.78), the moments do not exist when the shape parameter $\kappa \leq -1$. For the GPD in (3.85), when $\kappa > 1$, the MLEs do not exist and when $0.5 \leq \kappa \leq 1$, they may have numerical problems. Also, for $\kappa \leq -0.5$, the MOM and PWM do not exist because the second and higher order moments do not exist. Another, perhaps more serious problem with the MOM and PWM is that they can produce estimates of $\boldsymbol{\theta}$ not in the parameter space $\boldsymbol{\Theta}$ (see Chan and Balakrishnan (1995)).

In this section we describe the Elemental Percentile Method (EPM), proposed by Castillo and Hadi (1995a) for estimating the parameters and quantiles of $F(x; \boldsymbol{\theta}), \boldsymbol{\theta} \in \boldsymbol{\Theta}$. The method gives well-defined estimators for all values of $\boldsymbol{\theta} \in \boldsymbol{\Theta}$.

Simulation studies by Castillo and Hadi (1995b,c, 1997) indicate that no method is uniformly the best for all $\boldsymbol{\theta} \in \boldsymbol{\Theta}$, but this method performs well compared to all other methods.

5.4.1 Initial Estimates

This method obtains the estimates in two steps: First, a set of initial estimates based on selected subsets of the data are computed, then the obtained estimates are combined to produce final more-efficient and robust estimates of the parameters. The two steps are described below.

Elemental Percentile Estimates

Since $X = \{X_1, X_2, \ldots, X_n\}$ are iid random variables having a common cdf $F(x; \boldsymbol{\theta})$, then we have

$$F(x_{i:n}; \boldsymbol{\theta}) \cong p_{i:n}, \quad i = 1, 2, \ldots, n, \tag{5.67}$$

or, equivalently,

$$x_{i:n} \cong F^{-1}(p_{i:n}; \boldsymbol{\theta}), \; i = 1, 2, \ldots, n, \tag{5.68}$$

where $x_{i:n}$ are the order statistics and $p_{i:n}$ are empirical estimates of $F(x_i; \boldsymbol{\theta})$ or suitable plotting positions as defined in (5.1).

Let $I = \{i_1, i_2, \ldots, i_k\}$ be a set of indices of k distinct order statistics, where $i_j \in \{1, 2, \ldots, n\}$ and $j = \{1, 2, \ldots, k\}$. We refer to a subset of size k observations as an *elemental subset* and to the resultant estimates as *elemental estimates* of $\boldsymbol{\theta}$. For each observation in an elemental subset I, we set

$$x_{i:n} = F^{-1}(p_{i:n}; \boldsymbol{\theta}), \; i \in I, \tag{5.69}$$

where we have replaced the approximation in (5.68) by an equality. The set I is chosen so that the system in (5.69) contains k independent equations in k unknowns, $\boldsymbol{\theta} = \{\theta_1, \theta_2, \ldots, \theta_k\}$. An elemental estimate of $\boldsymbol{\theta}$ can then be obtained by solving (5.69) for $\boldsymbol{\theta}$.

Final Estimates

The estimates obtained from (5.69) depend on k distinct order statistics. For large n and k, the number of elemental subsets may be too large for the computations of all possible elemental estimates to be feasible. In such cases, instead of computing all possible elemental estimates, one may select a prespecified number, N, of elemental subsets either systematically, based on some theoretical considerations, or at random. For each of these subsets, an elemental estimate of $\boldsymbol{\theta}$ is computed. Let us denote these elemental estimates by $\hat{\theta}_{j1}, \hat{\theta}_{j2}, \ldots \hat{\theta}_{jN}, j = 1, 2, \ldots, k$. These elemental estimates can then be combined, using some suitable (preferably robust) functions, to obtain an overall final estimate of $\boldsymbol{\theta}$. Examples of robust functions include the *median* (MED) and the *α-trimmed mean* (TM_α), where α indicates the percentage of trimming. Thus, a final estimate of $\boldsymbol{\theta} = \{\theta_1, \theta_2, \ldots, \theta_k\}$, can be defined as

$$\hat{\theta}_j(\text{MED}) = \text{Median}(\hat{\theta}_{j1}, \hat{\theta}_{j2}, \ldots, \hat{\theta}_{jN}), \quad j = 1, 2, \ldots, k, \tag{5.70}$$

or

$$\hat{\theta}_j(\text{TM}_\alpha) = \text{TM}_\alpha(\hat{\theta}_{j1}, \hat{\theta}_{j2}, \ldots, \hat{\theta}_{jN}), \quad j = 1, 2, \ldots, k, \tag{5.71}$$

where $\text{Median}(y_1, y_2, \ldots, y_N)$ is the median of the set of numbers $\{y_1, y_2, \ldots, y_N\}$, and $\text{TM}_\alpha(y_1, y_2, \ldots, y_N)$ is the mean obtained after trimming the $(\alpha/2)\%$ largest and the $(\alpha/2)\%$ smallest order statistics of $y_1, y_2, \ldots, y_N$.

The MED estimators are very robust but inefficient. The TM_α estimators are less robust but more efficient than the MED estimators. The larger the trimming, the more robust and less efficient are the TM_α estimators.

The estimate of any desired quantile x_p can then be obtained by substituting the parameter estimates in (5.70) or (5.71) for $\boldsymbol{\theta}$ in

$$\hat{x}_p = F^{-1}(p; \hat{\boldsymbol{\theta}}), \quad 0 < p < 1. \tag{5.72}$$

5.4.2 Confidence Intervals

In some cases, the variance (and hence the standard error) of the resultant estimates may not be analytically readily available. In such cases, an estimate of the standard deviation can be easily obtained using sampling methods such as the *jackknife* and the *bootstrap* methods; see Efron (1979) and Diaconis and Efron (1974). Note that since the parameter and quantile estimates are well defined for all possible combinations of parameters and sample values, the standard error of these estimates (hence, confidence intervals for the corresponding parameter or quantile values) can be computed without difficulty. The bootstrap sampling (see Gomes and Oliveira (2001)) can be performed in two ways: the samples can be drawn with replacement directly from the data, or they can be drawn from the parametric cdf, $F(x; \hat{\boldsymbol{\theta}})$. However, it is preferable to use the *parametric bootstrap* to obtain the variance of the estimates of a particular method.

Accordingly, to obtain confidence intervals we simulate a large number of samples and obtain the corresponding estimates for each parameter θ_j. We use these estimates to obtain an empirical cdf for each parameter estimate $\hat{\theta}_j$. From each of these cdfs, we obtain the $(1-\alpha)100\%$ shortest confidence interval for θ_j as

$$\theta_j \in \left(\hat{\theta}_j(p), \hat{\theta}_j(1-\alpha+p)\right), \tag{5.73}$$

where $\hat{\theta}_j(p)$ is the pth quantile of the ecdf of $\hat{\theta}_j$.

Example 5.8 (EPM estimates for the exponential distribution). The exponential distribution has only one parameter θ, which means that for a sample of size n, there are n elemental subsets each contain only one observation. Thus, (5.69) becomes

$$x_{j:n} = F^{-1}(p_{j:n}; \theta) = -\theta \log(1 - p_{j:n}), \quad j = 1, 2, \ldots, n. \tag{5.74}$$

Solving for θ, we obtain

$$\hat{\theta}_j = \frac{-x_{j:n}}{\log(1 - p_{j:n})}, \quad j = 1, 2, \ldots, n. \tag{5.75}$$

The final estimates in (5.70) and (5.71) become

$$\hat{\theta}(\text{MED}) = \text{Median}(\hat{\theta}_1, \hat{\theta}_2, \ldots, \hat{\theta}_n), \tag{5.76}$$

and

$$\hat{\theta}(\text{TM}_\alpha) = \text{TM}_\alpha(\hat{\theta}_1, \hat{\theta}_2, \ldots, \hat{\theta}_n). \tag{5.77}$$

For example, assuming that the times between 48 consecutive telephone calls, given in Table 1.8, are exponentially distributed, the 48 initial estimates in (5.75) are shown in Table 5.1. Then the MED and TM_α of θ, for two specific values of α, are

$$\hat{\theta}(MED) = 1.006, \quad \hat{\theta}(\text{TM}_{25\%}) = 1.014, \quad \text{and} \quad \hat{\theta}(\text{TM}_{50\%}) = 1.007$$

minutes. We see that for this data set all estimators are very much the same. This should not be surprising because the exponential distribution is a one-parameter family and hence easy to estimate. ■

5.5 The Quantile Least Squares Method

The quantile least squares method estimates the parameters by minimizing the squares of the differences between the theoretical and the observed quantiles. Accordingly, the estimates are obtained by solving the following minimization problem:

$$\underset{\boldsymbol{\theta}}{\text{Minimize}} \quad \sum_{i=1}^{n} \left[x_{i:n} - F^{-1}\left(p_{i:n}; \boldsymbol{\theta}\right)\right]^2. \tag{5.78}$$

Confidence intervals can be obtained by simulating samples from the resulting population and using the bootstrap method.

Table 5.1: Telephone Data: Initial Estimates of the Exponential Parameter θ.

j	$\hat{\theta}_j$	j	$\hat{\theta}_j$	j	$\hat{\theta}_j$	j	$\hat{\theta}_j$
1	0.9700	13	1.0704	25	0.8266	37	1.0377
2	1.6798	14	1.0105	26	0.8198	38	1.0777
3	1.4245	15	0.9577	27	0.7992	39	1.0571
4	1.0569	16	1.0372	28	0.8026	40	1.0917
5	1.3007	17	0.9623	29	0.7700	41	1.0263
6	1.9139	18	0.9610	30	0.8761	42	0.9764
7	1.6218	19	0.9580	31	0.9586	43	1.0381
8	1.5147	20	0.9723	32	1.0108	44	0.9551
9	1.3797	21	1.0007	33	0.9649	45	0.8781
10	1.2267	22	0.9564	34	1.1066	46	0.8556
11	1.1013	23	0.8994	35	1.0696	47	1.0004
12	1.1748	24	0.8767	36	1.0400	48	0.8351

5.6 The Truncation Method

While dealing with extremes, we want to fit models to the tail, because only the tail of the distribution defines the domain of attraction. Then, it could be convenient to use the tail of interest alone to fit the models. To this end, we can fix a threshold value u and then consider only the sample values exceeding that threshold u.

5.7 Estimation for Multivariate Models

Several methods can be used for estimating the parameters of a multivariate distribution from data. Some of them are the following:

1. The maximum likelihood method. This is a standard and well-known method, but it may have some problems for particular families of distributions. For a detailed description of this method, see Coles (2001).

2. The weighted least squares cdf method.

3. The elemental percentile method.

4. The method based on least squares.

These are described below briefly.

5.7.1 The Maximum Likelihood Method

This is an extension of the method described for the univariate case in Section 5.1. Assume that $f(x_1, x_2, \ldots, x_n; \boldsymbol{\theta})$ is the pdf of a given population, where

$\boldsymbol{\theta}$ is a vector parameter. Then, the maximum likelihood method maximizes the loglikelihood $\ell(\boldsymbol{\theta}|x) = \sum_{i=1}^{n} \log f(x_1, x_2, \ldots, x_n; \boldsymbol{\theta})$ with respect to $\boldsymbol{\theta}$ possibly subject to some domain constraints for $\boldsymbol{\theta}$.

These estimates have good properties (see Section 5.1.2) and under some regularity conditions are asymptotically normal.

Example 5.9 (Bivariate Logistic Distribution). Let $(x_1, y_1), \ldots, (x_n, y_n)$ be a sample from a bivariate logistic distribution with joint cdf

$$F(x, y; \lambda, \delta, \sigma, \tau) = \frac{1}{1 + e^{-(x-\lambda)/\sigma} + e^{-(y-\delta)/\tau}}, \quad -\infty < x, y < \infty, \tag{5.79}$$

where $-\infty < \lambda, \delta < \infty$ are location parameters and $\sigma, \tau > 0$ are scale parameters. The joint pdf is given by

$$f(x, y; \lambda, \delta, \sigma, \tau) = \frac{2e^{-(x-\lambda)/\sigma-(y-\delta)/\tau}}{\sigma\tau\left[1 + e^{-(x-\lambda)/\sigma} + e^{-(y-\delta)/\tau}\right]^3}, \quad -\infty < x, y < \infty. \tag{5.80}$$

The likelihood function of λ, δ, σ, and τ is given by

$$\begin{aligned} L(\lambda, \delta, \sigma, \tau) &= \prod_{i=1}^{n} f(x_i, y_i; \lambda, \delta, \sigma, \tau) \\ &= \frac{2^n e^{-n(\bar{x}-\lambda)/\sigma-n(\bar{y}-\delta)/\tau}}{\sigma^n\tau^n \prod_{i=1}^{n}\left[1 + e^{-(x_i-\lambda)/\sigma} + e^{-(y_i-\delta)/\tau}\right]^3}, \end{aligned} \tag{5.81}$$

where $\bar{x}$ denotes the mean of $x_1, \ldots, x_n$. Letting $k = \log 2^n$, the loglikelihood function is

$$\begin{aligned} \ell(\lambda, \delta, \sigma, \tau) &= \log L(\lambda, \delta, \sigma, \tau) \\ &= k - \frac{n(\bar{x}-\lambda)}{\sigma} - \frac{n(\bar{y}-\delta)}{\tau} - n\log\sigma - n\log\tau \\ &\quad -3\sum_{i=1}^{n} \log\left[1 + e^{-(x_i-\lambda)/\sigma} + e^{-(y_i-\delta)/\tau}\right]. \end{aligned} \tag{5.82}$$

The partial derivatives of $\ell(\lambda, \delta, \sigma, \tau)$ with respect to λ, δ, σ, and τ are

$$\begin{aligned} \frac{\partial\ell(\lambda, \delta, \sigma, \tau)}{\partial\lambda} &= \frac{n}{\sigma} - \frac{3}{\sigma}\sum_{i=1}^{n} \frac{e^{-(x_i-\lambda)/\sigma}}{1 + e^{-(x_i-\lambda)/\sigma} + e^{-(y_i-\delta)/\tau}}, \\ \frac{\partial\ell(\lambda, \delta, \sigma, \tau)}{\partial\delta} &= \frac{n}{\tau} - \frac{3}{\tau}\sum_{i=1}^{n} \frac{e^{-(y_i-\delta)/\tau}}{1 + e^{-(x_i-\lambda)/\sigma} + e^{-(y_i-\delta)/\tau}}, \\ \frac{\partial\ell(\lambda, \delta, \sigma, \tau)}{\partial\sigma} &= \frac{n(\bar{x}-\lambda)}{\sigma^2} - \frac{n}{\sigma} - \frac{3}{\sigma^2}\sum_{i=1}^{n} \frac{(x_i-\lambda)e^{-(x_i-\lambda)/\sigma}}{1 + e^{-(x_i-\lambda)/\sigma} + e^{-(y_i-\delta)/\tau}}, \\ \frac{\partial\ell(\lambda, \delta, \sigma, \tau)}{\partial\tau} &= \frac{n(\bar{y}-\delta)}{\tau^2} - \frac{n}{\tau} - \frac{3}{\tau^2}\sum_{i=1}^{n} \frac{(y_i-\delta)e^{-(y_i-\delta)/\tau}}{1 + e^{-(x_i-\lambda)/\sigma} + e^{-(y_i-\delta)/\tau}}. \end{aligned} \tag{5.83}$$

Now, setting the partial derivatives all equal to zero and solving the system of equations in (5.83) for λ, δ, σ, and τ, we obtain the MLEs of λ, δ, σ, and τ. Also, the variance-covariance matrix of the MLEs may be approximated by the inverse of the Fisher information matrix.

■

5.7.2 The Weighted Least Squares CDF Method

Let (X, Y) be a bivariate random variable with cdf $F_{(X,Y)}(x, y; \boldsymbol{\theta})$, where $\boldsymbol{\theta} = (\theta_1, \ldots, \theta_k)$ is a possibly vector-valued parameter and $(x_1, y_1), \ldots, (x_n, y_n)$ a sample from F. Let

$$p^{xy} = \frac{m^{xy} - 0.5}{n},$$

where

$$m^{xy} = \text{number of points in the sample where } X \leq x \text{ and } Y \leq y.$$

The parameter $\boldsymbol{\theta}$ is estimated by

$$\underset{\boldsymbol{\theta}}{\text{Minimizing}} \quad \sum_{i=1}^{n} \frac{n}{p^{x_i y_i}(1 - p^{x_i y_i})} \left(F_{X,Y}(x_i, y_i; \boldsymbol{\theta}) - p^{x_i y_i}\right)^2 , \tag{5.84}$$

where the factors $\dfrac{n}{p^{x_i y_i}(1 - p^{x_i y_i})}$ are the weights that account for the variance of the different terms.

If one desires a tail fitting, the sum in (5.84) must be extended only to the corresponding tail, that is,

$$\underset{\boldsymbol{\theta}}{\text{Minimizing}} \quad \sum_{h(x_i, y_i) > a} \frac{n}{p^{x_i y_i}(1 - p^{x_i y_i})} \left[F_{X,Y}(x_i, y_i; \boldsymbol{\theta}) - p^{x_i y_i}\right]^2 , \tag{5.85}$$

where $h(x_i, y_i) > a$ defines the tail region.

5.7.3 The Elemental Percentile Method

Let $\boldsymbol{\theta} = (\theta_1, \theta_2, \ldots, \theta_k)$, that is, k be the number of parameters in $\boldsymbol{\theta}$. Consider a subset I_1 of k different sample points

$$I_1 = \{i_r | i_r \in \{1, 2, \ldots, n\},\ i_{r_1} \neq i_{r_2} \text{ if } r_1 \neq r_2,\ r = 1, 2, \ldots, k\}$$

and assume that the system of k equations in the k unknowns $(\theta_1, \theta_2, \ldots, \theta_k)$:

$$F_{(X,Y)}(x_{i_r}, y_{i_r}; \theta_{1r}, \theta_{2r}, \ldots, \theta_{kr}) = p^{x_{i_r} y_{i_r}}, \quad i_r \in I_1, \tag{5.86}$$

allows obtaining a set of elemental estimates $\{\hat{\theta}_1, \hat{\theta}_2, \ldots, \hat{\theta}_k\}$. Now we select m different sets $I_1, I_2, \ldots, I_m$ instead of just one I_1. Then, we can obtain m elemental estimates $\{\hat{\theta}_{1m}, \hat{\theta}_{2m}, \ldots, \hat{\theta}_{km}\}$. Finally, we can select an appropriate robust estimate $\hat{\theta}_j^*$ of $\theta_j, j = 1, 2, \ldots, k$. Two possibilities for such a final estimate are

1. The median,

$$\hat{\theta}_j^* = \text{Median}(\hat{\theta}_{j1}, \hat{\theta}_{j2}, \ldots, \hat{\theta}_{jm}), \quad j = 1, 2, \ldots, k. \tag{5.87}$$

2. The trimmed mean,

$$\hat{\theta}_j^* = \text{TM}_\alpha(\hat{\theta}_{j1}, \hat{\theta}_{j2}, \ldots, \hat{\theta}_{jm}), \quad j = 1, 2, \ldots, k, \tag{5.88}$$

where $\text{TM}_\alpha(y_1, y_2, \ldots, y_m)$ is the average of $y_1, y_2, \ldots, y_m$ after deleting (trimming) the smallest $(\alpha/2)100\%$ and the largest $(\alpha/2)100\%$ of the m estimates $y_1, y_2, \ldots, y_m$. The median function is very robust but inefficient. With an appropriate choice of the trimming percentage α such as $\alpha = 10$ or 20, the TM_α function is less robust but more efficient than the median.

5.7.4 A Method Based on Least Squares

The main idea of this method, that was proposed by Castillo, Sarabia, and Hadi (1997), is to write the predicted values as a function of the parameter $\boldsymbol{\theta}$, then an estimate of $\boldsymbol{\theta}$ is obtained by minimizing the sum of squared deviations between predicted and observed sample values.

Let X and Y be a bivariate random variable with cdf $F_{(X,Y)}(x, y; \boldsymbol{\theta})$. The marginal cdfs are denoted by $F_X(x; \boldsymbol{\theta})$ and $F_Y(y; \boldsymbol{\theta})$. Let

$$\begin{array}{lcl} p^x & = & \text{proportion of points in the sample where } (X \le x), \\ p^y & = & \text{proportion of points in the sample where } (Y \le y), \\ p^{xy} & = & \text{proportion of points in the sample where } (X \le x \text{ and } Y \le y). \end{array}$$

The idea is to use the joint and marginal cdfs as the basis for calculating the predicted values as functions of $\boldsymbol{\theta}$. We present here two possible methods:

1. Using $F_{(X,Y)}(x_i, y_i; \boldsymbol{\theta})$ and $F_X(x_i; \boldsymbol{\theta})$, we have

$$\left.\begin{array}{l} F_X(x_i; \boldsymbol{\theta}) = p^{x_i} \\ \\ F_{(X,Y)}(x_i, y_i; \boldsymbol{\theta}) = p^{x_i y_i} \end{array}\right\} \Rightarrow \left\{\begin{array}{l} \hat{x}_i(\boldsymbol{\theta}) = F_X^{-1}(p^{x_i}; \boldsymbol{\theta}), \\ \\ \hat{y}_i(\boldsymbol{\theta}) = F_{(X,Y)}^{-1}(p^{x_i y_i}; \hat{x}_i(\boldsymbol{\theta}), \boldsymbol{\theta}), \end{array}\right. \tag{5.89}$$

where $F_X^{-1}(p; \boldsymbol{\theta})$ is the inverse of $F_X(x_i; \boldsymbol{\theta})$ and $F_{(X,Y)}^{-1}(p; x_i, \boldsymbol{\theta})$ is the inverse of $F_{(X,Y)}(x_i, y_i; \boldsymbol{\theta})$ with respect to its second argument.

2. Using $F_{(X,Y)}(x_i, y_i; \boldsymbol{\theta})$ and $F_Y(y_i; \boldsymbol{\theta})$, we have

$$\left.\begin{array}{l} F_Y(y_i; \boldsymbol{\theta}) = p^{y_i} \\ \\ F_{(X,Y)}(x_i, y_i; \boldsymbol{\theta}) = p^{x_i y_i} \end{array}\right\} \Rightarrow \left\{\begin{array}{l} \hat{x}_i(\boldsymbol{\theta}) = F_{(X,Y)}^{-1}(p^{x_i y_i}; \hat{y}_i(\boldsymbol{\theta}), \boldsymbol{\theta}), \\ \\ \hat{y}_i(\boldsymbol{\theta}) = F_Y^{-1}(p^{y_i}; \boldsymbol{\theta}), \end{array}\right. \tag{5.90}$$

where $F_Y^{-1}(p; \boldsymbol{\theta})$ is the inverse of $F_Y(y_i; \boldsymbol{\theta})$ and $F_{(X,Y)}^{-1}(p; y_i, \boldsymbol{\theta})$ is the inverse of $F_{(X,Y)}(x_i, y_i; \boldsymbol{\theta})$ with respect to its first argument.

Taking the weighted average of (5.89) and (5.90), we obtain the new estimates

$$\begin{aligned}\hat{x}_i(\boldsymbol{\theta}) &= \beta F_X^{-1}(p^{x_i};\boldsymbol{\theta}) + (1-\beta)F_{(X,Y)}^{-1}\left(p^{x_iy_i}; F_Y^{-1}(p^{y_i};\boldsymbol{\theta}),\boldsymbol{\theta}\right),\\ \hat{y}_i(\boldsymbol{\theta}) &= \beta F_Y^{-1}(p^{y_i};\boldsymbol{\theta}) + (1-\beta)F_{(X,Y)}^{-1}\left(p^{x_iy_i}; F_X^{-1}(p^{x_i};\boldsymbol{\theta}),\boldsymbol{\theta}\right),\end{aligned} \tag{5.91}$$

where $0 \le \beta \le 1$ is an appropriately chosen weight.

Then, an estimator of $\boldsymbol{\theta}$ can now be obtained by minimizing

$$E = \sum_{i=1}^{n}\left([x_i - \hat{x}_i(\boldsymbol{\theta})]^2 + [y_i - \hat{y}_i(\boldsymbol{\theta})]^2\right) \tag{5.92}$$

with respect to $\boldsymbol{\theta}$.

With regard to the weight β in (5.91), we have two options

1. The data analyst selects an appropriate value of β. For example, taking $\beta = 0.5$, (5.91) reduces to

$$\begin{aligned}\hat{x}_i(\boldsymbol{\theta}) &= \frac{1}{2}\left(F_X^{-1}(p^{x_i};\boldsymbol{\theta}) + F_{(X,Y)}^{-1}\left(p^{x_iy_i}; F_Y^{-1}(p^{y_i};\boldsymbol{\theta}),\boldsymbol{\theta}\right)\right),\\ \hat{y}_i(\boldsymbol{\theta}) &= \frac{1}{2}\left(F_{(X,Y)}^{-1}\left(p^{x_iy_i}; F_X^{-1}(p^{x_i};\boldsymbol{\theta}),\boldsymbol{\theta}\right) + F_Y^{-1}(p^{y_i};\boldsymbol{\theta})\right),\end{aligned} \tag{5.93}$$

   which is the average of (5.89) and (5.90).

2. A value of β can be chosen optimally by minimizing (5.92) with respect to both $\boldsymbol{\theta}$ and β. Clearly, this option is better than the first one because $F_X^{-1}(p^{x_i};\boldsymbol{\theta})$ and $F_Y^{-1}(p^{y_i};\boldsymbol{\theta})$ are expected to be less variable than $F_{(X,Y)}^{-1}\left(p^{x_iy_i}; F_Y^{-1}(p^{y_i};\boldsymbol{\theta}),\boldsymbol{\theta}\right)$ and $F_{(X,Y)}^{-1}\left(p^{x_iy_i}; F_X^{-1}(p^{x_i};\boldsymbol{\theta}),\boldsymbol{\theta}\right)$ especially so near the tails of the probability distribution, hence they should be given different weights. Determining the weights optimally, however, requires more computations, especially in multivariate situations.

Note that the predictions in (5.89) and (5.93) are in some cases obtained by sequentially solving equations in a single unknown and that there is a unique solution which, when no closed solution exists, can be obtained by the bisection method.

A similar method can be used if we use the survival function, $S(x,y)$, instead of $F(x,y)$.

Example 5.10 (Standard Bivariate Logistic Distribution). The cdf of the bivariate logistic distribution $BL(\lambda,\sigma,\delta,\tau)$ is given by

$$F_{(X,Y)}(x,y;\boldsymbol{\theta}) = \left[1+\exp\left(-\frac{x-\lambda}{\sigma}\right)+\exp\left(-\frac{y-\delta}{\tau}\right)\right]^{-1}, \quad x,y\in(-\infty,\infty),$$

where $\boldsymbol{\theta} = (\lambda,\sigma,\delta,\tau)$, from which we get the marginal cdf

$$F_X(x;\boldsymbol{\theta}) = \left[1+\exp\left(-\frac{x-\lambda}{\sigma}\right)\right]^{-1}, \quad x\in(-\infty,\infty). \tag{5.94}$$

Setting $F_{(X,Y)}(\hat{x}, \hat{y}; \boldsymbol{\theta}) = p^{xy}$ and $F_X(\hat{x}; \boldsymbol{\theta}) = p^x$, we obtain the following system of equations in $\hat{x}$ and $\hat{y}$:

$$\begin{aligned} \alpha e^{-\hat{x}/\sigma} + \beta e^{-\hat{y}/\tau} &= -1 + 1/p^{xy}, \\ \alpha e^{-\hat{x}/\sigma} &= -1 + 1/p^x, \end{aligned} \tag{5.95}$$

where $\alpha = e^{\lambda/\sigma}$ and $\beta = e^{\delta/\tau}$, which has the following solution:

$$\begin{aligned} \hat{x} &= \lambda - \sigma \log(\tfrac{1}{p^x} - 1), \\ \hat{y} &= \delta - \tau \log(\tfrac{1}{p^{xy}} - \tfrac{1}{p^x}), \end{aligned} \tag{5.96}$$

provided that $p^{xy} \neq p^x$.

By symmetry, we can consider the estimators

$$\begin{aligned} \hat{x} &= \lambda - \sigma \log(\tfrac{1}{p^{xy}} - \tfrac{1}{p^y}), \\ \hat{y} &= \delta - \tau \log(\tfrac{1}{p^y} - 1), \end{aligned} \tag{5.97}$$

provided that $p^{xy} \neq p^y$.

Thus, we propose using the following equations to compute the predicted values, which are obtained by averaging (5.96) and (5.97) and replacing $\hat{x}$ and $\hat{y}$ by $\hat{x}_i$ and $\hat{y}_i$:

$$\begin{aligned} \hat{x}_i(\boldsymbol{\theta}) &= \lambda - \sigma r_i, \\ \hat{y}_i(\boldsymbol{\theta}) &= \delta - \tau s_i, \end{aligned} \tag{5.98}$$

where

$$\begin{aligned} r_i &= \begin{cases} \gamma \log(\frac{1}{p^{x_i}} - 1) + (1-\gamma) \log(\frac{1}{p^{x_i y_i}} - \frac{1}{p^{y_i}}), & if\ p^{x_i y_i} \neq p^{y_i}, \\ \log(\frac{1}{p^{x_i}} - 1), & if\ p^{x_i y_i} = p^{y_i}, \end{cases} \\ s_i &= \begin{cases} \gamma \log(\frac{1}{p^{y_i}} - 1) + (1-\gamma)(\log(\frac{1}{p^{x_i y_i}} - \frac{1}{p^{x_i}}), & if\ p^{x_i y_i} \neq p^{x_i}, \\ \log(\frac{1}{p^{y_i}} - 1), & if\ p^{x_i y_i} = p^{x_i}, \end{cases} \end{aligned} \tag{5.99}$$

where γ is the weight used for the solution (5.96). Note that when the sample size is finite, it is possible to have $p^{x_i y_i} = p^{x_i}$ or $p^{x_i y_i} = p^{y_i}$ for some sample values. Then, we minimize, with respect to λ, σ, δ, and τ,

$$\begin{aligned} E &= \sum_{i=1}^{n} \left[x_i - \hat{x}(\lambda, \sigma, \delta, \tau)\right]^2 + \left[y_i - \hat{y}(\lambda, \sigma, \delta, \tau)\right]^2 \\ &= \sum_{i=1}^{n} \left[(x_i - \lambda + \sigma r_i)^2 + (y_i - \delta + \tau s_i)^2\right]. \end{aligned} \tag{5.100}$$

Taking the derivative of E with respect to each of the parameters, we obtain

$$\begin{aligned}
\partial E/\partial \lambda &= -2\sum_{i=1}^{n}(x_i - \lambda + \sigma r_i),\\
\partial E/\partial \sigma &= 2\sum_{i=1}^{n}(x_i - \lambda + \sigma r_i)\, r_i,\\
\partial E/\partial \delta &= -2\sum_{i=1}^{n}(y_i - \delta + \tau s_i),\\
\partial E/\partial \tau &= 2\sum_{i=1}^{n}(y_i - \delta + \tau s_i)\, s_i.
\end{aligned} \tag{5.101}$$

Setting each of the above equations to zero, we obtain the system of equations

$$\begin{array}{llllll}
\lambda & n & - & \sigma & \sum_{i=1}^{n} r_i & = \sum_{i=1}^{n} x_i,\\
\lambda & \sum_{i=1}^{n} r_i & - & \sigma & \sum_{i=1}^{n} r_i^2 & = \sum_{i=1}^{n} x_i r_i,\\
\delta & n & - & \tau & \sum_{i=1}^{n} s_i & = \sum_{i=1}^{n} y_i,\\
\delta & \sum_{i=1}^{n} s_i & - & \tau & \sum_{i=1}^{n} s_i^2 & = \sum_{i=1}^{n} y_i s_i.
\end{array} \tag{5.102}$$

The solution of the above equations yield the estimators

$$\begin{aligned}
\hat{\lambda} &= \frac{\sum_{i=1}^{n} x_i r_i \sum_{i=1}^{n} r_i - \sum_{i=1}^{n} x_i \sum_{i=1}^{n} r_i^2}{\left(\sum_{i=1}^{n} r_i\right)^2 - n\sum_{i=1}^{n} r_i^2}, \quad
\hat{\sigma} = \frac{n\sum_{i=1}^{n} x_i r_i - \sum_{i=1}^{n} x_i \sum_{i=1}^{n} r_i}{\left(\sum_{i=1}^{n} r_i\right)^2 - n\sum_{i=1}^{n} r_i^2},\\
\hat{\delta} &= \frac{\sum_{i=1}^{n} y_i s_i \sum_{i=1}^{n} s_i - \sum_{i=1}^{n} y_i \sum_{i=1}^{n} s_i^2}{\left(\sum_{i=1}^{n} s_i\right)^2 - n\sum_{i=1}^{n} s_i^2}, \quad
\hat{\tau} = \frac{n\sum_{i=1}^{n} y_i s_i - \sum_{i=1}^{n} y_i \sum_{i=1}^{n} s_i}{\left(\sum_{i=1}^{n} s_i\right)^2 - n\sum_{i=1}^{n} s_i^2}.
\end{aligned} \tag{5.103}$$

■

Exercises

5.1 Consider a Bernoulli random variable with probability mass function

$$P(x;p) = \begin{cases} p, & \text{if } x = 1,\\ 1-p, & \text{if } x = 0, \end{cases} \tag{5.104}$$

and obtain:

(a) The maximum likelihood estimate of p.

(b) The moment estimate of p.

(c) The Fisher information matrix.

5.2 Consider a Poisson random variable with probability mass function

$$P(x;\lambda) = \frac{\lambda^x \exp\{-\lambda\}}{x!}, \quad x = 0, 1, \ldots, \quad \lambda > 0. \tag{5.105}$$

and obtain:

(a) The maximum likelihood estimate of λ.

(b) The moment estimate of λ.

(c) The Fisher information matrix.

5.3 Given the exponential density function

$$f(x;\lambda) = \lambda \exp\{-\lambda x\}, \quad x > 0, \quad \lambda > 0, \tag{5.106}$$

obtain:

(a) The maximum likelihood estimate of λ.

(b) The moment estimate of λ.

(c) The Fisher information matrix.

5.4 Let X be a logistic random variable with pdf

$$f(x) = \frac{\exp\{-(x-\alpha)/\beta\}}{\beta\left[1+\exp\{-(x-\alpha)/\beta\}\right]^2}, \quad -\infty < x, \alpha < \infty, \ \beta > 0.$$

Show that the information matrix is

$$\mathbf{I} = \begin{bmatrix} 1/(3\beta^2) & 0 \\ 0 & (3+\pi^2)/(9\beta^2) \end{bmatrix}.$$

5.5 Given the normal density

$$f(x;\mu,\sigma) = \frac{1}{\sigma\sqrt{2\pi}} \exp\left\{-\frac{1}{2\sigma^2}(x-\mu)^2\right\}, \tag{5.107}$$

obtain:

(a) The maximum likelihood estimates of μ and σ.

(b) The moment estimates of μ and σ.

(c) The Fisher information matrix.

5.6 Consider the family of densities:

$$f(x) = \begin{cases} 1 + \left(\frac{2-2a}{a^2}\right)x, & \text{if } 0 \le x \le a, \\ 0, & \text{otherwise.} \end{cases}$$

Obtain:

(a) The maximum likelihood estimate of a.

(b) The moment estimate of a.

(c) Given the sample data

$$\{0.1, 0.48, 0.3, 0.50, 0.52, 0.54, 0.35, 0.40, 0.45, 0.2\},$$

obtain the estimates of a using both methods.

5.7 For the two-parameter exponential distribution in Example 5.5, derive the MOM estimators of μ and θ. Also, derive their variances, compare them with those of the MLEs, and comment.

5.8 For the two-parameter exponential distribution in Example 5.5, derive the PWM estimators of μ and θ by considering S_{10}, S_{01} and S_{11}.

5.9 Obtain the moment parameter estimates for the Weibull and Gumbel families.

5.10 Obtain the elemental percentile estimates for the parameters of the Weibull distribution.

5.11 Obtain the quantile least squares method estimates for the parameters of the Gumbel distribution.

5.12 Obtain the maximum likelihood estimates for the bivariate normal distribution.

Chapter 6

Model Selection and Validation

Mathematical or statistical models are often initially specified based on knowledge in the field of study, statistical theory, the available data, and/or assumptions. The knowledge of the phenomena being studied could be theoretical or empirical. This knowledge is usually provided by experts in the field of study.

Sometimes, models are initially specified based on statistical or probabilistic arguments, but in many other situations the models are specified merely by assumptions. For example, if we wish to model a binomial experiment, then we are pretty sure that the corresponding random variable has a binomial distribution. On the other hand, if we wish to model the lifetime of an element, there are several candidate models available. Basically, any distribution that describes a nonnegative random variable could be used (e.g., exponential, Gamma, Weibull, Fréchet, Pareto, etc.).

If the knowledge in the field of study is lacking or limited, then statistical methods could be used to select a model based on the available data. Statistical methods themselves are often based on assumptions. All assumptions, regardless of their sources or reasons, should be validated before the finally chosen model is used in practice for decision-making.

Additionally, the initially specified models or family of models usually depend on unknown parameters. These parameters are estimated using the available data and estimation methods as discussed in general terms in Chapter 5 and applied to extremes models in Chapters 9, 10, and 11.

We should also keep in mind that, in all cases, mathematical and statistical models are simplifications and/or approximations of reality. For this and the above reasons, wise decision-making in the face of uncertainty requires that models must be chosen carefully and validated thoroughly before they are put to use in practice.

This chapter deals with the model selection and validation problems. It should be recognized, however, that model selection and validation are not easy

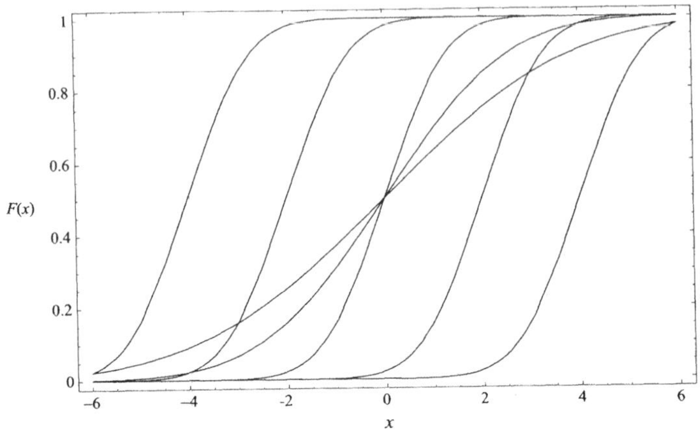

Figure 6.1: The cdfs corresponding to a normal family $N(\mu, \sigma^2)$ on arithmetic scales.

tasks and they require time and knowledge. Section 6.1 introduces probability plots and shows how to build several types of probability papers and how they can be used to select a parent distribution for the given sample. Section 6.2 deals with the problem of selecting models by hypothesis testing techniques. Section 6.3 discusses the problem of model validation using the Q-Q and the P-P plots.

6.1 Probability Paper Plots

One of the most commonly used graphical methods in statistics by engineers is the probability paper plot (PPP). In this section, the most important probability paper plots are described.[1]

If the cdfs of a given parametric family are drawn on arithmetic scales, one obtains a family of curves. For example, Figure 6.1 shows several cdfs corresponding to a normal family $N(\mu, \sigma^2)$.

The basic idea of PPP of a two-parameter family of distributions consists of changing the random variable X to $U = h(X)$ and the probability P to $V = g(P)$ in such a manner that the cdfs become a family of straight lines. In this way, when the cdf is drawn, a linear trend is an indication of the sample coming from the corresponding family.

It is important to realize that since the plane has two dimensions, probability plots of only two-parameter families are possible.

In practice, however, we may not know the exact cdf. In these cases, we use the *empirical* cumulative distribution function (ecdf) as an approximation

[1]Some of the material in this section is reprinted from the book *Extreme Value Theory in Engineering*, by E. Castillo, Copyright © Academic Press (1988), with permission from Elsevier.

to the true but unknown cdf. Let $x_1, x_2, \ldots, x_n$ be a random sample drawn from a two-parameter cdf, $F(x; a, b)$, where a and b are the parameters, and let $x_{1:n}, x_{2:n}, \ldots, x_{n:n}$ be the corresponding observed order statistics. Then, the *empirical cumulative distribution function* (ecdf) is defined as

$$F(x) = \begin{cases} 0, & \text{if } x < x_{1:n}, \\ i/n, & \text{if } x_{i:n} \leq x < x_{i+1:n}, \quad i = 1, \ldots, n-1, \\ 1, & \text{if } x \geq x_{n:n}. \end{cases} \tag{6.1}$$

This is a step (jump) function with steps $0, 1/n, 2/n, \ldots, 1$. There are several methods that can be used to smooth the ecdf (see Simonoff (1996)).

Due to the fact that in the order statistic $x_{i:n}$ the probability jumps from $(i-1)/n$ to i/n, for the two extreme cases of $i = 1$ and $i = n$ these jumps involve $p = 0$ and $p = 1$, respectively, that when we apply the scale transformation can transform to $g(0) = -\infty$ or $g(1) = \infty$, respectively. Thus, if one wants to plot this sample on a probability plot using the ecdf one runs into problems because the infinity points cannot be drawn. Thus, instead of plotting the set of points $\{(x_{i:n}, i/n) | i = 1, 2, \ldots, n\}$, one plots the set $\{(x_{i:n}, p_{i:n}) | i = 1, 2, \ldots, n\}$, where $p_{i:n}$ is one of the plotting positions given in (5.1)–(5.3). In this book, unless otherwise stated, we will use

$$p_{i:n} = i/(n+1), \quad i = 1, \ldots, n. \tag{6.2}$$

To obtain a probability plot, we look for a transformation that expresses the equation

$$p = F(x; a, b) \tag{6.3}$$

in the form of a straight line,

$$g(p) = g[F(x; a, b)] = ah(x) + b, \tag{6.4}$$

or, equivalently,

$$v = au + b, \tag{6.5}$$

where

$$u = h(x), \tag{6.6}$$

$$v = g(p). \tag{6.7}$$

The variable v is called the *reduced* variable. Thus, for the existence of a probability plot associated with a given family of cdfs, $F(x; a, b)$, it is necessary to have

$$F(x; a, b) = g^{-1}[ah(x) + b], \tag{6.8}$$

which is the key formula to derive probability plots.

Note that (6.8) is a functional equation with three unknowns F, h, and g, that when solved gives all possible families $F(x; a, b)$ of cdfs with probability plots, and the associated required transformations $h(x)$ and $g(p)$. If the family $F(x; a, b)$ is given, the functional equation gives the associated transformations $h(x)$ and $g(p)$. Using (6.8), one can also build the family $F(x; a, b)$ directly from the given transformations $h(x)$ and $g(p)$.

Definition 6.1 (Probability paper plot). *Let $x_1, \ldots, x_n$ be a sample drawn from $F(x; a, b)$. Let $x_{1:n}, x_{2:n}, \ldots, x_{n:n}$ be the corresponding order statistics and $p_{1:n}, p_{2:n}, \ldots, p_{n:n}$ be plotting positions such as those given in (5.1)–(5.3). The scatter plot of*

$$p_{i:n} \quad \textit{versus} \quad x_{i:n} \quad i = 1, 2, \ldots, n, \tag{6.9}$$

is called a probability paper plot (PPP), where the ordinate and abscissa axes are transformed by $v_i = g(p_{i:n})$ and $u_i = h(x_{i:n})$, as defined in (6.4), respectively.

If $x_1, x_2, \ldots, x_n$ have indeed been drawn from $F(x; a, b)$, the scatter of points in the PPP would exhibit a straight line trend with positive slope. However, due to the random character of samples, even in the case of the sample drawn from the given family, one should not expect that the corresponding graph will be an exact straight line. Thus, if the trend is approximately linear, it can be used as an evidence that the sample did come from the assumed family.

Probability Plot Bands. As we shall see in Chapter 7, the rth order statistic of a sample coming from a uniform distribution, $U(0, 1)$, is distributed as Beta$(r, n - r + 1)$. Using this result, one can obtain a $100(1 - \alpha)\%$ confidence band for the rth order statistic, which is given by

$$\left(F^{-1}_{Beta(r,n-r+1)}(\alpha/2), F^{-1}_{Beta(r,n-r+1)}(1 - \alpha/2)\right), \quad r = 1, 2, \ldots, n. \tag{6.10}$$

These bands can be plotted on a probability paper to indicate whether the plots are as expected or some deviations occur. This has been done in the plots appearing in the subsequent chapters, where we have used the 0.025 and 0.975 bands that correspond to $\alpha = 0.05$.

We should alert the reader here about the fact that these bands cannot be used for rejecting or accepting the hypothesis that the entire sample comes from the family associated with the probability paper for $\alpha = 0.05$. They are only an indication for each individual order statistic.

Return Periods. The PPP plot can be enhanced further by showing the return period[2] on the vertical axis on the right-hand side of the graph. Thus, from the PPP, the values of the random variable associated with given return periods can be easily found. The reader must be warned about the fact that the concept of return period depends on whether we are interested in minima or maxima values (see Section 7.5).

Once we have checked that the cdf belongs to the family, the PPP can also be used to estimate the parameters of the family, if desired. However, better methods for parameter and quantile estimation are given in Chapters 5 and 9.

The transformation required to construct the PPP for some families of distributions are derived below. Other distributions are left as an exercise for the reader. Table 6.1 summarizes the transformations associated with some of the most important PPPs. Unfortunately, not all two-parameter families possess a probability paper. For example, see Exercise 6.1.

[2]See Section 7.5 for a formal definition of the return period.

Table 6.1: Transformations Required for Different Probability Papers.

Probability Paper[a]	Reference Equation	Random Variable Scale u	Probability Scale v
Normal	(3.37)	x	$\Phi^{-1}(p)$
Log-normal	(3.43)	$\log x$	$\Phi^{-1}(p)$
Weibull_M	(3.60)	$-\log(\lambda - x)$	$-\log(-\log p)$
Gumbel_M	(3.63)	x	$-\log(-\log p)$
Fréchet_M	(3.69)	$\log(x - \lambda)$	$-\log(-\log p)$
Weibull_m	(3.57)	$\log(x - \lambda)$	$\log(-\log(1-p))$
Gumbel_m	(3.66)	x	$\log(-\log(1-p))$
Fréchet_m	(3.73)	$-\log(\lambda - x)$	$\log(-\log(1-p))$

[a] The symbols M and m indicate maximal and minimal, respectively.

6.1.1 Normal Probability Paper Plot

The pdf of the normal random variable, $N(\mu, \sigma^2)$, is given in (3.37). In this case, (6.3) can be written as

$$p = F(x; \mu, \sigma) = \Phi\left(\frac{x - \mu}{\sigma}\right), \tag{6.11}$$

where μ and σ are the mean and the standard deviation, respectively, and $\Phi(x)$ is the cdf of the standard normal distribution, which is given in (3.40). Equation (6.11) can in turn be written as

$$\Phi^{-1}(p) = \frac{x - \mu}{\sigma} = \frac{1}{\sigma}\, x - \frac{\mu}{\sigma}. \tag{6.12}$$

A comparison of (6.12) with (6.4) and (6.5) gives

$$v = g(p) = \Phi^{-1}(p), \quad u = h(x) = x, \quad a = \frac{1}{\sigma}, \quad b = -\frac{\mu}{\sigma},$$

and the family of straight lines becomes

$$v = \frac{1}{\sigma}\, u - \frac{\mu}{\sigma}. \tag{6.13}$$

Thus, in a normal PPP, the ordinate axis needs to be transformed by $v = \Phi^{-1}(p)$, whereas the abscissa axis need not be transformed.

Once the normality assumption has been checked, estimation of the parameters μ and σ is straightforward. In fact, setting $v = 0$ and $v = 1$ in (6.13), we obtain

$$\begin{aligned} v = 0 \quad &\Rightarrow \quad u = \mu, \\ v = 1 \quad &\Rightarrow \quad u = \mu + \sigma. \end{aligned} \tag{6.14}$$

Note that the probability scales on the normal probability paper have symmetry with respect to $p = 0.5$, that is, the distance $g(0.5 + \alpha) - g(0.5)$ on the reduced scale between the points $p = 0.5 + \alpha$ and $p = 0.5$ is identical to the distance $g(0.5) - g(0.5 - \alpha)$ between $p = 0.5$ and $p = 0.5 - \alpha$, for any valid value α. This is not true for other probability papers. This property allows distinguishing the normal PPP from other PPPs such as Gumbel, Weibull, and Fréchet.

Example 6.1 (Normal PPP). The upper panel in Figure 6.2 shows the normal PPP for the Houmb's wave heights data described in Section 1.4.5. The graph shows a clear linear trend, which means that the assumption that the data came from a normal distribution cannot be rejected. The parameters μ and σ can then be estimated as follows. Obtain the least squares line, that is, $v = 0.420u - 4.084$, and use (6.14) to get $\mu = 9.715$ and $\sigma = 2.379$. The same values can be obtained approximately from the graph as follows. Since $v = \Phi^{-1}(p)$, we have $p = \Phi(v)$. Thus, from the standard normal table (see Table A.1 in the Appendix) we see that for $v = 0$, $p = 0.5$ and for $v = 1$, $p = 0.8413$. From the graph in Figure 6.2 we see that for $p = 0.5$, $u = 9.7$, and for $p = 0.8413$ the corresponding value is $u = 12.1$; hence, from (6.14), we obtain $\mu \approx 9.7$ meters and $\sigma = u - \mu \approx 2.4$ meters.

The lower panel in Figure 6.2 shows the normal PPP for the wind data described in Section 1.4.1. The graph shows a clear curvature, which indicates that the normality assumption is not valid. An alternative model that fits these data better is given in Example 9.14. ■

6.1.2 Log-Normal Probability Paper Plot

The pdf of the log-normal random variable is given in (3.43). The case of the log-normal PPP can be obtained from the case of the normal PPP by taking into account the fact that X is log-normal if and only if its logarithm, $\log(X)$, is normal. Consequently, we transform X into $\log(X)$ and obtain a normal PPP. Thus, in addition to the transformation of the probability axis required for the normal probability paper plot, we need to transform the X scale to a logarithmic scale. The mean μ^* and the standard deviation σ^* of the log-normal distribution can then be estimated by (see (3.44))

$$\mu^* = e^{\mu+\sigma^2/2} \quad \text{and} \quad \sigma^{*2} = e^{2\mu}\left(e^{2\sigma^2} - e^{\sigma^2}\right), \tag{6.15}$$

where μ and σ are the values obtained according to (6.14).

Example 6.2 (Log-normal PPP). Let $x_1, x_2, \ldots, x_{40}$ represent the 40 observations in the Fatigue data, described in Section 1.4.10. Assuming that the data were drawn from a log-normal distribution, then $\log(x_i), i = 1, 2, \ldots, 40$, have a normal distribution. The normal PPP using $\log(x_i)$ is shown as the top panel in Figure 6.3. The scatter of points resemble a linear trend that after

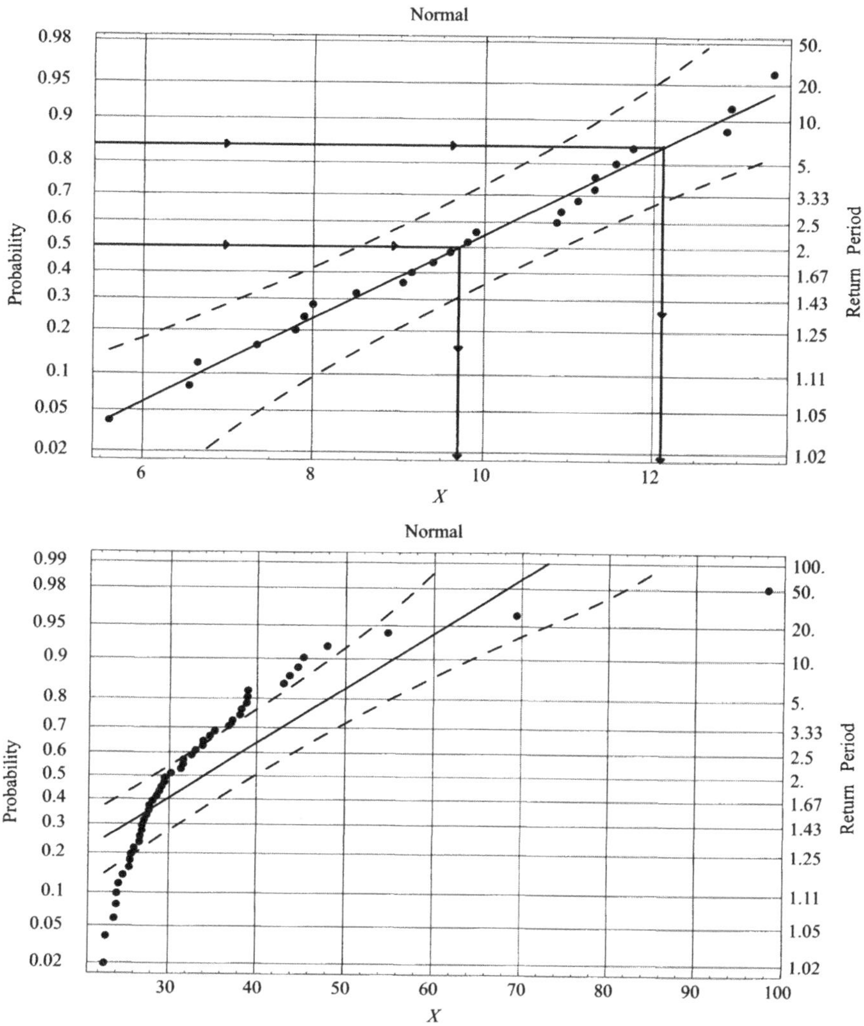

Figure 6.2: Normal probability paper plots. Top panel: Houmb's data. Lower panel: wind data.

fitting a least squares straight line gives $v = 2.0425u - 23.282$. As in Example 6.1, we find that $\mu = 11.4$ and $\sigma = 0.4896$. Consequently, according to (6.15) the mean and variance of the log-normal distribution are

$$\begin{aligned}
\mu^* &= \exp\left(11.4 + 0.4896^2/2\right) = 100596 \text{ cycles}, \\
\sigma^{*2} &= \exp\left(2 \times 11.4\right)\left[\exp\left(2 \times 0.4896^2\right) - \exp\left(0.4896^2\right)\right] \\
&= 2.7416 \times 10^9 \text{ cycles}^2.
\end{aligned}$$

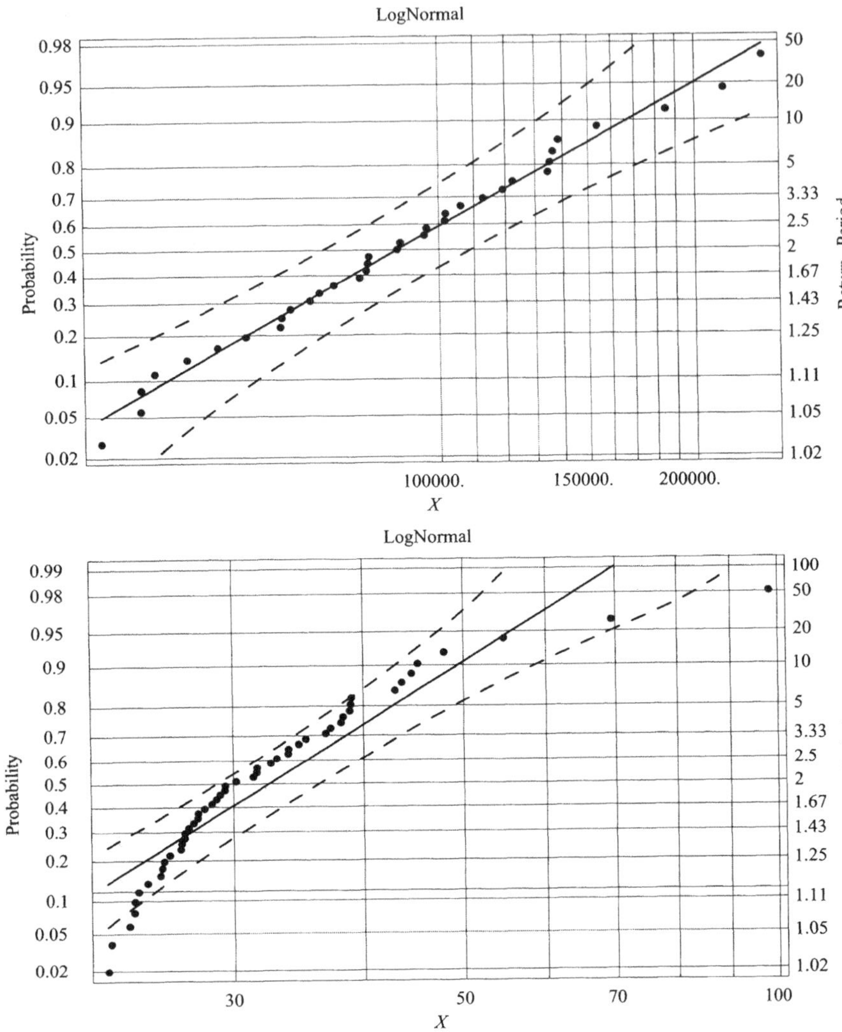

Figure 6.3: Log-normal probability paper plots. Top panel: Fatigue data. Lower panel: wind data.

Consider again the wind data, described in Section 1.4.1. The lower panel in Figure 6.3 shows the normal PPP of the logarithmic transformation of these data with a superimposed least squares line. The graph shows a curvature, which means that the wind data are not likely to have been drawn from a log-normal distribution. ■

6.1.3 Gumbel Probability Paper Plot

In this section, we derive the maximal and minimal Gumbel probability paper plots. The maximal Gumbel cdf is (see (3.63))

$$F(x;\lambda,\delta) = \exp\left[-\exp\left(\frac{\lambda - x}{\delta}\right)\right], \quad -\infty < x < \infty. \tag{6.16}$$

Let $p = F(x;\lambda,\delta)$. Then taking logarithms twice we get

$$-\log[-\log(p)] = \frac{1}{\delta}\,x - \frac{\lambda}{\delta}.$$

Upon comparison with (6.4), we get

$$\begin{aligned} v &= g(p) = -\log[-\log(p)], \quad (6.17)\\ u &= h(x) = x,\\ a &= 1/\delta, \quad (6.18)\\ b &= -\lambda/\delta, \end{aligned}$$

which shows that the transformation (6.17) transforms (6.16) to the family of straight lines

$$v = \frac{1}{\delta}\,u - \frac{\lambda}{\delta}.$$

Thus, in a maximal Gumbel PPP, the ordinate axis need to be transformed by $v = -\log[-\log(p)]$, whereas the abscissa axis need not be transformed.

Estimation of the two parameters λ and δ can be done by setting $v = 0$ and $v = 1$, and obtaining

$$\begin{aligned} v = 0 &\Rightarrow u = \lambda,\\ v = 1 &\Rightarrow u = \lambda + \delta. \end{aligned} \tag{6.19}$$

Once we have fitted a straight line to the data, the abscissas associated with the reduced variable, v, 0 and 1 are the values λ and $\lambda + \delta$, respectively.

The PPP for the minimal Gumbel is derived in a similar manner. The cdf is (see (3.66))

$$F(x) = 1 - \exp\left[-\exp\left(\frac{x - \lambda}{\delta}\right)\right], \quad -\infty < x < \infty, \tag{6.20}$$

from which it follows that

$$v = \frac{1}{\delta}\,u - \frac{\lambda}{\delta},$$

where

$$\begin{aligned} v &= g(p) = \log[-\log(1-p)], \quad (6.21)\\ u &= h(x) = x. \end{aligned}$$

Therefore, in a minimal Gumbel PPP, the ordinate axis need to be transformed by $v = \log[-\log(1-p)]$, whereas the abscissa axis need not be transformed. Once we have fitted a straight line to the data, the abscissas associated with the reduced variable, v, 0 and 1 are the values λ and $\lambda + \delta$, respectively.

Example 6.3 (Maximal Gumbel PPP). The yearly maximum wave heights in a given region measured in feet are given in Table 1.3. The data are plotted on a maximal Gumbel PPP in the top panel of Figure 6.4. Since the trend is clearly linear, the sample can be accepted as having come from the maximal Gumbel family of distributions.

Based on the fitted line in this figure, the following estimates are obtained. Since $v = 0.16504u - 1.84728$, for $v = 0$ and $v = 1$, we obtain $u = 11.193$ and $u = 17.252$, respectively. This gives $\lambda = 11.193$ feet and $\delta = 6.059$ feet. The same results can be obtained from the graph taking into account that, since $v = -\log[-\log(p)]$, we have $p = \exp[-\exp(-v)]$. Thus, for $v = 0$ and $v = 1$, we obtain $p = e^{-1} = 0.37$ and $e^{-1/e} = 0.692$. Using these values we obtain $\lambda = 11.19$ and $\lambda + \delta = 17.25$ (see Fig. 6.4).

From the vertical axis on right-hand side of the graph we can see that wave heights of 35, 25, and 13.4 feet have return periods of 50, 10, and 2 years, respectively.

Returning again to the wind data, described in Section 1.4.1, the lower panel in Figure 6.4 shows the maximal Gumbel PPP for these data with the corresponding least squares line. The curvature pattern of the points indicate that the maximal Gumbel distribution does not provide a good fit for these data. ■

6.1.4 Weibull Probability Paper Plot

The maximal Weibull cdf is (see (3.60))

$$p = F(x; \lambda, \beta, \delta) = \exp\left[-\left(\frac{\lambda - x}{\delta}\right)^{\beta}\right], \quad x \le \lambda, \beta > 0. \tag{6.22}$$

Letting $p = F(x; \lambda, \beta, \delta)$ and taking logarithms twice we get

$$-\log[-\log(p)] = -\beta \log\left(\frac{\lambda - x}{\delta}\right) = -\beta\log(\lambda - x) + \beta\log\delta.$$

A comparison with (6.4) gives

$$\begin{aligned} u &= h(x) = -\log(\lambda - x), \\ v &= g(p) = -\log[-\log(p)], \\ a &= \beta, \\ b &= \beta\log\delta. \end{aligned} \tag{6.23}$$

This shows that the transformation (6.23) transforms (6.22) to a family of straight lines

$$v = \beta u + \beta \log \delta. \tag{6.24}$$

Note that the v scale coincides with that for the maximal Gumbel PPP, but now the u scale is logarithmic. Note also that this is a three-parameter family,

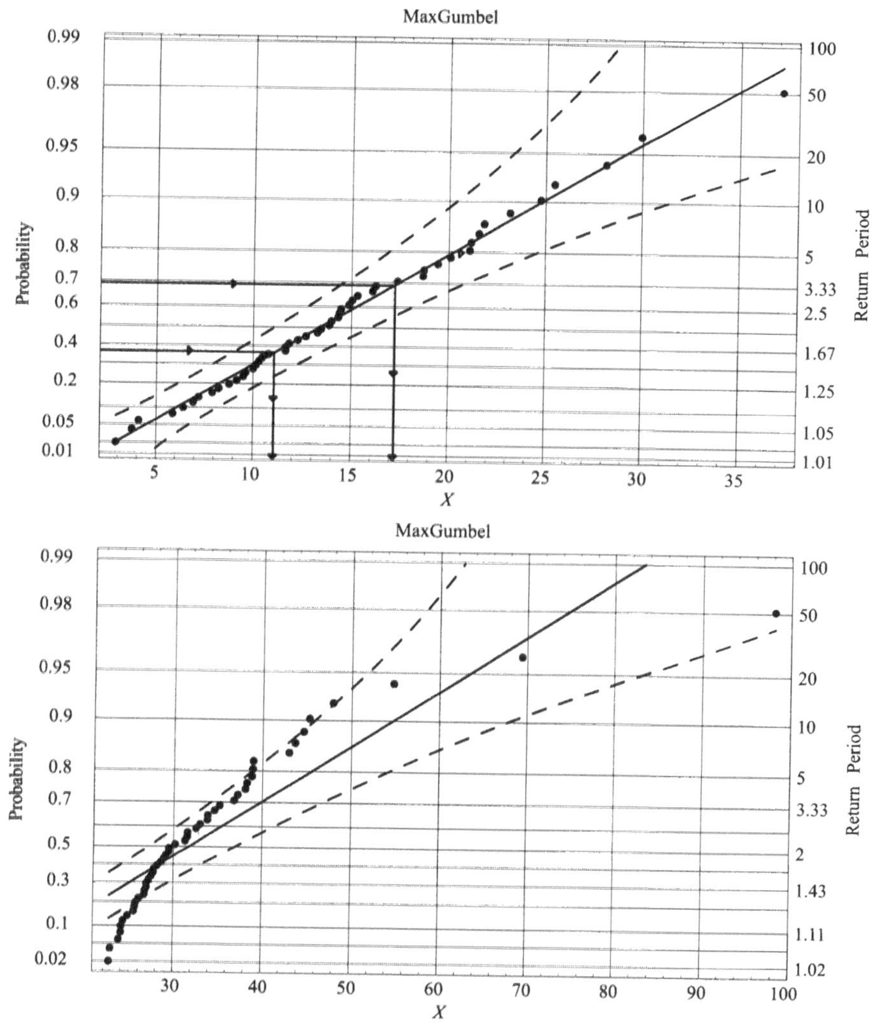

Figure 6.4: Maximal Gumbel probability paper plots. Top panel: wave heights data. Lower panel: wind data.

so we need to estimate λ to be able to construct the graph. For this purpose, we try successive values of $\lambda \geq x_{n:n}$ until we get a straight line, or select the upper end λ by other methods or physical reasons. Then, estimates of the remaining parameters β and δ are obtained as before by setting $u = 0$ and $u = 1$, to get

$$v = 0 \Rightarrow u = -\log \delta,$$
$$v = 1 \Rightarrow u = \frac{1}{\beta} - \log \delta.$$

The cdf of the minimal Weibull is (see (3.57))

$$F(x) = 1 - \exp\left[-\left(\frac{x-\lambda}{\delta}\right)^{\beta}\right], \quad x \geq \lambda, \beta > 0, \tag{6.25}$$

from which it follows that

$$v = \beta u - \beta \log \delta, \tag{6.26}$$

where

$$\begin{aligned} v &= g(p) = \log[-\log(1-p)], \\ u &= h(x) = \log(x - \lambda). \end{aligned} \tag{6.27}$$

Note that the v scale coincides with that for the minimal Gumbel PPP, but now the u scale is logarithmic.

Here also, we first need an estimate of λ. We try successive values of $\lambda \leq x_{1:n}$ until we get a straight line. Then estimates of the remaining parameters β and δ are obtained by setting $v = 0$ and $v = 1$, to get

$$\begin{aligned} v = 0 &\Rightarrow u = \log \delta, \\ v = 1 &\Rightarrow u = \frac{1}{\beta} + \log \delta. \end{aligned}$$

Example 6.4 (Maximal Weibull PPP). The data given in Table 1.5 are the yearly oldest age at death of men in Sweden during the period from 1905 to 1958. The data are plotted on a maximal Weibull probability paper for three estimates $\lambda = 106.8, 108$, and 115 years in Figure 6.5. Note that λ must be larger than $x_{n:n} = 106.5$. The fit first improves with increasing values of λ, then deteriorates afterwards. It is clear then that the $\lambda = 108$ (the middle panel) gives a better fit than the other values of λ. In this case, the least squares line becomes $v = 2.881u + 4.833$, so that (see (6.24)) $\beta = 2.881$ and $\delta = 0.187$.

The ages $108 - 1/0.31 = 104.77$ and $108 - 1/0.72 = 106.61$ years have return periods of 5 and 50 years, respectively. ∎

Example 6.5 (Minimal Weibull PPP). The data given in Table 1.4 are the yearly oldest age at death of women in Sweden during the period from 1905 to 1958. The data are plotted on a minimal Weibull probability paper for the three estimates $\lambda = 101, 98$, and 80 years in Figure 6.6. Note that λ must be smaller than $x_{1:n} = 101.5$. The fit first improves with decreasing values of λ, then deteriorates afterwards. It is clear then that the $\lambda = 98$ years (the middle panel) gives a better fit than the other two values of λ.

The surprising result is that a minimum family fits the maximum data well. This can be an indication of some dependence in the sample or that the number of data points is not sufficient to reveal the maximal domain of attraction. ∎

Note that the maximal probability papers magnify the right tails (see Fig. 6.5), and the minimal probability papers magnify the left tails (see Fig. 6.6).

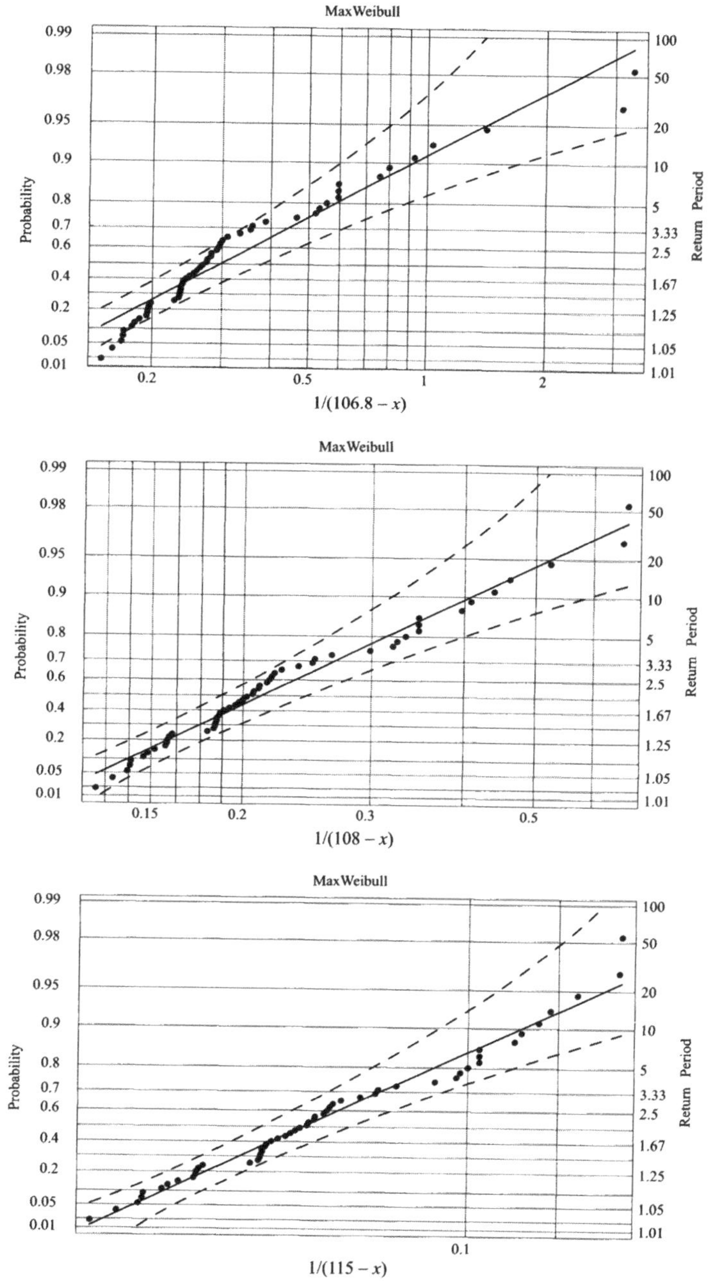

Figure 6.5: Maximal Weibull PPP for three estimates of λ for the Oldest Ages at Death data for men. The middle panel provides the best fit.

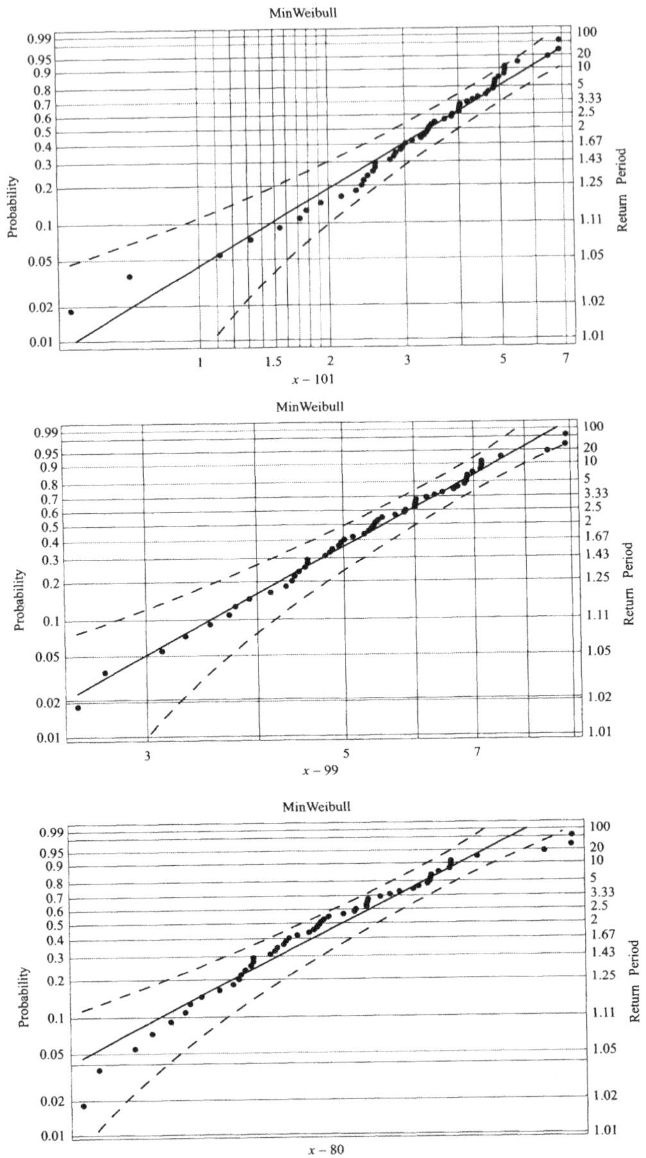

Figure 6.6: Minimal Weibull PPP for three estimates of λ for the Oldest Ages at Death data for women. The middle panel provides the best fit.

6.2 Selecting Models by Hypothesis Testing

The procedures described in the previous section are based on graphical displays. Another approach to the model selection problem is to formulate the problem into a hypothesis testing framework. To this end, consider a family of models $\mathcal{M}(\boldsymbol{\theta})$ corresponding to a parametric family of pdf $\{f(x;\boldsymbol{\theta})|\boldsymbol{\theta} \in \boldsymbol{\Theta}\}$, where $\boldsymbol{\theta}$ is a vector-valued parameter with dimension m. Let $\boldsymbol{\theta}$ be partitioned as $\boldsymbol{\theta} = (\boldsymbol{\theta}_1, \boldsymbol{\theta}_2)$, where m_1 and m_2 are the dimensions of $\boldsymbol{\theta}_1$ and $\boldsymbol{\theta}_2$, respectively, so that $m_1 + m_2 = m$. The family of models can then be written as $\mathcal{M}(\boldsymbol{\theta}_1, \boldsymbol{\theta}_2)$. Consider the restricted model $\mathcal{M}_0(\boldsymbol{\theta}_1^*, \boldsymbol{\theta}_2)$ resulting after fixing $\boldsymbol{\theta}_1 = \boldsymbol{\theta}_1^*$ in $\mathcal{M}(\boldsymbol{\theta}_1, \boldsymbol{\theta}_2)$. Now, we wish to test

$$H_0 : X \sim \mathcal{M}_0(\boldsymbol{\theta}_1^*, \boldsymbol{\theta}_2) \quad \text{versus} \quad H_1 : X \sim \mathcal{M}(\boldsymbol{\theta}_1, \boldsymbol{\theta}_2).$$

Let $\hat{\boldsymbol{\theta}}$ be the maximum likelihood estimate of $\boldsymbol{\theta}$ obtained from the unrestricted model $\mathcal{M}(\boldsymbol{\theta}_1, \boldsymbol{\theta}_2)$. Similarly, let $\hat{\boldsymbol{\theta}}_0$ be the maximum likelihood estimate of $\boldsymbol{\theta}$ obtained from the restricted model $\mathcal{M}_0(\boldsymbol{\theta}_1^*, \boldsymbol{\theta}_2)$. The corresponding log-likelihoods at their maxima values are denoted by $\ell(\hat{\boldsymbol{\theta}}_1, \hat{\boldsymbol{\theta}}_2)$ and $\ell(\hat{\boldsymbol{\theta}}_2|\boldsymbol{\theta}_1 = \boldsymbol{\theta}_1^*)$. Then, according to Theorem 5.2, we reject H_0 in favor of H_1 if

$$2\left[\ell(\hat{\boldsymbol{\theta}}_1, \hat{\boldsymbol{\theta}}_2) - \ell(\hat{\boldsymbol{\theta}}_2|\boldsymbol{\theta}_1 = \boldsymbol{\theta}_1^*)\right] > \chi^2_{m_1}(1-\alpha), \tag{6.28}$$

where $\chi^2_{m_1}(1-\alpha)$ is the $(1-\alpha)$ quantile of the χ^2 distribution with m_1 degrees of freedom.

Some recent references for tests for selection of extreme value models are Marohn (1998, 2000), Liao and Shimokawa (1999), and Galambos (2001).

Example 6.6 (Weibull versus Gumbel). Consider the wind data described in Section 1.4.1. If we use the maximum likelihood method to estimate a Gumbel model, we get the following parameter estimates and the log-likelihood optimal value:

$$\hat{\lambda} = 29.4484, \quad \hat{\delta} = 6.7284, \quad \text{and} \quad \ell = -179.291.$$

The standard errors of $\hat{\lambda}$ and $\hat{\delta}$ are 0.986 and 0.812, respectively.

If, alternatively, we use the maximum likelihood method to estimate a maximal GEVD, we obtain the following parameter estimates and the log-likelihood optimal value:

$$\hat{\lambda} = 28.095, \quad \hat{\delta} = 5.06472, \quad \hat{\kappa} = -0.419984, \quad \text{and} \quad \ell = -171.901.$$

The standard errors of $\hat{\lambda}$, $\hat{\delta}$, and $\hat{\kappa}$ are $0.839, 0.762$, and 0.150, respectively.

Since the interval estimate for $\hat{\kappa} = -0.419984 \pm 1.96 \times 0.149847$ does not contain the origin, we can conclude that the wind data does not come from a Gumbel parent, and that the GEVD is a better model.

Since

$$2\left[\ell(\hat{\boldsymbol{\theta}}_1, \hat{\boldsymbol{\theta}}_2) - \ell(\hat{\boldsymbol{\theta}}_2|\boldsymbol{\kappa}_1 = 0)\right] = -171.901 + 179.291 = 7.39$$

and $\chi_1^2(0.95) > 3.8415$ (see Table A.3 in the Appendix), we conclude once again that the GEVD is the right parent. ■

6.3 Model Validation

In this section, we give some methods for checking whether a fitted model is in agreement with the data. In other words, we measure the quality of the fitted model. We rely heavily on graphical displays rather than on formal test statistics. The pattern of points on a graph can be far more informative than the values of overall test statistics.

Let $x_1, x_2, \ldots, x_n$ be a sample from a given population with cdf $F(x)$. Let $x_{1:n}, x_{2:n}, \ldots, x_{n:n}$ be the corresponding order statistics and $p_{1:n}, p_{2:n}, \ldots, p_{n:n}$ be plotting positions such as those given in (5.1)–(5.3). Here we use $p_{i:n} = i/(n+1)$. Finally, let $\hat{F}(x)$ be an estimate of $F(x)$ based on $x_1, x_2, \ldots, x_n$. Thus, $\hat{F}^{-1}(p_{i:n})$ is the estimated quantile corresponding to the ith order statistic, $x_{i:n}$. Similarly, $\hat{F}(x_{i:n})$ is the estimated probability corresponding to $x_{i:n}$. Next we discuss the *Quantile-Quantile* (Q-Q) plots and the *Probability-Probability* (P-P) plots in model validation.

For other treatment of the validation problem, see Drees, de Haan, and Li (2003) and Fei, Lu, and Xu (1998).

6.3.1 The Q-Q Plots

Let $\hat{F}(x)$ be an estimate of $F(x)$ based on $x_1, x_2, \ldots, x_n$. The scatter plot of the points

$$\hat{F}^{-1}(p_{i:n}) \quad \text{versus} \quad x_{i:n}, \quad i = 1, 2, \ldots, n, \tag{6.29}$$

is called a Q-Q plot. Thus, the Q-Q plots show the estimated versus the observed quantiles. If the model fits the data well, the pattern of points on the Q-Q plot will exhibit a 45-degree straight line. Note that all the points of a Q-Q plot are inside the square $[\hat{F}^{-1}(p_{1:n}), \hat{F}^{-1}(p_{n:n})] \times [x_{1:n}, x_{n:n}]$.

6.3.2 The P-P Plots

Let $x_1, x_2, \ldots, x_n$ be a sample from a given population with estimated cdf $\hat{F}(x)$. The scatter plot of the points

$$\hat{F}(x_{i:n}) \quad \text{versus} \quad p_{i:n}, \quad i = 1, 2, \ldots, n, \tag{6.30}$$

is called a P-P plot. If the model fits the data well, the graph will be close to the 45-degree line. Note that all the points in the P-P plot are inside the unit square $[0, 1] \times [0, 1]$.

Example 6.7 (P-P and Q-Q plots). As an example of the P-P and Q-Q plots, we fit the GEVD in the tail of interest using the maximum likelihood method to two of the data sets in Chapter 1. Figures 6.7 and 6.8 show the P-P and Q-Q plots for the wind and Epicenter data sets, respectively. As can be seen from the straight line pattern in Figure 6.7, the GEVD fits the wind data very well. The pattern of points in Figure 6.8 indicates that the GEVD does not fit the Epicenter data. ■

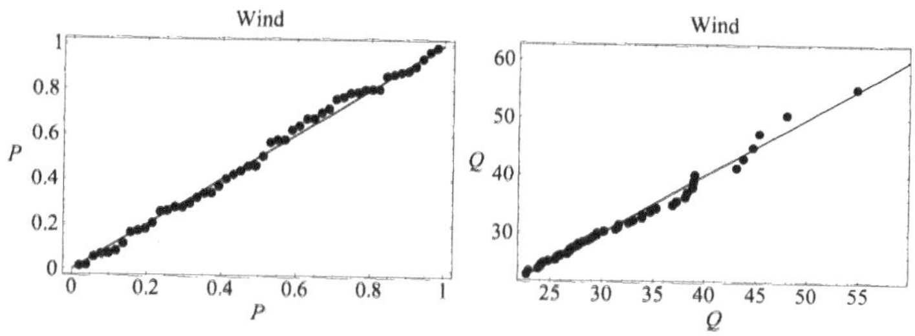

Figure 6.7: P-P and Q-Q plots for the wind data.

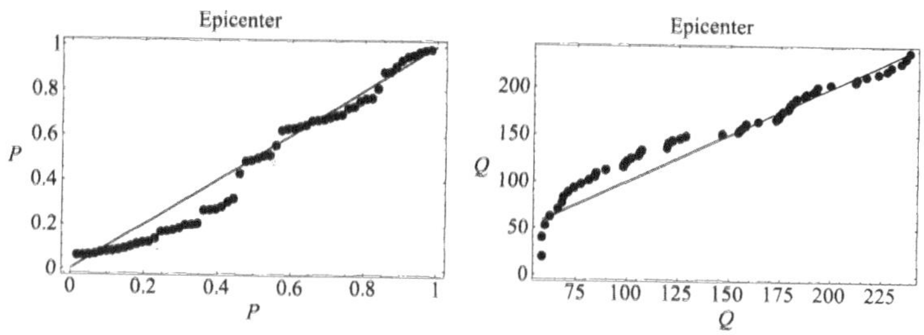

Figure 6.8: P-P and Q-Q plots for the Epicenter data.

Exercises

6.1 Show that no PPP can be constructed for the GPD distribution given in (3.2.17).

6.2 Can a PPP be constructed for the maximal GEVD given in Section 3.2.16 for given κ?

6.3 Can a PPP be constructed for the maximal GEVD given in Section 3.2.16 for given δ?

6.4 Find the transformation needed to draw the PPP for the maximal Fréchet domain of attraction.

6.5 Draw the wind data described in Section 1.4.1 on a PPP for the maximal Fréchet domain of attraction. Does the maximal Fréchet provide a good fit for the data?

6.6 Repeat Exercise 6.5 using the Precipitation data described in Section 1.4.11.

6.7 Find the transformation needed to draw the PPP for the minimal Fréchet domain of attraction.

6.8 Draw wind data described in Section 1.4.1 on a PPP for the minimal Fréchet domain of attraction. Does the minimal Fréchet provide a good fit for the data?

6.9 Repeat Exercise 6.8 using the Oldest Ages at Death data given in Table 1.4 and described in Section 1.4.4.

6.10 Repeat the test of Gumbel versus GEVD in Example 6.6 for the wave heights data described in Section 1.4.3.

6.11 Plot the wave heights data described in Section 1.4.3 on a maximal Gumbel probability plot and obtain the wave heights associated with return periods of 50 and 100 years.

6.12 An engineer is interested in constructing a probability paper plot for the location-scale exponential distribution. Can you do this construction?

Part IV

Exact Models for Order Statistics and Extremes

Chapter 7

Order Statistics

7.1 Order Statistics and Extremes

Let $X_1, X_2, \ldots, X_n$ be a sample of size n drawn from a common pdf $f(x)$ and cdf $F(x)$. Arrange $(X_1, \ldots, X_n)$ in an increasing order of magnitude and let $X_{1:n} \leq \cdots \leq X_{n:n}$ be the ordered values. The rth element of this sequence, $X_{r:n}$, is called the rth *order statistic* in the sample. The first and last order statistics, $X_{1:n}$ and $X_{n:n}$, are the *minimum* and *maximum* of $(X_1, \ldots, X_n)$, respectively. The minimum and maximum order statistics are called the *extremes*. For more detailed presentations of order statistics, see, for example, the books by Arnold, Balakrishnan, and Nagaraja (1992), Balakrishnan and Cohen (1991), and David and Nagaraja (2003). The two-volume handbook on order statistics prepared by Balakrishnan and Rao (1998a,b) also serve as a useful reference for readers.

Order statistics are very important in practice and especially the minimum, $X_{1:n}$, and the maximum, $X_{n:n}$, because they are the critical values used in engineering, physics, medicine, etc.; see, for example, Castillo and Hadi (1997).

The sample size itself may be known or random. In Section 7.2, we discuss the distributions of order statistics when $(X_1, X_2, \ldots, X_n)$ is an independently and identically distributed (iid) sample of known size n drawn from a common cdf $F(x)$. Section 7.3 discusses the case when the sample size is random. Section 7.4 deals with design values based on exceedances. Section 7.5 discusses return periods of random variables. The assumption of independence is then relaxed in Section 7.6.

7.2 Order Statistics of Independent Observations

In this section, we deal with order statistics for the particular case of independent observations.[1]

[1]Some of the material in this section is reprinted from the book *Extreme Value Theory in Engineering*, by E. Castillo, Copyright © Academic Press (1988), with permission from Elsevier.

7.2.1 Distributions of Extremes

Let X be a random variable with pdf $f(x)$ and cdf $F(x)$. Let $(x_1, x_2, \ldots, x_n)$ be an iid sample drawn from $F(x)$. Then, using (4.14), their joint pdf is

$$f(x_1, x_2, \ldots, x_n) = \prod_{i=1}^{n} f(x_i), \tag{7.1}$$

and their cdf is

$$F(x_1, x_2, \ldots, x_n) = \prod_{i=1}^{n} F(x_i). \tag{7.2}$$

The cdf of the maximum order statistic, $X_{n:n}$, is

$$\begin{aligned} F_{max}(x) &= \Pr(X_{n:n} \le x) \\ &= \Pr(\max(X_1, X_2, \ldots, X_n) \le x) \\ &= \Pr(\text{all } X_i \le x) \\ &= \prod_{i=1}^{n} \Pr(X_i \le x) = \prod_{i=1}^{n} F(x) = [F(x)]^n. \end{aligned} \tag{7.3}$$

Thus, the cdf of the maximum can be obtained by raising the original cdf to the nth power. The pdf of $X_{n:n}$ is obtained by differentiating (7.3) with respect to x, that is,

$$f_{max}(x) = nf(x)[F(x)]^{n-1}. \tag{7.4}$$

Figure 7.1 shows the pdf and cdf of the maximum order statistic, $X_{n:n}$, in samples of sizes $1, 2, 10, 100, 1000$, and 10000 drawn from the normal $N(0,1)$, maximal Gumbel $G_M(0,1)$, and minimal Gumbel $G_m(0,1)$ distributions, respectively. Note that the curves for $n = 1$, shown with thicker, darker lines, are of the parent's pdf $f(x)$ and cdf $F(x)$. Observe that in most cases the pdf and cdf of the maximum order statistics change location and scale with increasing n. An exception is the minimal Gumbel case, where the scale remains constant. In particular, the pdf moves to the right with decreasing scale as n increases, and the cdf moves to the right and with increasing slope as n increases. As can be inferred from the graphs, at $n = \infty$, $X_{n:n}$ becomes a degenerate random variable. This fact can actually be obtained directly from (7.3).

The cdf of the minimum order statistic, $X_{1:n}$, can be obtained in a similar way as follows:

$$\begin{aligned} F_{min}(x) &= \Pr(X_{1:n} \le x) \\ &= \Pr(\min(X_1, X_2, \ldots, X_n) \le x) \\ &= 1 - \Pr(\min(X_1, X_2, \ldots, X_n) > x) \\ &= 1 - \Pr(\text{all } X_i > x) = 1 - \prod_{i=1}^{n} \Pr(X_i > x) \\ &= 1 - \prod_{i=1}^{n}(1 - F(x)) = 1 - [1 - F(x)]^n = 1 - [S(x)]^n, \end{aligned} \tag{7.5}$$

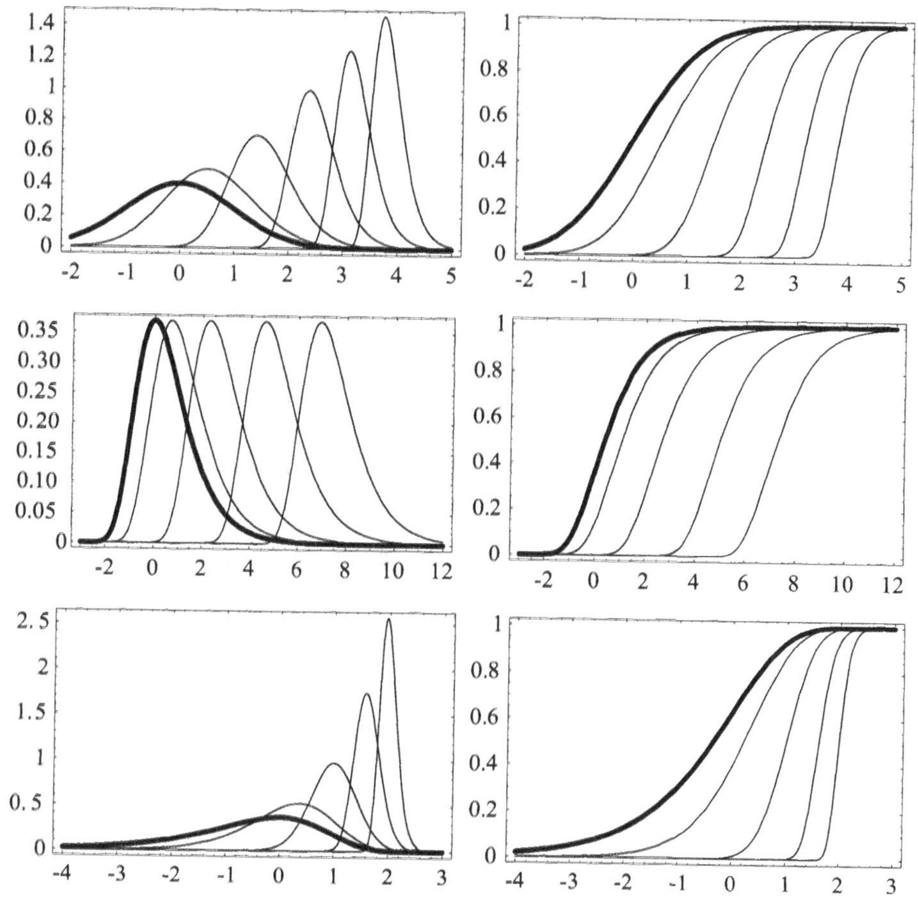

Figure 7.1: The pdf and cdf of the maxima of a sample of sizes $1, 2, 10, 100, 1000$, and 10000 drawn from the normal $N(0,1)$, maximal Gumbel $G_M(0,1)$, and minimal Gumbel $G_m(0,1)$ distributions, respectively.

where $S(x)$ is the survival function defined in (3.92). Thus, the cdf of the minimum order statistic is one minus the survival function raised to the power n.

This implies that

$$S_{min}(x) = 1 - F_{min}(x) = S^n(x)$$

or, in other words, the survival function of the minimum is the survival function of the parent raised to the nth power.

The pdf of $X_{1:n}$ is obtained by differentiating (7.5) with respect to x as

$$f_{min}(x) = nf(x)[1 - F(x)]^{n-1}. \tag{7.6}$$

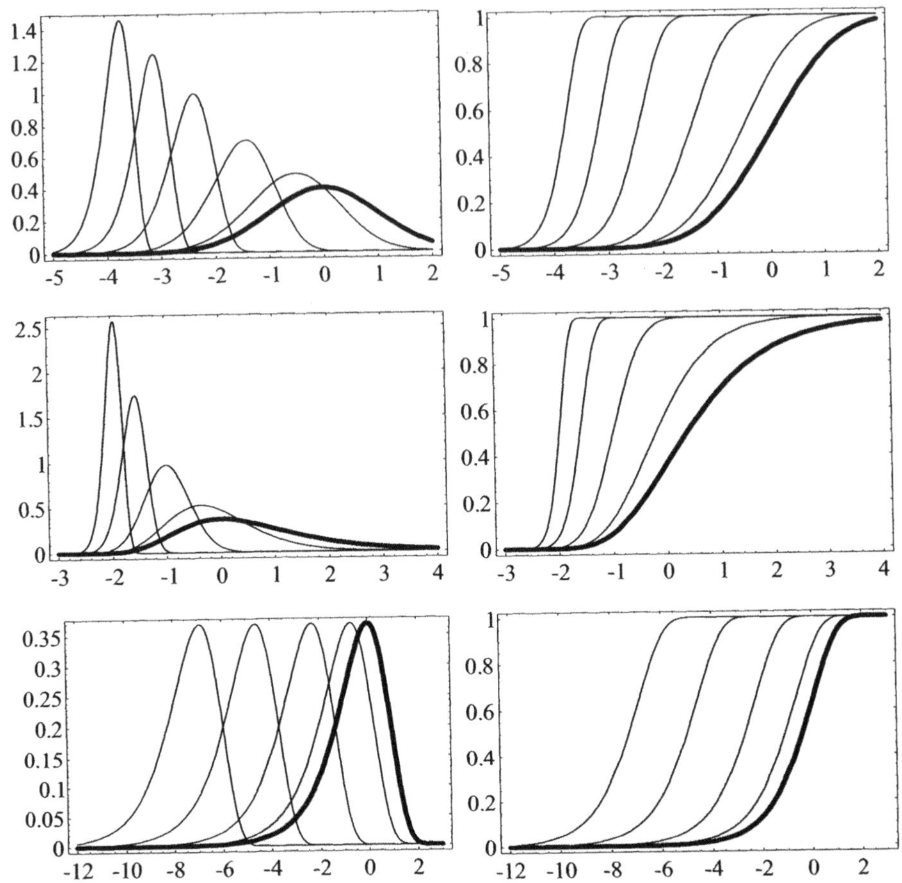

Figure 7.2: The pdf and cdf of the minima of a sample of sizes $1, 2, 10, 100, 1000$, and 10000 drawn from the normal $N(0,1)$, maximal Gumbel $G_M(0,1)$, and minimal Gumbel $G_m(0,1)$ distributions, respectively.

Figure 7.2 shows the pdf and cdf of the minimum order statistic, $X_{1:n}$, in samples of sizes $1, 2, 10, 100, 1000$, and 10000 drawn from the normal $N(0,1)$, maximal Gumbel $G_M(0,1)$, and minimal Gumbel $G_m(0,1)$ distributions, respectively. Note that the curves for $n = 1$, shown with thicker, darker lines, are of the parent's pdf $f(x)$ and cdf $F(x)$. Compare the pdf and cdf in Figure 7.2 to those in Figure 7.1 and note the similarities as well as differences. Note also that all the graphs in Figure 7.2 can be obtained from the graphs in Figure 7.1 after one or two symmetric transformations. The pdf and cdf of the minimum order statistics can also change location and scale with increasing n, but they move in the opposite direction. The pdf of the minimal Gumbel moves to the left but with fixed scale, whereas the pdfs of the normal and maximal Gumbel

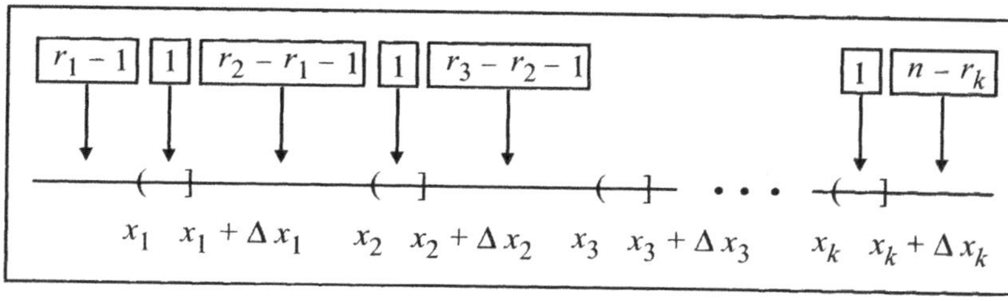

Figure 7.3: An illustration of the multinomial experiment used to determine the joint pdf of a subset of k order statistics.

move to the left with decreasing scale as n increases, and their cdfs move to the left with increasing slope as n increases. It can also be inferred from the graphs that at $n = \infty$, $X_{1:n}$ becomes a degenerate random variable. This can also be obtained directly from (7.5).

7.2.2 Distribution of a Subset of Order Statistics

Let $X_{r_1:n}, \ldots, X_{r_k:n}$, where $r_1 < \ldots < r_k$, be a subset of k order statistics in a random sample of size n drawn from a population with pdf $f(x)$ and cdf $F(x)$. The joint pdf of this subset is obtained as follows. Consider the event $x_j < X_{r_j:n} \leq x_j + \Delta x_j, 1 \leq j \leq k$, for small values of Δx_j (see Fig. 7.3). That is, k values in the sample belong to the intervals $(x_j, x_j + \Delta x_j)$ for $1 \leq j \leq k$ and the rest are distributed in such a way that exactly $(r_j - r_{j-1} - 1)$ belong to the interval $(x_{j-1} + \Delta x_{j-1}, x_j)$ for $1 \leq j \leq k$, where $\Delta x_0 = 0$, $r_0 = 0$, $r_{k+1} = n + 1$, $x_0 = -\infty$ and $x_{k+1} = \infty$.

Consider the following multinomial experiment with the $2k+1$ possible outcomes associated with the $2k+1$ intervals illustrated in Figure 7.3. We obtain a sample of size n from the population and determine to which of the intervals they belong. Since we assume independence and replacement, the numbers of elements in each interval is a multinomial random variable with parameters:

$$\begin{gathered}\{n; f(x_1)\Delta x_1, \ldots, f(x_k)\Delta x_k, [F(x_1) - F(x_0)], \\ [F(x_2) - F(x_1)], \ldots, [F(x_{k+1}) - F(x_k)]\},\end{gathered}$$

where the parameters are n (the sample size) and the probabilities associated with the $2k+1$ intervals. Consequently, we can use the results for multinomial random variables to obtain the joint pdf of the k order statistics and obtain (see (4.9))

$$\begin{aligned} f_{r_1,\ldots,r_k:n}(x_1, \ldots, x_k) &= n! \prod_{i=1}^{k} f(x_i) \prod_{j=1}^{k+1} \frac{[F(x_j) - F(x_{j-1})]^{r_j - r_{j-1} - 1}}{(r_j - r_{j-1} - 1)!}, \\ & \quad x_1 \leq x_2 \leq \ldots \leq x_k. \end{aligned} \tag{7.7}$$

7.2.3 Distribution of a Single Order Statistic

The pdf and cdf of the rth order statistic in a sample of size n are

$$
\begin{aligned}
f_{r:n}(x) &= f(x)\frac{[F(x)]^{r-1}[1-F(x)]^{n-r}}{B(r,n-r+1)}, &(7.8)\\
F_{r:n}(x) &= I_{\beta}(F(x);r,n-r+1), &(7.9)
\end{aligned}
$$

where $B(r,n-r+1)$ is the Beta function defined in (3.33) and $I_{\beta}(p;a,b)$ is the incomplete Beta function defined in (3.36).

The pdf of $X_{r:n}$ can be obtained from (7.7) by setting

$$
k=1,\;\; x_1=x,\;\; r_0=0,\;\; r_1=r,\;\; r_2=n+1,\;\; x_0=-\infty,\;\; x_2=\infty,
$$

and obtaining

$$
\begin{aligned}
f_{r:n}(x) &= n!f(x)\frac{[F(x)-F(-\infty)]^{r-0-1}}{(r-0-1)!}\frac{[F(\infty)-F(x)]^{n+1-r-1}}{(n+1-r-1)!}\\
&= n!f(x)\frac{[F(x)]^{r-1}}{(r-1)!}\frac{[1-F(x)]^{n-r}}{(n-r)!}\\
&= f(x)\frac{[F(x)]^{r-1}[1-F(x)]^{n-r}}{B(r,n-r+1)}. &(7.10)
\end{aligned}
$$

The cdf can be obtained as follows:

$$
\begin{aligned}
F_{r:n}(x) &= \Pr(X_{r:n}\le x) = 1-F_{m_n(x)}(r-1)\\
&= \sum_{k=r}^{n}[F(x)]^k[1-F(x)]^{n-k}\\
&= \int_0^{F(x)} u^{r-1}(1-u)^{n-r}du\\
&= I_{\beta}(F(x);r,n-r+1), &(7.11)
\end{aligned}
$$

where $m_n(x)$ is the number of sample elements such that $X \le x$ and $I_{\beta}(p;a,b)$ is the incomplete Beta function defined in (3.36).

Figures 7.4–7.7 show the pdf and cdf of the order statistics in a sample of size five drawn from a uniform $U(0,1)$, exponential $E(1)$, gamma $G(3,1)$, and normal $N(0,1)$ distributions, respectively. Figure 7.8 shows the pdf and cdf of the last five order statistics in a sample of size 20 drawn from the exponential $E(1)$ distribution. The parent pdf and cdf are also shown on these graphs with a darker curve.

Figures 7.9 and 7.10 show the pdf and cdf of the first and last five order statistics in a sample of size 20 drawn from the normal $N(0,1)$ distributions, respectively. The parent pdf and cdf are also shown on these graphs with a darker curve.

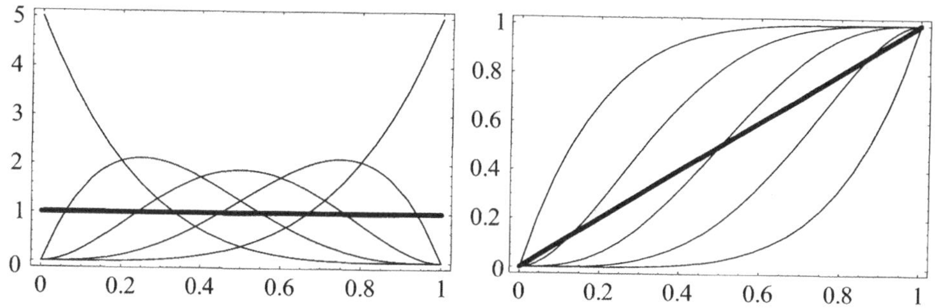

Figure 7.4: The pdf and cdf of the five order statistics of a sample of size five from the uniform $U(0, 1)$ distribution.

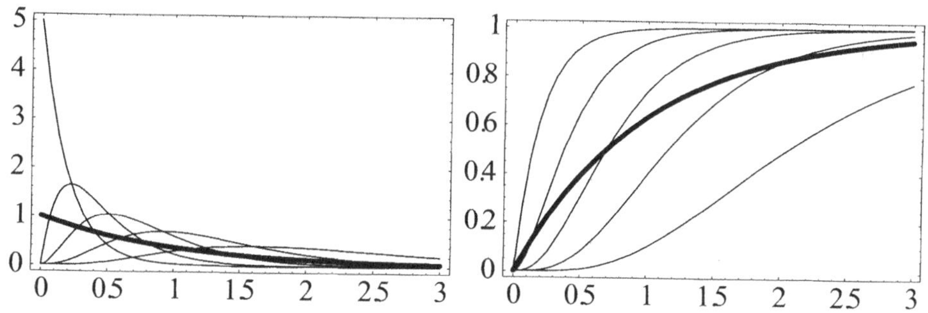

Figure 7.5: The pdf and cdf of the five order statistics of a sample of size five from the exponential $E(1)$ distribution.

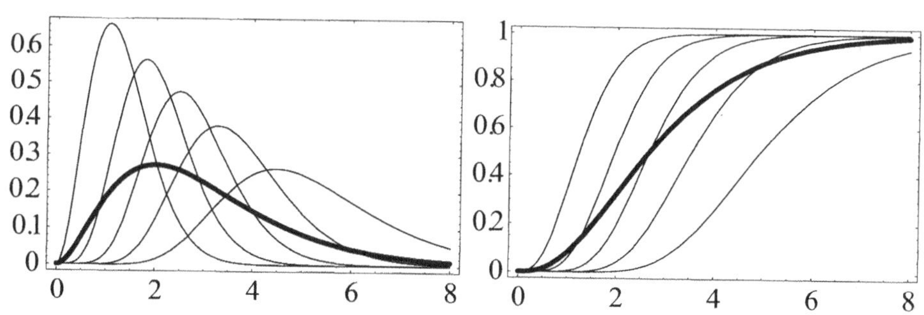

Figure 7.6: The pdf and cdf of the five order statistics of a sample of size five from the gamma $G(3, 1)$ distribution.

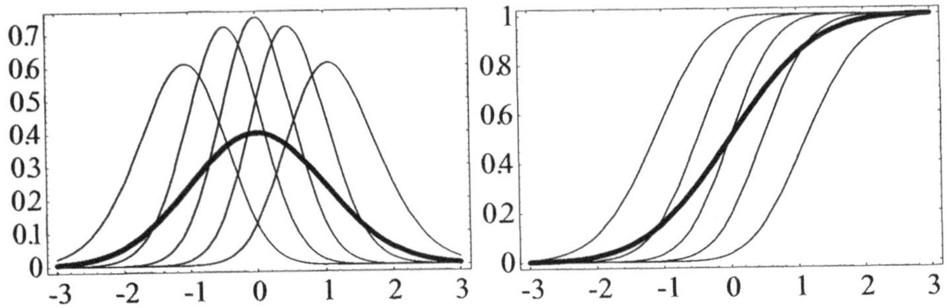

Figure 7.7: The pdf and cdf of the five order statistics of a sample of size five from the normal $N(0,1)$ distribution.

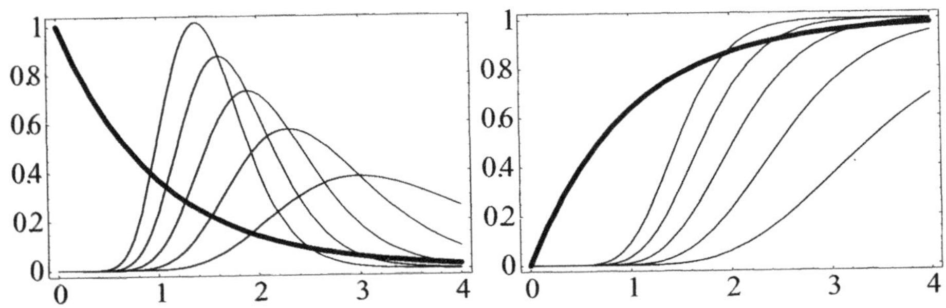

Figure 7.8: The pdf and cdf of the last five order statistics of a sample of size 20 from the exponential $E(1)$ distribution.

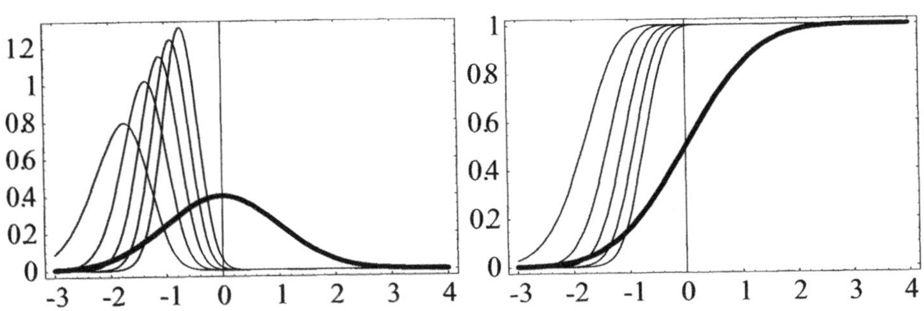

Figure 7.9: The pdfs of the first five order statistics in a sample of size 20 drawn from the normal $N(0,1)$ distribution.

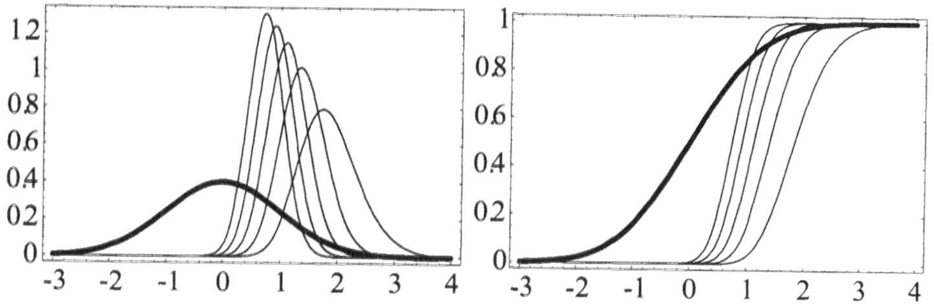

Figure 7.10: The pdfs of the last five order statistics in a sample of size 20 drawn from the normal $N(0,1)$ distribution.

Example 7.1 (Uniform parent). From (7.8), the density of the order statistic $X_{r:n}$ from a uniform $U(0,1)$ parent is

$$f_{r:n}(x) = \frac{x^{r-1}(1-x)^{n-r}}{B(r, n-r+1)}, \quad 0 < x < 1, \tag{7.12}$$

that is, a Beta distribution $B(r, n-r+1)$. ∎

Example 7.2 (Wave heights). Longuet-Higgins (1952) uses the Rayleigh distribution to model the wave heights, X, in a given location. The pdf is

$$f(x) = \frac{x}{\mu_0}\, e^{-x^2/2\mu_0}, \quad \mu_0 > 0, x > 0,$$

where μ_0 is the zero moment of its energy spectrum. The pdf of the rth order statistic in a sample of size 1000 is

$$f_{X_{r:1000}}(x) = r\binom{1000}{r}\left[1 - e^{-x^2/(2\mu_0)}\right]^{r-1} e^{-x^2(1000-r)/(2\mu_0)}(x/\mu_0)e^{-x^2/(2\mu_0)}$$

and the corresponding cdf is

$$F_{X_{r:1000}}(x) = \sum_{k=r}^{1000}\binom{1000}{k}\left[1 - e^{-x^2/(2\mu_0)}\right]^{k} e^{-x^2(1000-k)/(2\mu_0)}.$$

For example, if it is known that $\mu_0 = 4$ m^2 and that a rigid breakwater fails when $x = 9.6$ m, then the survival probability after a series of 1000 waves is

$$p = F_{X_{1000:1000}}(9.6) = 0.99.$$

Similarly, if a flexible breakwater fails after five waves with height above $x = 7$ m, its survival probability to 1000 waves is

$$p = F_{X_{996:1000}}(7) = 0.92.$$

∎

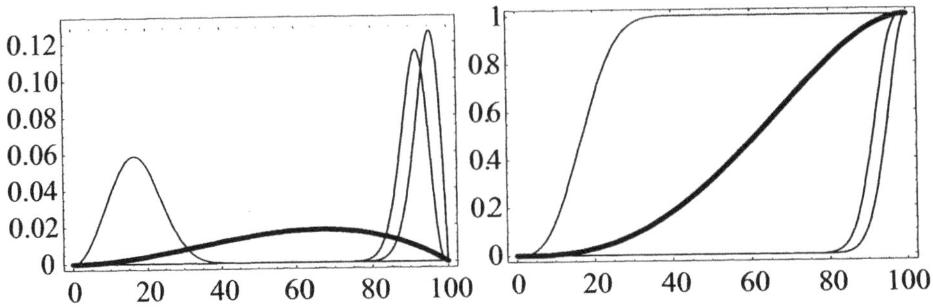

Figure 7.11: The pdf and cdf of the best, second best and the worst students in a class of 40 students corresponding to Example 7.4.

Example 7.3 (Floods). Suppose that the maximum yearly water level X, in meters, in a given section of a river follows a Gumbel distribution with cdf

$$F(x) = \exp\left\{-\exp\left[\left(\frac{14-x}{2}\right)\right]\right\}.$$

Suppose also that a value of $x = 6.0$ m leads to floods. The probability of having a maximum of two floods in a period of 50 years is $F_{X_{49:50}}(4) = 0.77$. ∎

Example 7.4 (Exam scores). The score, X, in a final exam has the following pdf and cdf:

$$\begin{aligned} f(x) &= \frac{12x^2(100-x)}{10^8}, \quad 0 \le x \le 100, \\ F(x) &= \begin{cases} 0, & \text{if } x < 0, \\ \dfrac{12x^3(\frac{100}{3} - \frac{x}{4})}{10^8}, & \text{if } 0 \le x < 100, \\ 1, & \text{if } x \ge 100. \end{cases} \end{aligned} \tag{7.13}$$

The pdf and cdf of the best, second best and the worst students in a class of 40 students are shown in Figure 7.11. These correspond to the 40th, 39th, and 1st order statistics, respectively, of a sample of size 40. Note how the probabilities change depending on the student's performance. ∎

7.2.4 Distributions of Other Special Cases

The distribution of a subset of order statistics in (7.7) leads to several special cases:

Joint distribution of maximum and minimum

Setting $k = 2$, $r_1 = 1$, and $r_2 = n$ in (7.7), we obtain the joint pdf of the maximum and the minimum of a sample of size n, which becomes

$$f_{1,n:n}(x_1, x_2) = n(n-1)f(x_1)f(x_2)[F(x_2) - F(x_1)]^{n-2}, \quad x_1 \le x_2.$$

Joint distribution of two consecutive order statistics

Setting $k = 2, r_1 = i$ and $r_2 = i+1$ in (7.7), we get the joint density of the statistics of orders i and $i+1$ as

$$f_{i,i+1:n}(x_1, x_2) = \frac{n! f(x_1) f(x_2) F^{i-1}(x_1)[1 - F(x_2)]^{n-i-1}}{(i-1)!(n-i-1)!}, \quad x_1 \le x_2.$$

Joint distribution of any two order statistics

The joint distribution of the statistics of orders r and s $(r < s)$ is

$$f_{r,s:n}(x_r, x_s) = \frac{n! f(x_r) f(x_s) F^{r-1}(x_r)[F(x_s) - F(x_r)]^{s-r-1}[1 - F(x_s)]^{n-s}}{(r-1)!(s-r-1)!(n-s)!},$$
$$x_r \le x_s.$$

Figures 7.12 and 7.13 show the pdfs of the pairs $\{(1,9), (2,8), (3,7), (4,5)\}$ of order statistics in a random sample of size $n = 9$ from uniform $U(0,1)$ and normal $N(0,1)$ parents, respectively.

Joint distribution of all order statistics

The joint density of all order statistics can be obtained from (7.7) by setting $k = n$, as

$$f_{1,\ldots,n:n}(x_1, \ldots, x_n) = n! \prod_{i=1}^{n} f(x_i), \quad x_1 \le \ldots \le x_n.$$

Joint distribution of the last k order statistics

The joint density of the last k order statistics can be obtained from (7.7) by setting $r_1 = n-k+1, r_2 = n-k+2, \ldots, r_{k-1} = n-1, r_k = n$, as

$$f_{n-k+1,\ldots,n:n}(x_{n-k+1}, \ldots, x_n) = n! \prod_{i=n-k+1}^{n} f(x_i) \frac{F(x_{n-k+1})^{n-k}}{(n-k)!},$$
$$x_{n-k+1} \le \cdots \le x_n.$$

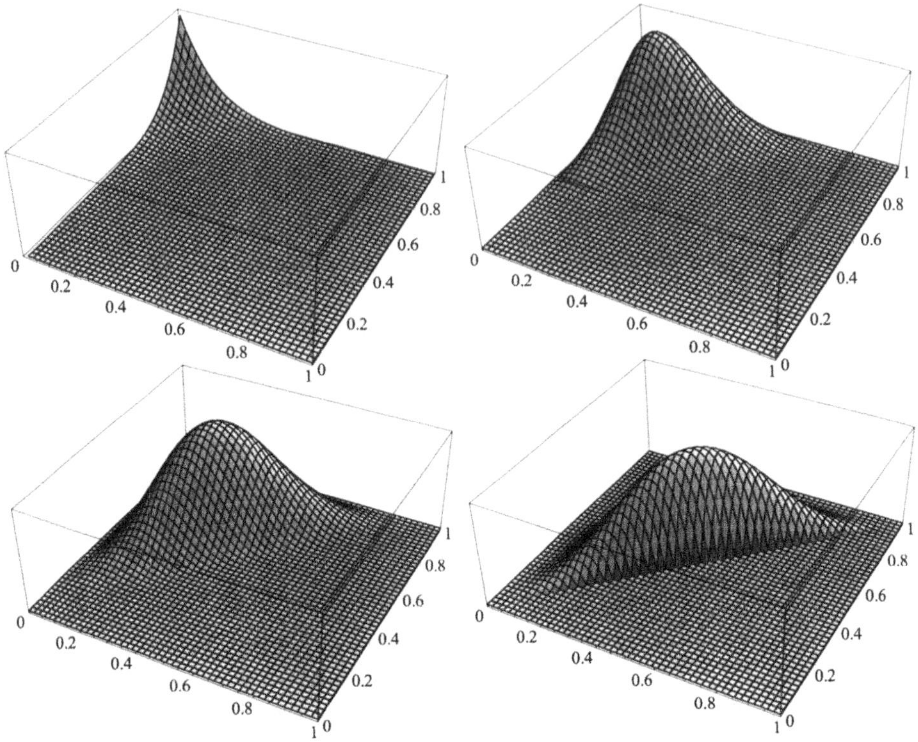

Figure 7.12: The pdfs of the pairs $\{(1,9),(2,8),(3,7),(4,5)\}$ of order statistics in a random sample of size $n = 9$ from a uniform $U(0,1)$ parent.

Joint distribution of the first k order statistics

The joint density of the first k order statistics can be obtained from (7.7) by setting $r_1 = 1, r_2 = 2, \ldots, r_k = k$, as

$$f_{1,2,\ldots,k:n}(x_{(1)}, \ \cdots \ , x_{(k)}) = n! \prod_{i=1}^{k} f(x_{(i)}) \frac{[1 - F(x_k)]^{n-k}}{(n-k)!}, \qquad x_{(1)} \le x_{(2)} \le \cdots \le x_{(k)}.$$

7.3 Order Statistics in a Sample of Random Size

The above derivations of the pdf and cdf of the order statistics is based on the assumption that the sample size n is known. However, in some practical situations, the sample size is random. For example, the number of earthquakes that will occur in a given region next year is a random variable and we wish to know the pdf of the earthquake with the maximum intensity during the year. Here, the sample size is not known in advance. In such cases, where the sample

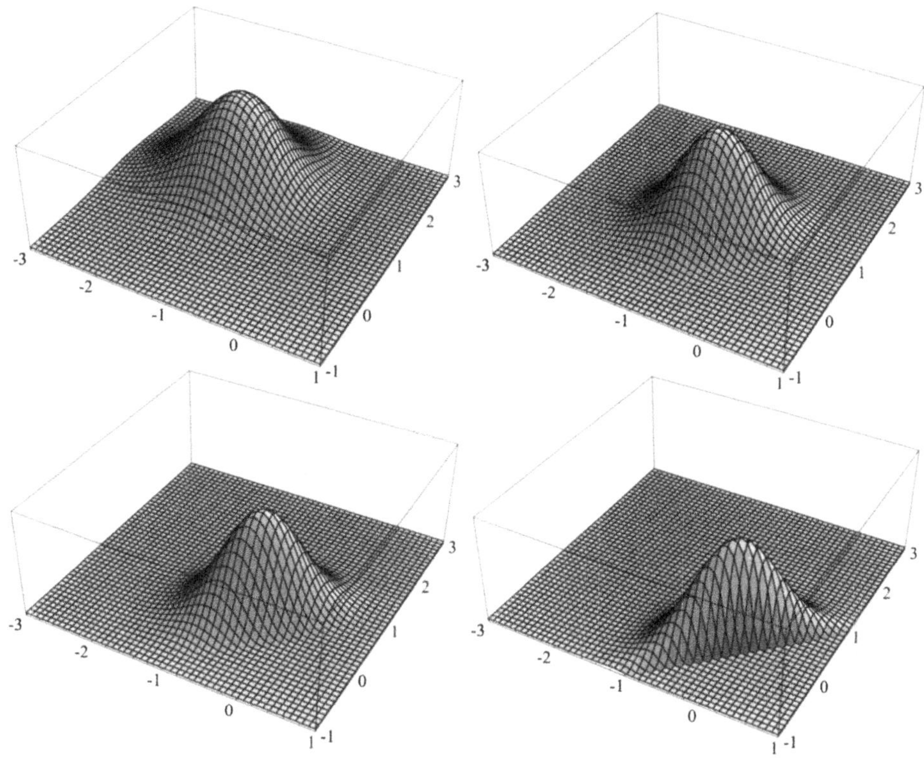

Figure 7.13: The pdfs of the pairs $\{(1,9),(2,8),(3,7),(4,5)\}$ of order statistics in a random sample of size $n = 9$ from a normal $N(0,1)$ parent.

size is random, the previous formulas for the distributions of order statistics are no longer valid.

Let the sample size N be a discrete random variable with pdf

$$\Pr(N = n), \quad n = 1, 2, \ldots,$$

and let $f_X(x;n)$ and $F_X(x;n)$ be the pdf and cdf of the statistic X under consideration (e.g., order statistic, difference of order statistics, etc.) for a fixed sample size n. Then, the *total probability theorem* permits writing the pdf and cdf of X as

$$g_X(x) = \sum_n \Pr(N = n)\, f_X(x;n),$$

and

$$G_X(x) = \sum_n \Pr(N = n)\, F_X(x;n),$$

where the sum is extended to all possible values of the random variable N.

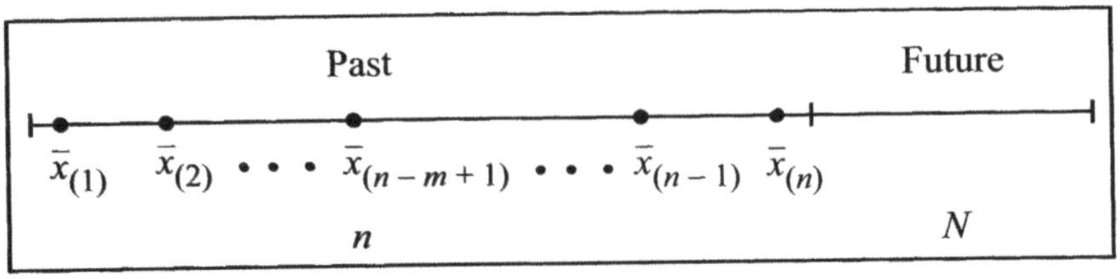

Figure 7.14: An illustration of the design values based on exceedances.

Example 7.5 (Earthquakes). The occurrence of earthquakes in a given region follows a truncated Poisson process of intensity λ earthquakes/year:

$$\Pr(N = n) = \frac{e^{-\lambda t}(\lambda t)^n}{n!(1 - e^{-\lambda t})}, \quad n = 1, 2, \ldots \tag{7.14}$$

and the corresponding intensity, X, is a random variable with cdf $H(x)$.

Then, the maximum intensity of an earthquake during a period of duration t has cdf:

$$\begin{aligned} F_{X_{\max}}(x) &= \sum_{n=1}^{\infty} \frac{e^{-\lambda t}(\lambda t)^n}{n!(1 - e^{-\lambda t})} H^n(x) \\ &= \frac{\exp\{-\lambda t[1 - H(x)]\} - \exp\{-\lambda t\}}{1 - \exp\{-\lambda t\}}, \quad x \geq 0. \end{aligned}$$

■

7.4 Design Values Based on Exceedances

An important practical problem related to exceedances is to determine the probability of having in the future N observations, r exceedances of the mth largest observation occurring in the past n observations (see Fig. 7.14). If p_m is the probability of exceeding the mth largest observation occurred in the past n experiments, then the probability of r exceedances in the next N observations is

$$P(r|n, m, N, p_m) = \binom{N}{r} p_m^r (1 - p_m)^{N-r}. \tag{7.15}$$

The problem is that p_m is random, that is,

$$p_m = 1 - F(X_{n-m+1:n}) = F(X_{m:n}) = U_{m:n}, \tag{7.16}$$

where $X_{n-m+1:n}$ is the $(n - m + 1)$th order statistic of the past n observations, $U_{m:n}$, is the mth order statistic in a sample of size n from a uniform $U(0, 1)$ and the last two equalities must be interpreted in the sense of having identical cdf.

Note that (7.16) is independent of $F(x)$ and has a clear intuitive meaning. Note also that p_m is random (it depends on the previous sample, that is random) with pdf

$$f(p_m) = \frac{p_m^{m-1}(1-p_m)^{n-m}}{B(m, n-m+1)}, \quad 0 \le p_m \le 1. \tag{7.17}$$

Consequently, the total probability theorem leads to

$$\begin{aligned} P(r|n,m,N) &= \int_0^1 P(r|n,m,N,p_m) f(p_m) dp_m \\ &= \binom{N}{r} \int_0^1 \frac{p_m^{r+m-1}(1-p_m)^{N-r+n-m}}{B(m,n-m+1)} dp_m \\ &= \binom{N}{r} \frac{B(r+m, N-r+n-m+1)}{B(m,n-m+1)}. \end{aligned}$$

The mean number of exceedances, $\bar{r}(n,m,N)$, taking into account that the mean of the binomial variable in (7.15) is Np_m, the total probability theorem, and that the mean of the mth order statistic $U_{m:n}$ is $m/(n+1)$, becomes

$$\bar{r}(n,m,N) = \int_0^1 N p_m f(p_m) dp_m = N\mu_{U_{m:n}} = \frac{Nm}{n+1} \tag{7.18}$$

and its variance

$$\sigma^2(n,m,N) = \frac{Nm(n-m+1)(N+n+1)}{(n+1)^2(n+2)}. \tag{7.19}$$

This variance attains a minimum value when $m = (n+1)/2$. However, the coefficient of variation decreases with m, as can be expected.

Example 7.6 (Temperatures). The maximum temperature in a given location during the last 40 years was 42°C. Determine the mean value of the number of exceedances of 42°C of the annual maxima in the next 30 years.

From (7.18) and (7.19), we obtain

$$\bar{r}(40,1,30) = 30/41 = 0.732,$$

$$\sigma^2(40,1,30) = \frac{30 \times 40 \times 71}{41^2 \times 42} = 1.207.$$

■

Example 7.7 (Floods). The yearly maximum floods, in a given cross-section of a river, measured in m^3/sec, during the last 60 years are shown in Table 1.2. We wish to select a flood design value for having a mean of four exceedances in the next 20 years. According to (7.18), we have

$$\bar{r}(60,m,20) = \frac{20m}{61} \approx 4 \Longrightarrow m = 12,$$

which shows that the 12th largest order statistic in the sequence must be selected, that is, 50.17 m^3/sec. ■

7.5 Return Periods

If $F(x)$ is the cdf of the yearly maxima of a random variable X, the return period, τ_x, of the event $\{X > x\}$ is $1/[1 - F(x)]$ years. Similarly, if $F(x)$ is the cdf of the yearly minima of a random variable X, the return period τ_x of the event $\{X < x\}$ is $1/F(x)$ years. Then, we have

$$\tau_x = [1 - F(x)]^{-1} \tag{7.20}$$

for exceedances, and

$$\tau_x = [F(x)]^{-1} \tag{7.21}$$

for shortfalls.

Consider a time interval of small duration τ (unit time) and the Bernoulli experiment consisting of determining whether or not the event $\{X > x\}$ occurs in such interval. Consider now a sequence of time intervals of the same duration and the corresponding Bernoulli experiments. The number of required Bernoulli experiments for the event to occur for the first time is a Pascal or geometric random variable $Ge(p)$, which means $1/p$ is the return period measured in τ units. For this to be valid, p must be small, to guarantee that no more than one event occurs in each time interval (a small probability of having more than one occurrence of the event).

Note that if an engineering work fails if and only if the event A occurs, then the mean life of the engineering work is the return period of A.

The importance of return period in engineering is due to the fact that many design criteria use return periods, that is, an engineering work is designed to withstand, on average, return periods of 50, 100, or 500 years.

In addition, the probability of occurrence of the event A before the return period is (see the geometric distribution)

$$F(\tau) = 1 - (1 - p)^{\tau} = 1 - (1 - p)^{1/p}, \tag{7.22}$$

which for $\tau \to \infty$ ($p \to 0$) tends to 0.63212.

Remark 7.1 *If instead of recording yearly maxima of X, the maxima for a period of k years were recorded, then τ_x in (7.20) and (7.21) would be measured in periods of k years.*

Example 7.8 (Return period of a flood). The cdf of the yearly maximum flood (in m^3/sec) in a given cross-section of a river is given by

$$F(x) = \exp\left[-\exp\left(-\frac{x - 38.5}{7.8}\right)\right].$$

Then, the return periods of floods of 60 and 70 m^3/sec are

$$\tau_{60} = \frac{1}{1 - F(60)} = 16.25 \text{ years}$$

and

$$\tau_{70} = \frac{1}{1 - F(70)} = 57.24 \text{ years}$$

This means that yearly maximum floods of 60 and 70 m^3/sec occur, on average, once every 16.25 and 57.24 years, respectively. ∎

Example 7.9 (Design wave height for a breakwater). If a breakwater is to be designed to withstand a mean useful life of 50 years, and the yearly maximum wave height h (in feet) is known, from previous experience, to be

$$F(h) = \exp\left[-\exp\left(-\frac{h-15}{4}\right)\right]. \tag{7.23}$$

Then, the design wave height must satisfy the equation

$$\frac{1}{1 - F(h)} = 50,$$

which leads to a design wave height of $h = 30.61$ feet. ∎

Let $F_k(x)$ and $F_r(x)$ be the cdf of the maximum of a random variable X which is observed during periods of k and r years, respectively, where $r = sk$, and assume that the maximum in different years are independent random variables.

Then, we have

$$F_r(x) = F_k^s(x). \tag{7.24}$$

The return periods $\tau_x^{(k)}$ and $\tau_x^{(r)}$ of a value x of X, measured in periods of k and r years, respectively, are

$$\tau_x^{(k)} = \frac{1}{1 - F_k(x)} \quad \text{and} \quad \tau_x^{(r)} = \frac{1}{1 - F_r(x)}. \tag{7.25}$$

Then, using (7.25) we get

$$\tau_x^{(r)} = \frac{1}{1 - F_r(x)} = \frac{1}{1 - F_k^s(x)} = \frac{1}{1 - (1 - S_k(x))^s} \approx \frac{1}{sS_k(x)} \tag{7.26}$$

and, since

$$\tau_x^{(k)} = \frac{1}{1 - F_k(x)} = \frac{1}{1 - (1 - S_k(x))} = \frac{1}{S_k(x)}, \tag{7.27}$$

we get

$$\tau_x^{(k)} = s\tau_x^{(r)}. \tag{7.28}$$

For example, if r and k refer to one year and one month periods, respectively, then $s = 12$, and the return period measured in months is 12 times the return period measured in years.

Note that this approximation, based on the geometric distribution, is valid only if the function $F(x)$ at x is very small, so that the approximation

$$[1 - F(x)]^s \approx 1 - sF(x)$$

is valid. For that to hold, the length of the unit period considered must be small enough for the probability of one event to be small, and the probability for two or more events to be negligible. Otherwise, the model is invalid.

Example 7.10 (Return periods based on different data durations). Assume that the yearly maximum wave height at a given location has the Gumbel distribution in (7.23), where the return period of a wave of height $h = 39.855$ feet is 500 years. If the maximum yearly waves are assumed to be independent, the maximum wave height in a decade has the following cdf

$$F_{10}(h) = F^{10}(h) = \exp\left[-\exp\left(-\frac{h - 15 - 4\log 10}{4}\right)\right].$$

The return period of $h = 39.855$ feet is

$$\begin{aligned} \tau_{10} &= \frac{1}{1 - F(30.61)} \\ &= \frac{1}{1 - \exp\left[-\exp\left(-(30.61 - 15 - \log 10)/4\right)\right]} \\ &= 50.4588 \text{ decades,} \end{aligned}$$

which is approximately 50 decades. ∎

7.6 Order Statistics of Dependent Observations

So far in this chapter, we have assumed that the sample is an iid sample. However, in many practical situations, this assumption is not valid, and some kind of dependence must be assumed. Think, for example, of the amount of rain in a given location on consecutive days; it is obvious that the chances of rain after a rainy day are much higher than the chances of rain after a dry day. As another example, consider the sequences of waves reaching a breakwater. If the height of a wave is high, the chance of the following wave to be high is much higher than that for a wave following a low wave. Therefore, dependence is very frequent in practice. Actually, it can be argued that dependent observations are more likely in real practice than independent ones. Consequently, dependence models are required in order to be able to solve real problems. In this section, we relax the independence assumption and present the distribution of order statistics for the case of dependent samples. Sometimes, we cannot obtain exact results and we need to work with inequalities (see Galambos and Simonelli (1996)).

7.6.1 The Inclusion-Exclusion Formula

If A and B are given events of a probability space, it is well known from probability theory that

$$\Pr(A \cup B) = \Pr(A) + \Pr(B) - \Pr(A \cap B).$$

This formula can be easily generalized to n events. Let $C_1, C_2, \ldots, C_n$ be n arbitrary events. Then, the following inclusion-exclusion formula holds:

$$\begin{aligned}\Pr\left(\bigcup_{i=1}^{n} C_i\right) &= \sum_{i=1}^{n}\Pr(C_i) - \sum_{1\le i_1<i_2\le n}\Pr(C_{i_1}\cap C_{i_2}) \\ &+ \sum_{1\le i_1<i_2<i_3\le n}\Pr(C_{i_1}\cap C_{i_2}\cap C_{i_3}) - \ldots + (-1)^{n+1} \\ &\times \sum_{1\le i_1<i_2<\ldots<i_n\le n}\Pr(C_{i_1}\cap C_{i_2}\cap\ldots\cap C_{i_n}). \qquad (7.29)\end{aligned}$$

Note that the last term on the right-hand side of (7.29) is an intersection of all n events and therefore can be simply written as $\Pr(C_1 \cap C_2 \cap \ldots \cap C_n)$. Now, let

$$S_{k,n} = \sum_{1\le i_1<i_2<\ldots<i_k\le n}\Pr(C_{i_1}\cap C_{i_2}\cap\ldots\cap C_{i_k}), \quad S_{0,n}=1, \qquad (7.30)$$

and recall that

$$\Pr\left(\bigcup_{i=1}^{n} C_i\right) = 1 - \Pr\left(\bigcap_{i=1}^{n} \overline{C}_i\right),$$

where $\overline{C}_i$ is the complement of C_i. Then (7.29) leads to

$$\Pr\left(\bigcap_{i=1}^{n} \overline{C}_i\right) = 1 - \Pr\left(\bigcup_{i=1}^{n} C_i\right) = \sum_{i=1}^{n}(-1)^i S_{i,n}. \qquad (7.31)$$

If $m_n(x)$ and $\underline{m}_n(x)$ are the number of C_i and $\overline{C}_i$ $(i = 1, 2, \ldots, n)$ that occur in the sample, respectively, according to Galambos (1978), we have

$$\Pr(m_n(x) = t) = \sum_{i=0}^{n-t}(-1)^i\binom{i+t}{t}S_{i+t,n} \qquad (7.32)$$

and

$$\Pr(\underline{m}_n(x) = t) = \sum_{i=0}^{n-t}(-1)^i\binom{i+t}{t}\overline{S}_{i+t,n}, \qquad (7.33)$$

where

$$\overline{S}_{k,n} = \sum_{1\le i_1<i_2<\ldots<i_k\le n}\Pr(\overline{C}_{i_1}\cap \overline{C}_{i_2}\cap\ldots\cap \overline{C}_{i_k}), \qquad (7.34)$$

$$S_{0,n} = 0, \quad \text{and} \quad \overline{S}_{0,n} = 1.$$

7.6.2 Distribution of a Single Order Statistic

By selecting the adequate sets $C_i, i = 1, 2, \ldots, n$, one can use the above formulas to obtain the cdf of any order statistic. Letting

$$C_i(x) = \{X_i \le x\}, \quad \text{and} \quad \overline{C}_i(x) = \{X_i > x\},$$

we have

$$F_{X_{r:n}}(x) = \Pr[X_{r:n} \le x] = \Pr[m_n(x) \ge r] = \Pr[\underline{m}_n(x) \le n - r], \tag{7.35}$$

$$\begin{aligned} F_{X_{n-r+1:n}}(x) &= \Pr[X_{n-r+1:n} \le x] = 1 - \Pr[X_{n-r+1:n} > x] \\ &= 1 - \Pr[\underline{m}_n(x) \ge r] = 1 - \Pr[m_n(x) \le n - r], \end{aligned} \tag{7.36}$$

where $m_n(x)$ and $\underline{m}_n(x)$ satisfy the relation

$$m_n(x) + \underline{m}_n(x) = n.$$

Now combining (7.35) and (7.36) with (7.32) and (7.33), one gets

$$F_{X_{r:n}}(x) = \sum_{t=0}^{n-r}\sum_{i=0}^{n-t}(-1)^i\binom{i+t}{t}\overline{S}_{i+t,n}(x) = \sum_{t=r}^{n}\sum_{i=0}^{n-t}(-1)^i\binom{i+t}{t}S_{i+t,n}(x) \tag{7.37}$$

and

$$\begin{aligned} F_{X_{n-r+1:n}}(x) &= 1 - \sum_{t=r}^{n}\sum_{i=0}^{n-t}(-1)^i\binom{i+t}{t}\overline{S}_{i+t,n}(x) \\ &= 1 - \sum_{t=0}^{n-r}\sum_{i=0}^{n-t}(-1)^i\binom{i+t}{t}S_{i+t,n}(x). \end{aligned} \tag{7.38}$$

Thus, (7.37) and (7.38) for the extremes become

$$F_{X_{n:n}}(x) = \sum_{i=0}^{n}(-1)^i\overline{S}_{i,n}(x) = S_{n,n}(x) = F(x, x, \ldots, x) \tag{7.39}$$

and

$$F_{X_{1:n}}(x) = 1 - \sum_{i=0}^{n}(-1)^i S_{i,n}(x) = 1 - \overline{S}_{n,n}(x) = 1 - S(x, x, \ldots, x), \tag{7.40}$$

where $F(x_1, x_2, \ldots, x_n)$ is the n-dimensional cdf of $(X_1, \ldots, X_n)$ and

$$S(x_1, x_2, \ldots, x_n) = \Pr(X_1 > x_1, X_2 > x_2, \ldots, X_n > x_n) \tag{7.41}$$

is the n-dimensional *survival* function, which are the generalizations of (7.3) and (7.5).

Example 7.11 (Multivariate Mardia's distribution). Consider a system made up of n equal serial components whose lifetimes jointly follow the multivariate Mardia's distribution with a survival function

$$S(x_1, x_2, \ldots, x_n) = \left[\sum_{j=1}^{n}\exp\left(\frac{x_j}{a}\right) - n + 1\right]^{-a}.$$

Then, since the components are serial, then the lifetime of the system is the lifetime of its weakest component. Thus, the cdf of the system lifetime, according to (7.40), is

$$F(x) = 1 - \left[n \exp\left(\frac{x}{a}\right) - n + 1\right]^{-a}.$$

■

Example 7.12 (Marshall-Olkin distribution). Consider again a system made up of n equal serial components whose lifetimes jointly follow the multivariate Marshall-Olkin distribution with a survival function

$$\begin{aligned} S(x_1, x_2, \ldots, x_n) &= \exp\left(-\sum_{i=1}^{n} \lambda x_i - \sum_{i<j} \lambda^2 \max(x_i, x_j)\right) \times \ldots \\ &\times \exp\left(- \sum_{1 \le i_1 < \ldots < i_{n-1} \le n} \lambda^{n-1} \max(x_{i_1}, x_{i_2}, \ldots, x_{i_{n-1}})\right) \\ &\times \exp\left(-\lambda^n \max(x_1, x_2, \ldots, x_n)\right). \end{aligned}$$

Letting $x_1 = \ldots = x_k = x$ and $x_{k+1} = \ldots = x_n = 0$, the k-dimensional survival function becomes

$$\alpha_k(x) = S(x, x, \ldots, x) = \exp\left\{\left[-\sum_{r=1}^{n} \binom{n}{r} \lambda^r + \sum_{r=1}^{n-k} \binom{n-k}{r} \lambda^r\right] x\right\},$$

from which we can write

$$\overline{S}_{k,n}(x) = \binom{n}{k} \alpha_k(x) = \binom{n}{k} \exp\left\{\left[-\sum_{r=1}^{n} \binom{n}{r} \lambda^r + \sum_{r=1}^{n-k} \binom{n-k}{r} \lambda^r\right] x\right\},$$

and, according to (7.37), the cdf of the rth order statistic is

$$\begin{aligned} F_{X_{r:n}}(x) &= \sum_{t=0}^{n-r} \sum_{i=0}^{n-t} (-1)^i \binom{i+t}{t} \binom{n}{i+t} \\ &\times \exp\left\{\left[-\sum_{r=1}^{n} \binom{n}{r} \lambda^r + \sum_{r=1}^{n-i-t} \binom{n-i-t}{r} \lambda^r\right] x\right\}. \end{aligned}$$

■

Exercises

7.1 Consider the random variable X with cdf

$$F(x) = \begin{cases} 0, & \text{if } x < 0, \\ x^k, & \text{if } 0 \le x < 1, \\ 1, & \text{if } x \ge 1, \end{cases}$$

where $k > 0$, and assume that an independent and identically distributed sample is obtained from X.

Obtain:

(a) The pdf of the order statistic $X_{j:n}$.

(b) The joint density of all order statistics.

(c) The joint density of $X_{i:n}$ and $X_{j:n}$.

(d) The joint distribution of the maximum and minimum, directly and using the previous result.

(e) The joint density of $X_{i:n}$, $X_{i+1:n}$ and $X_{i+2:n}$ for $i < n-1$.

7.2 Let $X_i \sim \text{Rayleigh}(\delta_i), i = 1, 2, \ldots, k$. Show that

$$Y = \min(X_1, X_2, \ldots, X_k)$$

is Rayleigh(δ), where

$$\delta = \frac{\prod_{i=1}^{k} \delta_i}{\sqrt{\sum_{i=1}^{k} \prod_{i_1 \neq i_2 \neq \ldots \neq i_{k-1}} \delta_{i_1}^2 \delta_{i_2}^2 \cdots \delta_{i_{k-1}}^2}}.$$

7.3 Given a sample of size five coming from a uniform $U(0,1)$ distribution, obtain:

(a) The joint density of the order statistics $X_{2:5}, X_{3:5}, X_{4:5}$.

(b) The density of the range $X_{5:5} - X_{1:5}$.

(c) The density of $Z = (X_{5:5} + X_{1:5})/2$.

(d) The density of the median $X_{3:5}$.

(e) Compare the last two densities and draw some conclusions.

7.4 A family of random variables is said to be stable with respect to max (min) operations if and only if the maximum (minimum) of a set of independent random variables in such a family belongs also to that family. Discuss:

(a) The stability of the Weibull family with respect to min operations. Can the location, scale, and shape parameters be all different for stability?

(b) The stability of the reversed Weibull family with respect to max operations.

(c) The stability of the Weibull family with respect to max operations.

7.5 The yearly maximum floods, in a given cross-section of a river, measured in m^3/sec, during the last 60 years are shown in Table 1.2. How many exceedances of 46 m^3/sec are expected during the next 10 years?

7.6 The time required by a car to go between two given cities is a random variable with cdf

$$F(x) = \begin{cases} 0, & \text{if } x < 3, \\ 1 - \exp(3 - x), & \text{if } x \geq 3. \end{cases}$$

(a) What is the pdf of the time required by the slowest (fastest) car in a group of four cars?

(b) If we work with a large number of cars, propose a distribution for the time associated with the slowest (fastest) car.

(c) What is the exact distribution for the time associated with the k fastest car?

7.7 The grade obtained by a randomly selected student in a statistics course has pdf

$$f(x) = \begin{cases} x/50, & \text{if } 0 \leq x \leq 10, \\ 0, & \text{otherwise.} \end{cases}$$

In a class with 100 students, determine:

(a) The pdf of the grade associated with the best student in the class.

(b) The pdf of the grade associated with the third best student in the class.

(c) The pdf of the grade associated with the worst student in the class.

7.8 Repeat Example 7.10 with a Weibull distribution and obtain the relation between the return periods obtained for the yearly and century cdfs.

7.9 Let $Z_{1:n}, Z_{2:n}, \ldots, Z_{n:n}$ be the order statistics from n independent and identically distributed standard exponential random variables with pdf

$$f(x) = e^{-x}, \quad 0 \leq x < \infty.$$

(a) Show that the normalized spacings

$$S_1 = nZ_{1:n}, Z_2 = (n-1)(Z_{2:n} - Z_{1:n}), \ldots, S_n = Z_{n:n} - Z_{n-1:n},$$

are statistically independent and that they are all distributed as standard exponential variables.

(b) Using this independence result, show that

$$\begin{aligned} E(Z_{r:n}) &= \sum_{i=1}^{r} \frac{1}{n-i+1}, \\ \operatorname{Var}(Z_{r:n}) &= \sum_{i=1}^{r} \frac{1}{(n-i+1)^2}, \\ \operatorname{Cov}(Z_{r:n}, Z_{s:n}) &= \operatorname{Var}(Z_{r:n}) \end{aligned}$$

for $1 \leq r < s \leq n$.

7.10 Using the results in Exercise 7.9, considering the MLE $\hat{\theta}$ of θ in Example 5.2, show that $2n\hat{\theta}/\theta$ has a central χ^2 distribution with $2n$ degrees of freedom.

7.11 Once again, by using the results in Exercise 7.9 and considering the MLEs $\hat{\mu}$ and $\hat{\theta}$ of μ and θ in Example 5.5, show that $P_1 = 2n(\hat{\mu} - \mu)/\theta$ and $P_2 = 2n\hat{\theta}/\theta$ are independently distributed as central χ^2 with 2 and $2n-2$ degrees of freedom, respectively. As a consequence, prove that $(n-1)(\hat{\mu}-\mu)/\hat{\theta}$ has a central F distribution with $(2, 2n-2)$ degrees of freedom (see Section 3.2.12).

7.12 Using the results in Exercise 7.9, establish that the MLE $\hat{\theta}$ of θ in Example 5.2 is also the Best Linear Unbiased Estimator of θ (that is, an estimator which is a linear function of order statistics, unbiased and has the minimum most variance); see Balakrishnan and Rao (1997).

7.13 Let $X_1, \ldots X_n$ be independent random variables with X_i being distributed as exponential with pdf

$$f_i(x) = \theta_i e^{-\theta_i x}, \quad x \geq 0,\ \theta_i > 0,\ i = 1, \ldots, n.$$

(a) Obtain the cdf and pdf of $X_{1:n}$.

(b) What is the distribution of $X_{r:n}$?

(c) Obtain the cdf and pdf of $X_{n:n}$.

(d) What is the joint density function of $X_{1:n}$ and $X_{n:n}$?

Chapter 8

Point Processes and Exact Models

There are two types of models related to extremes: finite (exact) and asymptotic models. Asymptotic models are used when no exact models are available. Chapters 9 and 11 deal with asymptotic models. In this chapter, we present some exact models related to extremes. Section 8.1 discusses point processes in general that are useful for understanding the limiting distributions of extremes and exceedances. Sections 8.2 to 8.6 discuss different particular models such as the Poisson flaws model, the mixture models, competing risk models, Poissonian storms model, etc.

8.1 Point Processes

Point processes are stochastic processes whose outcomes are points in n-dimensional space. Point processes are defined when two pieces of information are given: (a) the occurrence of points and (b) the probability of occurrence. Once the point process is defined, the following probabilities can be easily calculated: (a) the probability of the occurrence of a given number of events, (b) the probability of the time between successive events, (c) the probability of the time to the kth event to be smaller than a certain value t, etc. These probabilities are all essential for engineering design.

Point processes are very good tools for modeling engineering problems. Consider, for example, the occurrence of very high waves or winds, earthquakes, storms, failures, etc. They occur at random times in the one-dimensional space of time. Similarly, earthquakes and cracks in a concrete dam occur at random locations in the earth or in the domain of $\mathbb{R}^3$ defined by the dam structure, respectively. Thus, point processes help to calculate the probability of having k of these events during a given period of time or in a given region.

An example of a point process is the Poisson homogeneous model that was described earlier in Chapter 2. In that case, the point process is unidimensional and has intensity $\lambda(x) = \lambda$, that is, the intensity function is constant (homogeneous). However, there are other point processes that are not homogeneous.

A direct and simple way of defining the statistical properties of a point process is by means of its intensity function, which is a function $\lambda(\mathbf{x})$, that when integrated over a given region A gives the expected value of the number of events occurring in that region, that is,

$$\Lambda(A) = \int_A \lambda(\mathbf{x})d\mathbf{x}, \tag{8.1}$$

where $\Lambda(A)$ is the expected number of events occurring in the region A. This function allows defining the nonhomogeneous Poisson process.

Definition 8.1 (Nonhomogeneous Poisson process). *A point process is said to be a nonhomogeneous Poisson process with intensity $\lambda(\mathbf{x})$ if*

1. *The number of events, $N(A)$, occurring in A is a Poisson random variable, that is,*
$$N(A) \sim Po(\Lambda(A)).$$

2. *The number of events $N(A)$ and $N(B)$ occurring in nonoverlapping regions A and B, respectively, are independent random variables.*

An important problem of a nonhomogeneous Poisson process defined on a set $B \subset \mathbb{R}^m$ is the estimation of its intensity function $\lambda(\mathbf{x})$. This problem can be dealt with as parametric and nonparametric. To simplify the problem, we assume that we have a parametric family $\lambda(\mathbf{x};\boldsymbol{\theta})$ and that we wish to estimate $\boldsymbol{\theta}$. To this end, we may use the maximum likelihood method based on some observations, that is, some points $(\mathbf{x}_1, \mathbf{x}_2, \ldots, \mathbf{x}_n)$, where the event has occurred. For simplicity, we assume that they are not coincident (note that the probability of this coincidence is zero). Then, the likelihood is the probability of observing such information given the model. To facilitate the derivation of the likelihood function consider n spheres $S_i, i = 1, 2, \ldots, n$, centered at each of the points $\mathbf{x}_i$ (see Fig. 8.1). The radius r_i of S_i is small enough so that the spheres do not overlap. Then it is clear that we have observed one event in each of the S_i spheres, and no event in the set $Q \equiv B - \cup_{i=1}^n S_i$.

Since

$$\begin{aligned} \Lambda(S_i;\boldsymbol{\theta}) &= \int_{S_i} \lambda(\mathbf{x};\boldsymbol{\theta})d\mathbf{x} \approx \lambda(\mathbf{x}_i;\boldsymbol{\theta})\Omega(S_i), \\ \Lambda(Q;\boldsymbol{\theta}) &= \int_Q \lambda(\mathbf{x};\boldsymbol{\theta})d\mathbf{x}, \end{aligned}$$

where $\Omega(S_i)$ is the hypervolume of S_i. According to the Poisson assumption, their probabilities are

$$\begin{aligned} p_i &= \Pr[N(S_i) = 1] = \exp(-\Lambda(S_i))\Lambda(S_i) \approx \lambda(\mathbf{x}_i;\boldsymbol{\theta})\Omega(S_i), \\ q &= \Pr[N(Q) = 0] = \exp(-\Lambda(Q;\boldsymbol{\theta})). \end{aligned}$$

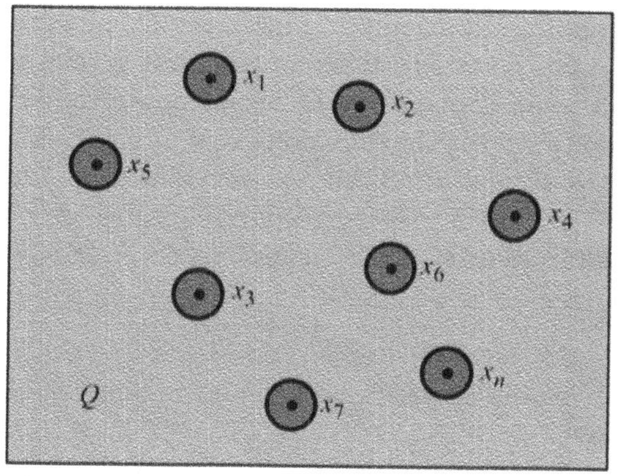

Figure 8.1: Illustration of the derivation of the likelihood function for a Poisson process.

The above approximation follows because S_i has a near zero radius, hence a near zero volume. Finally, since they are nonoverlapping regions, they are independent events. Then the likelihood function becomes

$$L(\boldsymbol{\theta}; \mathbf{x}) \propto \exp(-\Lambda(Q; \boldsymbol{\theta})) \prod_{i=1}^{n} \lambda(\mathbf{x}_i; \boldsymbol{\theta}), \tag{8.2}$$

where the constant $\prod_{i=1}^{n} \Omega(S_i)$ has been removed, hence the proportionality symbol, $\propto$, in (8.2).

Example 8.1 (Storms). Storms are not equally likely during the spring, summer, fall, and winter times, so that one can think of a function $\lambda(t)$, where t ranges from 1 to 365 days, which takes larger values at times where the chances of storms are higher. If $\lambda(t)$ is chosen such that

$$\int_0^{365} \lambda(t) dt = \eta,$$

where η is the expected number of storms during one year, then this function λ can be interpreted as the intensity of a nonhomogeneous Poisson process. Then, the expected number of storms during a period starting at time t_0 and ending at time t_1 is $\int_{t_0}^{t_1} \lambda(t) dt$. ∎

Example 8.2 (Poissonian Storm Model). In modeling the occurrence of storms, failures of marine structures, for example, are likely to occur due to exceptional events such as waves of considerable heights. Then, the Poisson model for occurrence of storms can be appropriate. However, one is interested not only in the number of storms occurring during a given period of time but also in the corresponding duration t and intensity H_s of storms, where H_s is the significant wave height, a statistical parameter that is related to storm severity. Then, a Poissonian point process can be an adequate solution.

Assume, for example, that the duration t and intensity H_s of a storm are random, that is, the two-dimensional random variable (t, H_s) has probability density function $q(t; H_s)$, and that the corresponding rates are $r(t, H_s)$ events/year. Then, the Poissonian intensity function becomes

$$\lambda(t, H_s) = r(t, H_s)q(t; H_s).$$

As a particular example, based on real data from the Gijón buoy, Castillo et al. (2004) propose

1. A constant rate $r(t, H_s) = r$.
2. A normal $N(a + bt, \sigma^2)$, density for $H_s|t$.
3. An exponential density with mean d for the duration t, that is,

$$\lambda(t, H_2) = r\frac{1}{d}\exp\left(-\frac{t}{d}\right) \times \frac{1}{\sigma\sqrt{2\pi}}\exp\left[-\frac{1}{2}\left(\frac{H_s - a - bt}{\sigma}\right)^2\right], \quad t \geq 0,$$

where a and b are regression parameters, r is the yearly rate (storms/year), and $a, b \in \mathbb{R}$ and $\sigma \in \mathbb{R}^+$. ■

Theorem 8.1 (Exceedances as a Poisson process). *Let $\{X_n\}$ be a sequence of independent and identically distributed (iid) random variables with cdf $F(x)$. Assume that the sequence $\{x_n\}$ of real numbers satisfies the condition*

$$\lim_{n\to\infty} n[1 - F(x_n)] = \tau, \quad 0 \leq \tau \leq \infty. \tag{8.3}$$

Then

$$\lim_{n\to\infty} \Pr[S(x_n) = r] = \frac{\exp(-\tau)\tau^r}{r!}, \quad r \geq 0, \tag{8.4}$$

where $S(x_n)$ is the number of X_i $(i = 1, 2, \ldots, n)$ exceeding x_n, and the right-hand side of (8.4) must be taken as 1 or 0 depending on whether $\tau = 0$ or $\tau = \infty$, respectively.

The proof of this theorem is just the well known Poisson limit for the binomial distribution if $0 < \tau < \infty$ and some trivial considerations for the two extreme cases $\tau = 0$ and $\tau = \infty$.

The practical importance of this theorem lies in the fact that the number of exceedances approximately follows a Poisson distribution, in the sense of (8.4) if (8.3) holds.

Definition 8.2 (Convergence of a sequence of point processes). *Consider a sequence $\{N_i; i = 1, 2, \ldots\}$ of point processes on a set A. This sequence is said to converge in distribution to the point process N, if $(N_n(A_1), N_n(A_2),$*

..., $N_n(A_k)$) converges in distribution to $(N(A_1), N(A_2), \dots, N(A_k))$ for all possible values of k and bounded sets $A_1, A_2, \dots, A_k$ such that the probabilities of boundary events in A_i are null for $i = 1, 2, \dots, k$. When the sequence $\{N_n\}$ converges to N, we write $N_n \to N$.

The practical importance of this convergence is that one can calculate probabilities using the statistical properties of the limit process for large enough values of n.

8.2 The Poisson-Flaws Model

Let $X(t)$ be the lifetime (time until failure) of a longitudinal element of length ℓ (e.g., wire, cable). We make the following assumptions:

1. The element contains a random number $N(\ell)$ of flaws. These flaws are distributed along the length of the longitudinal element according to a nonzero Poisson process of intensity $\lambda(\tau) > 0$. This means that elements have at least one flaw, that is, there are no infinitely resistent elements (elements with infinite lifetime). Consequently, the number of flaws in a piece of length ℓ is a truncated Poisson($\mu(\ell)$), where

$$\mu(\ell) = \int_0^\ell \lambda(\tau) d\tau.$$

2. The $X_1, X_2, \dots, X_{N(\ell)}$ lifetimes associated with the $N(\ell)$ flaws are assumed to be independent random variables with survival function $S(x; \boldsymbol{\theta})$, where $\boldsymbol{\theta}$ is a vector of parameters.

3. The lifetime of the longitudinal element is governed by the weakest link principle, that is,

$$X(\ell) = \min(X_1, X_2, \dots, X_{N(\ell)}). \tag{8.5}$$

With these assumptions, the survival function $S(x)$ of $X(\ell)$ assuming $N(\ell)$ flaws is $\Pr(X(\ell) > x) = S(x; \boldsymbol{\theta})^{N(\ell)}$ and taking into account that $N(\ell)$ is random and using the total probability theorem we get

$$\begin{aligned} S(x) = \Pr(X(\ell) > x) &= \frac{1}{1 - e^{-\mu(\ell)}} \sum_{N(\ell)=1}^{\infty} \frac{e^{-\mu(\ell)}[\mu(\ell)]^{N(\ell)}}{N(\ell)!} [S(x; \boldsymbol{\theta})]^{N(\ell)} \\ &= \frac{e^{-\mu(\ell)}}{1 - e^{-\mu(\ell)}} \sum_{N(\ell)=1}^{\infty} \frac{[\mu(\ell) S(x; \boldsymbol{\theta})]^{N(\ell)}}{N(\ell)!} \\ &= \frac{e^{-\mu(\ell)}}{1 - e^{-\mu(\ell)}} [\exp\{\mu(\ell) S(x; \boldsymbol{\theta})\} - 1] \qquad (8.6) \\ &= \frac{\exp\{\mu(\ell)[S(x; \boldsymbol{\theta}) - 1]\} - e^{-\mu(\ell)}}{1 - e^{-\mu(\ell)}}. \qquad (8.7) \end{aligned}$$

The identity in (8.6) follows because

$$\sum_{n=1}^{\infty} \frac{y^n}{n!} = e^y - 1.$$

Note that since $\lim_{x\to 0} S(x;\boldsymbol{\theta}) = 1$ and $\lim_{x\to\infty} S(x;\boldsymbol{\theta}) = 0$, $S(x)$ is a proper survival function. Note also that this would not be the case if a Poisson process (instead of a nonzero Poisson process) would have been assumed, since then a nonzero probability of zero flaws would exist, leading to the possibility of the existence of infinitely resistent elements.

This model can be extended in two ways. First, to include nonhomogeneous Poisson flaws, that is, Assumption 1 above can be replaced by assuming a Poisson process with varying intensity function $\mu(\ell)$. In this case, (8.7) becomes

$$\begin{aligned} S(x;\boldsymbol{\theta}, A(\ell)) &= \Pr(X(\ell) > x) \\ &= \exp(-A(\ell))\left[\exp(A(\ell)S(x;\boldsymbol{\theta})) - 1\right], \end{aligned} \tag{8.8}$$

where

$$A(\ell) = \int_0^t \mu(u)du. \tag{8.9}$$

Second, random varying intensity functions $\mu(\ell)$ or $A(\ell)$ can be assumed. In this case, (8.7) becomes

$$S(x;\boldsymbol{\theta}) = E_A\left[\exp(-A(\ell))\left\{\exp(A(\ell)S(x;\boldsymbol{\theta})) - 1\right\}\right], \tag{8.10}$$

where the expectation E_A is taken with respect to the random variable A. Some similar models are described by Taylor (1994).

Example 8.3 (Poisson-Flaws Model). Consider a cable material such that the number of flaws in it is Poissonian with intensity (see Fig. 8.2)

$$\lambda(t) = 4 + \sin(2t),$$

that is, the number of flaws in a cable of length T is a Poissonian random variable with mean

$$\mu(T) = \int_0^T [4 + \sin(2t)]dt = 1/2 + 4T - \cos(2T)/2.$$

Assume also that the survival function $S(x)$ of a flaw is Weibull $W_m(0, 1, 2)$, that is,

$$S(x) = e^{-x^2} \tag{8.11}$$

with $\lambda = 0, \delta = 1$ and $\beta = 2$. Then, according to (8.7) the survival function of a cable of length T is

$$S(x) = \frac{A^B - A}{1 - A},$$

where $A = \exp(1/2 + 4T - \cos(2T)/2)$ and $B = e^{-x^2} - 1$. Figure 8.3 shows the flaw survival function $S(x)$ (the thick line) and the survival functions $S(x)$ associated with cables of lengths 1, 2, 4, 8, and 16 m. ∎

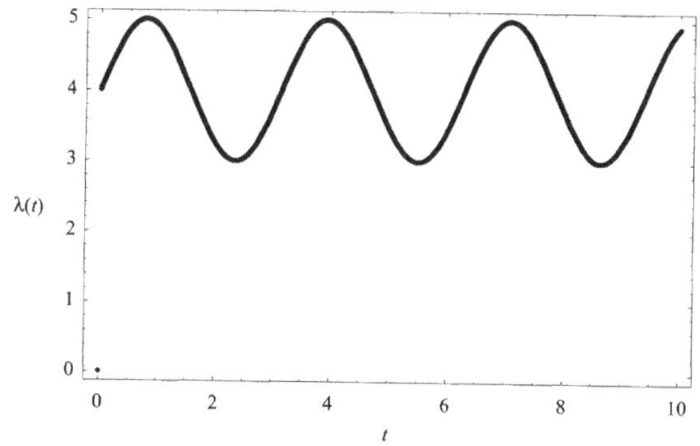

Figure 8.2: Intensity function $\lambda(t)$ of the flaws in a cable material.

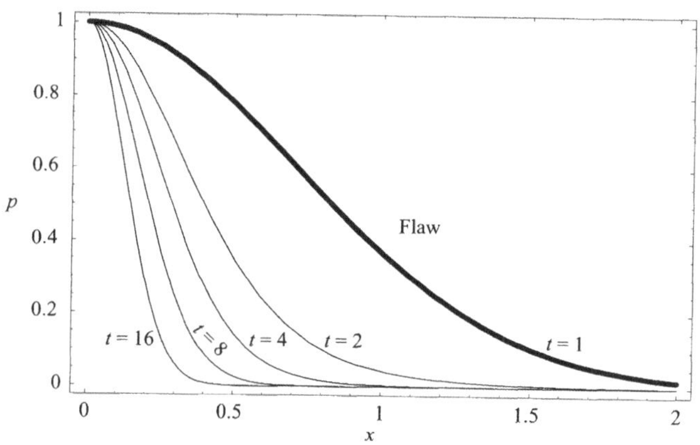

Figure 8.3: Survival functions of a typical flaw and cables of lengths $1, 2, 4, 8$, and 16 m.

8.3 Mixture Models

In mixture models, we assume that the model can be chosen at random, with given probability, from a given set of possible models. Here, we make the following assumptions:

1. The model has a survival function in the family of survival functions

$$\{S(x;\boldsymbol{\theta});\boldsymbol{\theta}\in\Theta\},$$

   where $\boldsymbol{\theta}$ is a vector of parameters, and Θ is the parameter space.

2. The probability of having a survival function $S(x;\boldsymbol{\theta})$ governed by the cdf $F(\boldsymbol{\theta})$.

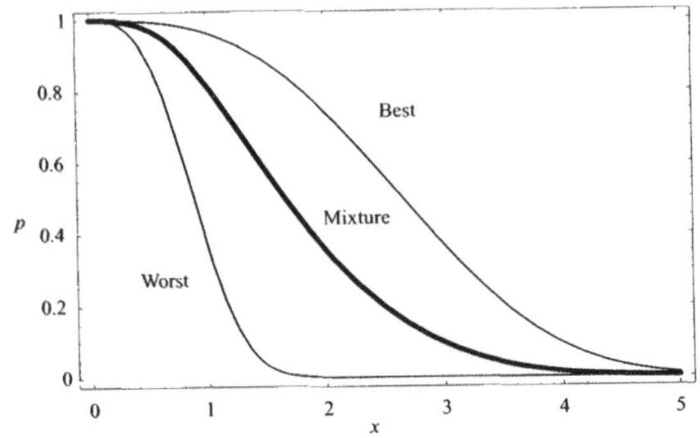

Figure 8.4: Mixture survival function of a family of survival functions, where best and worst elements are indicated.

Then, the mixture model leads to a survival function of the type

$$S(x; F) = \int_{\boldsymbol{\theta} \in \Theta} S(x; \boldsymbol{\theta}) dF(\boldsymbol{\theta}). \tag{8.12}$$

Note that (8.10) is an example of a mixture model.

Example 8.4 (Mixture model). Consider a family $\{S(x; \delta) | 1 \leq \delta \leq 3\}$ of survival functions of the Weibull type

$$S(x; \delta) = e^{-x^3/\delta}, \quad x \geq 0,$$

and assume that $\delta \sim U(1, 3)$. Then, the mixture survival function becomes

$$S(x) = \int_1^3 \frac{e^{-x^3/\delta}}{2} \, d\delta.$$

The resulting $H(x)$ function together with the best and worst survival function in the given family are shown in Figure 8.4.

■

8.4 Competing Risk Models

The competing risk model considers elements with several alternative and simultaneous modes of failure, and failure is said to occur as soon as the first failure takes place, that is, the following assumptions are made:

1. A set of different failure modes are considered with survival functions $\{S_i(x), i \in I\}$.

2. The failure of an element is said to occur based on the weakest link principle, that is, the survival function for the element is

$$S(x; I) = \prod_{i \in I} S_i(x).$$

Example 8.5 (Minimum temperatures). Assume that the minimum temperatures in month i is a Weibull $W_m(\lambda, \delta_i, \beta), i = 1, 2, \ldots, 12$, and consider the problem of minimum temperatures in a given period of long duration n years. Then, the cdf of the minimum temperatures in a period of n years is

$$\begin{aligned} F_n(x) &= 1 - \prod_{i=1}^{12n} \left(1 - F_{W_m(\lambda,\delta_i,\beta)}(x)\right) = 1 - \exp\left[-\sum_{i=1}^{12n} \left(\frac{x-\lambda}{\delta_i}\right)^{\beta}\right] \\ &= 1 - \exp\left[-n\left(\frac{x-\lambda}{\delta_0}\right)^{\beta}\right], \end{aligned}$$

where

$$\frac{1}{\delta_0^{\beta}} = \sum_{i=1}^{12} \frac{1}{\delta_i^{\beta}}.$$

Then

$$\lim_{n \to \infty} F_n\left(\delta_0 \left(\frac{x}{n}\right)^{1/\beta} + \lambda\right) = 1 - \exp(-x),$$

which shows the asymptotic property of the model. In this case 12 models, one per each month, compete to give the minimum temperature.

For complete years or large n, in fact, the model is identical or practically equivalent to assuming that minimum temperature distribution for all months are identical and Weibull $W_m(\lambda, \delta_0, \beta)$. ∎

8.5 Competing Risk Flaws Models

This model is a generalization of the model described in Section 8.2. Let $S(t)$ be the strength of a longitudinal element of length t. We make the following assumptions:

1. The element contains r different types of flaws that cause stress concentrations, such that they are distributed along the length of the longitudinal element. These flaws include internal flaws such as voids, inclusions, and weak grain boundaries, and external or surface flaws such as notches or cracks.

2. The number of existing flaws of type s $(s = 1, 2, \ldots, r)$ in the element is a random number $N_s(t)$, which depends on the length t.

3. $N_s(t)$ is a discrete positive random variable with probability mass function $P_s(x, t)$.

4. The $X_{s1}, X_{s2}, \ldots, X_{sN_s(t)}$ strengths of the $N_s(t)$ flaws of type s are assumed to be independent random variables with survival function $S_s(x; \boldsymbol{\theta}_s)$, where $\boldsymbol{\theta}_s \in \Theta_s$ is a vector of parameters in the parameter space Θ_s.

5. The strengths of different types of flaws are assumed to be independent.

6. The longitudinal element strength is governed by the weakest link principle, that is,

$$X(t) = \min_{s=1,2,\ldots,r} \min(X_{s1}, X_{s2}, \ldots, X_{sN_s(t)}). \tag{8.13}$$

With these assumptions, the survival function of $X(t)$ taking into account the total probability theorem becomes

$$\begin{aligned} S(x; \boldsymbol{\theta}) &= \Pr(X(t) > x) \\ &= \prod_{s=1}^{r} \left(\sum_{N_s(t)=1}^{\infty} P_s(N_s(t), t) S_s^{N_s(t)}(x; \boldsymbol{\theta}_s) \right), \end{aligned} \tag{8.14}$$

where $\boldsymbol{\theta} = (\boldsymbol{\theta}_1, \boldsymbol{\theta}_2, \ldots, \boldsymbol{\theta}_r)$. Note that since $\lim_{x \to 0} S(x; \boldsymbol{\theta}) = 1$ and $\lim_{x \to \infty} S(x; \boldsymbol{\theta}) = 0$, $S(x; \boldsymbol{\theta})$ is a proper survival function. Note also that this would not be the case if $\Pr(N_s(t) = 0) > 0$, $s = 1, 2, \ldots, r$.

If, in addition, the $P_s(N_s(t), t)$ is assumed to be random, then the resulting survival function would be

$$S(x; \boldsymbol{\theta}) = E_P \left\{ \prod_{s=1}^{r} \left(\sum_{N_s(t)=1}^{\infty} P_s(N_s(t), t) S_s^{N_s(t)}(x; \boldsymbol{\theta}_s) \right) \right\}. \tag{8.15}$$

where the $E(\cdot)$ is the expected value. Note that the model in Section 8.2 is a particular case for $r = 1$.

8.6 Poissonian Storm Model

In this section, we present a Poissonian storm model with a given storm description. The model is based on the following assumptions:

1. A storm, in a simplified form, can be defined by a pair (t, α), where t is its duration and α is its severity. This means that storms with different durations and intensity levels are considered.

2. The storms (t, α) are generated according to a Poisson point process with intensity $\lambda(t, \alpha)$ events/year.

3. The statistical parameter κ at time τ associated with failure during a storm of duration τ and severity α is assumed to vary according to a given law $\kappa = h(\tau, \alpha)$.

4. Failures occur only during storms and the probability of failure at time τ has probability $p(h(\tau,\alpha);\boldsymbol{\theta})$, dependent on the value $h(\tau,\alpha)$ of κ and a vector of parameters $\boldsymbol{\theta}$.

5. Failures occurring at times τ_1 and τ_2 during a storm are independent for $\tau_1 \neq \tau_2$.

6. One storm can cause only one failure. Note that in the hypothetical case of the occurrence of several failures in a storm, only one failure must be considered, because it is not possible to repair during storms.

With these assumptions the number of (t,α)-storms, that is, storms with duration t in the interval $(t, t+dt)$ and intensity α in the interval $(\alpha, \alpha+d\alpha)$ occurring during the useful life D (years) of the structure being considered is a Poisson random variable with parameter $(D\lambda(t,\alpha)dtd\alpha)$ random variable, and the number X of (t,α)-storms causing failure is a Poisson random variable withy parameter $(D\lambda(t,\alpha)p_S(t,\alpha;\boldsymbol{\theta})dtd\alpha)$ random variable, where $p_S(t,\alpha;\boldsymbol{\theta})$ is the probability of failure during a (t,α)-storm, and is given by

$$p_S(t,\alpha;\boldsymbol{\theta}) = 1 - \exp\left\{\int_0^t \log([1 - p(h(\tau,\alpha);\boldsymbol{\theta})]\, d\tau\right\}. \tag{8.16}$$

To understand (8.16), we state first a discrete version.

1. The probability of no failure during the storm time unit j is

$$1 - p_m(h(\tau,\alpha_j);\boldsymbol{\theta}).$$

2. A storm with duration t leads to no failure if all its storm time units produce no failure, and this event has probability

$$1 - p_S(t,\alpha;\boldsymbol{\theta}) = \prod_{j=1}^{t}(1 - p(h(\tau,\alpha_j);\boldsymbol{\theta})).$$

Thus, taking logarithms one gets

$$\log(1 - p_S(t,\alpha;\boldsymbol{\theta})) = \sum_{j=1}^{t} \log\left[1 - p(h(\tau,\alpha_j);\boldsymbol{\theta})\right].$$

Passing now to the continuous case, that is, replacing sums by integrals, we get

$$\log(1 - p_S(t,\alpha;\boldsymbol{\theta})) = \int_0^t \log([1 - p(h(\tau,\alpha);\boldsymbol{\theta})]\, d\tau,$$

from which (8.16) is obtained.

Since storms do not have a fixed duration t and intensity α, but random numbers, with probability density function $q(t;\alpha)$, then the resulting number of storms causing failures X in the lifetime is Poissonian, that is,

$$X \sim P_0\left(D\int_{\alpha_{min}}^{\alpha_{max}}\int_{t_{min}}^{t_{max}} \lambda(t,\alpha)\left[q(t,\alpha)p_S(t,\alpha;\boldsymbol{\theta})\right] dt\; d\alpha\right),$$

and the expected number of storms causing failure is

$$E[X_\alpha] = D\int_{\alpha_{min}}^{\alpha_{max}}\int_{t_{min}}^{t_{max}} \lambda(t,\alpha)\left[q(t,\alpha)p_S(t,\alpha;\boldsymbol{\theta})\right] dt\; d\alpha.$$

Exercises

8.1 Propose a Poisson model for obtaining the final scores of two teams after a soccer match. Make the model dependent on the quality of the teams and the need they have for winning the match.

8.2 Propose a nonhomogeneous Poisson model for the numbers of cars arriving at a traffic light.

8.3 A set of different failure modes are considered with survival functions $\{S_i(x;\theta), i \in I\}$. The competing risk model, developed in Section 8.4, states that the failure of an element is said to occur based on the weakest link principle, that is, the survival function for the element is

$$S(x) = \prod_{i\in I} S(x;\theta_i).$$

Discuss the existence of a family of survival functions that is stable with respect to this model.

8.4 The mixture model described in Section 8.3 leads to a survival function of the type

$$S(x;F) = \int_{\boldsymbol{\theta}\in\Theta} S(x;\boldsymbol{\theta})dF(\boldsymbol{\theta}). \tag{8.17}$$

Discuss the conditions under which S and F make the parametric family $S(x;\boldsymbol{\theta})$ closed, that is,

$$S(x;\theta(F)) = \int_{\boldsymbol{\theta}\in\Theta} S(x;\boldsymbol{\theta})dF(\boldsymbol{\theta}).$$

8.5 Let X_1 and X_2 be the random strengths associated with the edge and surface flaws, respectively, for a given piece of material. Assume that they are independent random variables with pdf $f_{X_1}(x_1;\boldsymbol{\theta}_1)$ and $f_{X_2}(x_2;\boldsymbol{\theta}_2)$, respectively. The piece of material is tested up to failure that occurs when the minimum edge or surface strength is reached, that is, with $\min(X_1, X_2)$. Then, the observed cdf is given by

$$F(x; \boldsymbol{\theta}_1, \boldsymbol{\theta}_2) = 1 - [1 - F_{X_1}(x_1; \boldsymbol{\theta}_1)][1 - F_{X_2}(x_2; \boldsymbol{\theta}_2)]. \tag{8.18}$$

If the additional information of the type of failure (edge or surface) that has occurred for each sample data is available, then

(a) Derive the conditional pdfs $h_1(x; \boldsymbol{\theta}_1, \boldsymbol{\theta}_2)$ and $h_2(x; \boldsymbol{\theta}_1, \boldsymbol{\theta}_2)$ of the strengths given the type of failure.

(b) Show that if

$$f_{X_1}(x_1) = \lambda_1 \exp(-\lambda_1 x) \quad \text{and} \quad f_{X_2}(x_2) = \lambda_2 \exp(-\lambda_2 x), \tag{8.19}$$

then

$$h_1(x; \boldsymbol{\theta}_1, \boldsymbol{\theta}_2) = (\lambda_1 + \lambda_2) e^{-(\lambda_1 + \lambda_2)x} \tag{8.20}$$

and

$$h_2(x; \boldsymbol{\theta}_1, \boldsymbol{\theta}_2) = (\lambda_1 + \lambda_2) e^{-(\lambda_1 + \lambda_2)x}. \tag{8.21}$$

8.6 Use the nonhomogeneous Poisson process theory in Section 8.1 to derive the limit GEVD and GPD distributions that appear in Chapters 5 and 6 (see Coles (2001)).

8.7 Discuss the assumptions of the Poisson-flaws model, indicating which ones are more critical. Suggest some changes to overcome these limitations and extend its applicability to real problems. What type of real problems are candidates for this model?

8.8 Discuss the assumptions of the Poissonian storm model and suggest some changes to extend its applicability to real problems. What type of real problems are candidates for this model? Can the Poisson assumption be improved? How?

8.9 Discuss the applicability of the competing risk flaws models to engineering design when different failure modes can occur. Discuss their assumptions and suggest how its limitations can be overcome.

8.10 Give a list of real engineering examples where the mixed model is justified. Indicate how to obtain the $S(x; \boldsymbol{\theta})$ and the F function appearing in (8.12) for some particular cases.

Explain clearly when this and the competing risk models must be used.

Part V

Asymptotic Models for Extremes

Chapter 9

Limit Distributions of Order Statistics

The exact distributions of order statistics from independent and dependent samples have been discussed in Chapter 7 (see Sections 7.2 and 7.6). A superficial analysis of these formulas could lead to the conclusion that they solve most of the problems that arise in real-life practice. Unfortunately, this is not true, because in many practical problems these expressions are not very useful. This happens, for example, in the following cases:

1. When the sample size is very large or goes to infinity.

2. When the cdf, $F(x)$, of the parent population is unknown.

3. When the sample size is unknown.

In this chapter, we discuss the limit distributions of order statistics, that is, when $n \to \infty$. These results are useful when the sample size n is large.

This chapter is organized as follows. The limit distributions of order statistics from independent observations are given in Section 9.1. Various methods for the estimation of the parameters and quantiles of the resulting limit distributions are presented in Sections 9.2 and 9.3. The use of probability paper plots for the limit distributions are discussed and guidelines for selecting a domain of attraction from data are provided in Section 9.4. The Q-Q and P-P plots for model validation are given in Section 9.5. Section 9.6 presents the hypothesis testing approach to the model selection problems. Finally, the case where the observations are dependent is dealt with in Section 9.7.

9.1 The Case of Independent Observations

Among order statistics, the minimum and the maximum are the most relevant to engineering applications. Thus, we start with these order statistics. Next, we deal with other order statistics, but mainly with high- and low-order statistics because of their relevance in practical problems.

9.1.1 Limit Distributions of Maxima and Minima

We have seen that the cdf of the maximum $X_{n:n}$ and minimum $X_{1:n}$ of a sample of size n drawn from a population with cdf $F(x)$ are

$$H_n(x) = \Pr[X_{n:n} \leq x] = [F(x)]^n$$

and

$$L_n(x) = \Pr[X_{1:n} \leq x] = 1 - [1 - F(x)]^n.$$

When n tends to infinity, we have

$$\lim_{n\to\infty} H_n(x) = \lim_{n\to\infty} [F(x)]^n = \begin{cases} 1, & \text{if } F(x) = 1, \\ 0, & \text{if } F(x) < 1, \end{cases}$$

and

$$\lim_{n\to\infty} L_n(x) = \lim_{n\to\infty} 1 - [1 - F(x)]^n = \begin{cases} 0, & \text{if } F(x) = 0, \\ 1, & \text{if } F(x) > 0. \end{cases}$$

This means that the limit distributions are degenerate (they take values 0 and 1 only).

To avoid degeneracy, we look for linear transformations such that the limit distributions

$$\lim_{n\to\infty} H_n(a_n + b_n x) = \lim_{n\to\infty} [F(a_n + b_n x)]^n = H(x), \quad \forall\, x, \tag{9.1}$$

and

$$\lim_{n\to\infty} L_n(c_n + d_n x) = \lim_{n\to\infty} 1 - [1 - F(c_n + d_n x)]^n = L(x), \quad \forall\, x, \tag{9.2}$$

are not degenerate, where a_n, b_n, c_n, and d_n are constants, depending on n.

Since the maximum cdf moves to the right and can change slope as n increases (see Fig. 9.1), a translation a_n and a scale change b_n, both depending on n, are used to keep it fixed and with the same shape as that of $H(x)$. A similar treatment can be done with minima instead of maxima, using constants c_n and d_n. When this is possible, we say that $F(x)$ belongs to the domain of attraction of the limit distribution.

Figure 9.1 illustrates this and shows the cdf of the maxima in samples of sizes $1, 2, 10, 100, 1000$, and 10000 drawn from a gamma $G(3, 1)$, normal $N(0, 1)$, and exponential $E(1)$ distributions, respectively. Note that the left-most of the curves on each of the graphs is the parent cdf (where $n = 1$).

For a summary, see Smith (1990), and for a different approach to extremes, see Gomes and de Haan (1999).

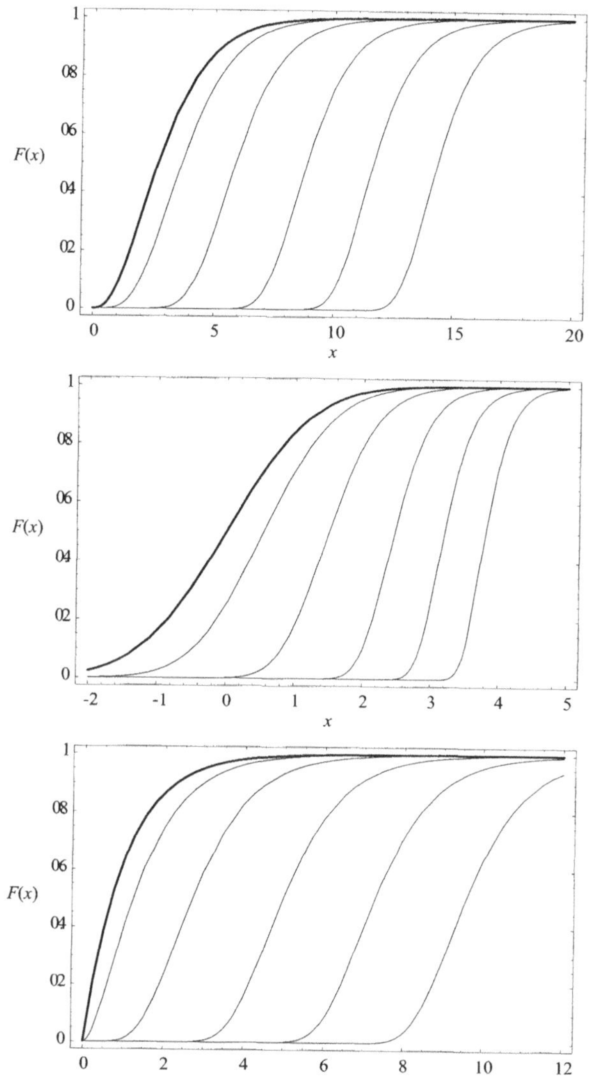

Figure 9.1: The cdf of the maxima in samples of sizes $1, 2, 10, 100, 1000$, and 10000 drawn from a gamma $G(3,1)$, normal $N(0,1)$, and exponential $E(1)$ distributions, respectively.

Definition 9.1 (Domain of attraction of a given distribution). *A given distribution, $F(x)$, is said to belong to the maximal domain of attraction of $H(x)$, if (9.1) holds for at least one pair of sequences $\{a_n\}$ and $\{b_n > 0\}$. Similarly, if $F(x)$ satisfies (9.2), we say that it belongs to the minimal domain of attraction of $L(x)$.*

The problem of limit distributions can then be stated as follows:

1. Find conditions under which (9.1) and (9.2) are satisfied.

2. Give rules for building the sequences $\{a_n\}, \{b_n\}, \{c_n\}$ and $\{d_n\}$.

3. Find the possible distributions for $H(x)$ and $L(x)$.

The answer to the third problem is given by the following theorems (see Fisher and Tippett (1928), Galambos (1987), and Tiago de Oliveira (1958)).

The surprising result is that only one parametric family is possible as a limit for maxima and only one for minima.

Theorem 9.1 (Feasible limit distribution for maxima). *The only nondegenerate family of distributions satisfying (9.1) is*

$$H_\kappa(x; \lambda, \delta) = \exp\left\{-\left[1-\kappa\left(\frac{x-\lambda}{\delta}\right)\right]^{1/\kappa}\right\}, \quad 1-\kappa\left(\frac{x-\lambda}{\delta}\right) \geq 0, \kappa \neq 0, \tag{9.3}$$

where the support is $x \leq \lambda + \delta/\kappa$, if $\kappa > 0$, or $x \geq \lambda + \delta/\kappa$, if $\kappa < 0$. The family of distributions for the case $\kappa = 0$ is obtained by taking the limit of (9.3) as $\kappa \to 0$ and getting

$$H_0(x; \lambda, \delta) = \exp\left[-\exp\left(\frac{\lambda - x}{\delta}\right)\right], \quad -\infty < x < \infty. \tag{9.4}$$

The distributions in (9.3) and (9.4) are called the von-Mises family of distributions for maxima or the maximal generalized extreme value distributions *(GEVDs), which we denote by* $\mathrm{GEVD}_M(\lambda, \delta, \kappa)$.

Note that for $\kappa > 0$, the distribution is limited on the right-hand side (the tail of interest), that is, it has the finite upper end $\lambda + \delta/\kappa$. Otherwise, it is unlimited on the right. Note also that for $\kappa < 0$, the distribution is limited on the left.

The corresponding p-quantiles can be obtained by inverting H_κ and H_0 in (9.3) and (9.4) and obtaining

$$x_p = \begin{cases} \lambda + \delta\left[1 - (-\log p)^\kappa\right]/\kappa, & \text{if } \kappa \neq 0, \\ \lambda - \delta \log(-\log p), & \text{if } \kappa = 0. \end{cases} \tag{9.5}$$

Theorem 9.2 (Feasible limit distribution for minima). *The only nondegenerate family of distributions satisfying (9.2) is*

$$L_\kappa(x; \lambda, \delta) = 1 - \exp\left\{-\left[1+\kappa\left(\frac{x-\lambda}{\delta}\right)\right]^{1/\kappa}\right\}, \quad 1+\kappa\left(\frac{x-\lambda}{\delta}\right) \geq 0, \kappa \neq 0, \tag{9.6}$$

where the support is $x \geq \lambda - \delta/\kappa$, if $\kappa > 0$, or $x \leq \lambda - \delta/\kappa$, if $\kappa < 0$. The family of distributions for the case $\kappa = 0$ is obtained by taking the limit of (9.6) as $\kappa \to 0$ and getting

$$L_0(x; \lambda, \delta) = 1 - \exp\left[-\exp\left(\frac{x-\lambda}{\delta}\right)\right], \quad -\infty < x < \infty. \tag{9.7}$$

The distributions in (9.6) and (9.7) are called the von-Mises family of distributions for minima or the minimal GEVDs, which we denote by $\text{GEVD}_m(\lambda, \delta, \kappa)$.

Note that for $\kappa > 0$, the distribution is limited on the left-hand side (the tail of interest), that is, it has the finite lower end $\lambda - \delta/\kappa$. Otherwise it is unlimited on the left. Note also that for $\kappa < 0$, the distribution is limited on the right.

The corresponding p-quantiles can be obtained by inverting L_κ and L_0 in (9.6) and (9.7) and getting

$$x_p = \begin{cases} \lambda - \delta\left[1 - (-\log(1-p))^\kappa\right]/\kappa, & \text{if } \kappa \neq 0, \\ \\ \lambda + \delta \log(-\log(1-p)), & \text{if } \kappa = 0. \end{cases} \tag{9.8}$$

The following theorem allows one to obtain the $GEVD_m$ from the $GEVD_M$. It also shows how a problem involving minima can be converted to a problem involving maxima just by changing the sign of the random variable involved.

Theorem 9.3 (Obtaining $GEVD_m$ from $GEVD_M$). *If the random variable $X \sim GEVD_m(\lambda, \delta, \kappa)$, then $Y = -X \sim GEVD_M(-\lambda, \delta, \kappa)$, and if the random variable $X \sim GEVD_M(\lambda, \delta, \kappa)$, then $Y = -X \sim GEVD_m(-\lambda, \delta, \kappa)$.*

Proof. Let $Y = -X$, then we have

$$\begin{aligned} F_Y(y) &= \Pr[Y \leq y] = \Pr[-X \leq y] = \Pr[X \geq -y] \\ &= 1 - \Pr[X < -y] = 1 - F_{GEVD_m(\lambda,\delta,\kappa)}(-y) \\ &= \exp\left\{-\left[1 + \kappa\left(\frac{-y-\lambda}{\delta}\right)\right]^{1/\kappa}\right\} \\ &= \exp\left\{-\left[1 - \kappa\left(\frac{y-(-\lambda)}{\delta}\right)\right]^{1/\kappa}\right\} \\ &= F_{GEVD_M(-\lambda,\delta,\kappa)}(y), \end{aligned}$$

which shows that $Y = -X \sim GEVD_M(-\lambda, \delta, \kappa)$. The converse can be established similarly. ∎

Example 9.1 (Deriving $L_\kappa(x; \lambda, \delta)$ from $H_\kappa(x; \lambda, \delta)$). Suppose that $X \sim GEVD_M(\lambda, \delta, \kappa)$, and let $Y = -X$. Then, by Theorem 9.3, $F_Y(x) = 1 - F_X(-x)$, and we have

$$\begin{aligned} F_Y(x) &= 1 - F_X(-x) = 1 - F_{GEVD_M(\lambda,\delta,\kappa)}(-x) \\ &= 1 - H_\kappa(-x; \lambda, \delta) = 1 - \exp\left\{-\left[1 + \kappa\left(\frac{x+\lambda}{\delta}\right)\right]^{1/\kappa}\right\}. \end{aligned}$$

Next, changing sign of λ one gets:

$$L_\kappa(x;\lambda,\delta) = 1 - \exp\left\{-\left[1+\kappa\left(\frac{x-\lambda}{\delta}\right)\right]^{1/\kappa}\right\},$$

which is the desired result. ■

For the sake of simplicity in what follows, $H_\kappa(x)$ and $L_\kappa(x)$ will be used instead of $H_\kappa(x;\lambda,\delta)$ and $L_\kappa(x;\lambda,\delta)$.

9.1.2 Weibull, Gumbel, and Fréchet as GEVDs

The maximal GEVD family in (9.3) and (9.4),

$$H_\kappa(x;\lambda,\delta) = \begin{cases} \exp\left\{-\left[1-\kappa\left(\frac{x-\lambda}{\delta}\right)\right]^{1/\kappa}\right\}, & 1-\kappa\left(\frac{x-\lambda}{\delta}\right) \geq 0, \\ & \kappa \neq 0, \\ \exp\left[-\exp\left(\frac{\lambda-x}{\delta}\right)\right], & -\infty < x < \infty, \\ & \kappa = 0, \end{cases} \tag{9.9}$$

includes the well-known Weibull, Gumbel, and Fréchet families for maxima, as special cases:

Maximal Weibull or Reversed Weibull:

$$H(x) = \begin{cases} \exp\left[-\left(\frac{\lambda-x}{\delta}\right)^\beta\right], & \text{if } x \leq \lambda, \beta > 0 \\ 1, & \text{otherwise.} \end{cases} \tag{9.10}$$

Gumbel or Maximal Gumbel:

$$H(x) = \exp\left[-\exp\left(\frac{\lambda-x}{\delta}\right)\right], \quad -\infty < x < \infty. \tag{9.11}$$

Fréchet or Maximal Fréchet:

$$H(x) = \begin{cases} 0, & \text{if } x < \lambda, \\ \exp\left[-\left(\frac{\delta}{x-\lambda}\right)^\beta\right], & \text{if } x \geq \lambda, \beta > 0. \end{cases} \tag{9.12}$$

Note that $\beta > 0$ in (9.10) and (9.12), and that $\delta > 0$ in (9.10)–(9.12) because it is a scale parameter.

The graphs of the Weibull, Gumbel, and Fréchet distributions in Figure 9.2 show that the Weibull and Fréchet distributions converge to the Gumbel

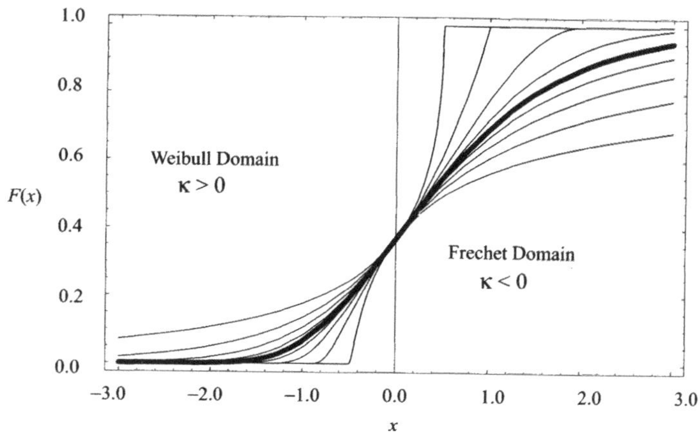

Figure 9.2: The Weibull and Fréchet distributions converge to the Gumbel distribution.

distribution. In other words, the Gumbel distribution can be approximated as much as desired by Weibull or Fréchet families letting $\beta \to \infty$.

Similarly, the GEVD family of distributions for minima,

$$L_\beta(x;\lambda,\delta) = \begin{cases} 1 - \exp\left\{-\left[1+\kappa\left(\dfrac{x-\lambda}{\delta}\right)\right]^{1/\kappa}\right\}, & 1+\kappa\left(\dfrac{x-\lambda}{\delta}\right) \geq 0, \\ & \text{if } \kappa \neq 0, \\ 1 - \exp\left[-\exp\left(\dfrac{x-\lambda}{\delta}\right)\right], & -\infty < x < \infty, \\ & \text{if } \kappa = 0, \end{cases} \tag{9.13}$$

includes the well-known Weibull, Gumbel, and Fréchet families for minima, as special cases:

Weibull or Minimal Weibull:

$$L(x) = \begin{cases} 0, & \text{if } x < \lambda, \\ 1 - \exp\left[-\left(\dfrac{x-\lambda}{\delta}\right)^\beta\right], & \text{otherwise.} \end{cases} \tag{9.14}$$

Minimal Gumbel or Reversed Gumbel:

$$L(x) = 1 - \exp\left[-\exp\left(\frac{x-\lambda}{\delta}\right)\right], \quad -\infty < x < \infty. \tag{9.15}$$

Minimal Fréchet or Reversed Fréchet:

$$L(x) = \begin{cases} 1 - \exp\left[-\left(\dfrac{\delta}{\lambda - x}\right)^\beta\right], & \text{if } x \leq \lambda, \\ 1, & \text{otherwise.} \end{cases} \tag{9.16}$$

Table 9.1: The Relationship Between the Parameters of the GEVDs in (9.9) and (9.13), and the Weibull, Gumbel, and Fréchet Distributions, Where the Asterisks Refer to the Parameters of the GEVDs.

Distribution	Maximal		Minimal	
Weibull	$\lambda = \lambda^* + \delta/\kappa^*$	$\lambda^* = \lambda - \delta$	$\lambda = \lambda^* - \delta^*/\kappa^*$	$\lambda^* = \lambda + \delta$
	$\delta = \delta^*/\kappa^*$	$\delta^* = \delta/\beta$	$\delta = \delta^*/\kappa^*$	$\delta^* = \delta/\beta$
	$\beta = 1/\kappa^*$	$\kappa^* = 1/\beta$	$\beta = 1/\kappa^*$	$\kappa^* = 1/\beta$
Gumbel	$\lambda = \lambda^*$	$\lambda^* = \lambda$	$\lambda = \lambda^*$	$\lambda^* = \lambda$
	$\delta = \delta^*$	$\delta^* = \delta$	$\delta = \delta^*$	$\delta^* = \delta$
Fréchet	$\lambda = \lambda^* + \delta^*/\kappa^*$	$\lambda^* = \lambda + \delta$	$\lambda = \lambda^* - \delta^*/\kappa^*$	$\lambda^* = \lambda - \delta$
	$\delta = -\delta^*/\kappa^*$	$\delta^* = \delta/\beta$	$\delta = -\delta^*/\kappa^*$	$\delta^* = \delta/\beta$
	$\beta = -1/\kappa^*$	$\kappa^* = -1/\beta$	$\beta = -1/\kappa^*$	$\kappa^* = -1/\beta$

Note that $\beta > 0$ in (9.14) and (9.16), and that $\delta > 0$ in (9.14)–(9.16). The relationship between the parameters of the GEVDs in (9.9) and (9.13), on one hand, and the Weibull, Gumbel, and Fréchet, on the other hand, are given in Table 9.1. The cdf and some other characteristics of the Weibull, Gumbel, and Fréchet are given in Tables 9.2–9.4, respectively.

Expressions (9.1) and (9.2) together with the previous theorems allow, for sufficiently large values of n, replacing $[F(a_n + b_n x)]^n$ by $H_\kappa(x)$ or $[F(x)]^n$ by $H_\kappa((x - a_n)/b_n)$ or, what is equivalent, for large values of x, replacing $F(x)$ by $H_\kappa^{1/n}[(x - a_n)/b_n]$, which belongs to the GEVD$_M$ family. The practical importance of this result is that for any continuous cdf $F(x)$, only the GEVD family is possible as a limit. Consequently, for extremes, the infinite degrees of freedom when choosing $F(x)$, reduce to selecting the parameters κ, a_n, and b_n.

9.1.3 Stability of Limit Distributions

Definition 9.2 (Stable family). *A parametric family of cumulative distribution functions, $\{F(x;\theta); \theta \in \Theta\}$, is said to be stable with respect to maxima if $[F(x;\theta)]^n \in \{F(x;\theta); \theta \in \Theta\}$, that is, $[F(x;\theta)]^n = F(x;\theta(n))$, where $\theta(n)$ is a parameter that depends on n. Similarly, a parametric family of survival functions, $\{S(x;\theta); \theta \in \Theta\}$, is said to be stable with respect to minima if $[S(x;\theta)]^n = S(x;\theta(n))$.*

Table 9.2: Some Characteristics of Weibull Distributions.

	Maximal Weibull	Minimal Weibull
CDF	$F(x) = \exp\left[-\left(\frac{\lambda-x}{\delta}\right)^\beta\right]$	$F(x) = 1 - \exp\left[-\left(\frac{x-\lambda}{\delta}\right)^\beta\right]$
Mean	$\lambda - \delta\Gamma\left(1+\frac{1}{\beta}\right)$	$\lambda + \delta\Gamma\left(1+\frac{1}{\beta}\right)$
Median	$\lambda - \delta\ 0.693^{1/\beta}$	$\lambda + \delta\ 0.693^{1/\beta}$
Mode	$\lambda - \delta\left(\frac{\beta-1}{\beta}\right)^{1/\beta}, \quad \beta > 1$ $\lambda, \quad \beta \le 1$	$\lambda + \delta\left(\frac{\beta-1}{\beta}\right)^{1/\beta}, \quad \beta > 1$ $\lambda, \quad \beta \le 1$
Variance	$\delta^2\left[\Gamma\left(1+\frac{2}{\beta}\right) - \Gamma^2\left(1+\frac{1}{\beta}\right)\right]$	$\delta^2\left[\Gamma\left(1+\frac{2}{\beta}\right) - \Gamma^2\left(1+\frac{1}{\beta}\right)\right]$
Quantile	$\lambda - \delta(-\log p)^{1/\beta}$	$\lambda + \delta(-\log(1-p))^{1/\beta}$

Table 9.3: Some Characteristics of the Gumbel Distributions.

	Maximal Gumbel	Minimal Gumbel
CDF	$F(x) = \exp\left[-\exp\left(\frac{\lambda-x}{\delta}\right)\right]$	$F(x) = 1 - \exp\left[-\exp\left(\frac{x-\lambda}{\delta}\right)\right]$
Mean	$\lambda + 0.57772\delta$	$\lambda - 0.57772\delta$
Median	$\lambda + 0.3665\delta$	$\lambda - 0.3665\delta$
Mode	λ	λ
Variance	$\frac{\pi^2\delta^2}{6}$	$\frac{\pi^2\delta^2}{6}$
Quantile	$\lambda - \delta(\log(-\log p))$	$\lambda + \delta(\log(-\log(1-p)))$

Table 9.4: Some Characteristics of the Fréchet Distributions.

	Maximal Fréchet	Minimal Fréchet
CDF	$F(x) = \exp\left[-\left(\frac{\delta}{x-\lambda}\right)^{\beta}\right]$	$F(x) = 1 - \exp\left[-\left(\frac{\delta}{\lambda-x}\right)^{\beta}\right]$
Mean	$\lambda + \delta\Gamma\left(1 - 1/\beta\right), \quad \beta > 1$	$\lambda - \delta\Gamma\left(1 - 1/\beta\right), \quad \beta > 1$
Median	$\lambda + \delta \; 0.693^{-1/\beta}$	$\lambda - \delta \; 0.693^{-1/\beta}$
Mode	$\left(1+\frac{1}{\beta}\right)^{\frac{-1}{\beta}} \left[\lambda\left(1+\frac{1}{\beta}\right)^{\frac{1}{\beta}} + \delta\right]$	$\left(1+\frac{1}{\beta}\right)^{\frac{-1}{\beta}} \left[\lambda\left(1+\frac{1}{\beta}\right)^{\frac{1}{\beta}} - \delta\right]$
Variance	$\delta^2\left[\Gamma\left(1-\frac{2}{\beta}\right) - \Gamma^2\left(1-\frac{1}{\beta}\right)\right],$ $\beta > 2$	$\delta^2\left[\Gamma\left(1-\frac{2}{\beta}\right) - \Gamma^2\left(1-\frac{1}{\beta}\right)\right],$ $\beta > 2$
Quantiles	$\lambda + \delta(-\log p)^{-1/\beta}$	$\lambda - \delta(-\log(1-p))^{-1/\beta}$

Theorem 9.4 (Asymptotic stability with respect to extremes). *The asymptotic families of distributions for maxima (minima) (Weibull, Gumbel, and Fréchet) are stable with respect to maxima (minima) operations. In other words, the maxima (minima) of iid samples drawn from such families belong to them.*

Proof.

$$H_\kappa^n(x;\lambda,\delta) = \left[\exp\left\{-\left[1-\kappa\left(\frac{x-\lambda}{\delta}\right)\right]^{1/\kappa}\right\}\right]^n \tag{9.17}$$

$$= \exp\left\{-n\left[1-\kappa\left(\frac{x-\lambda}{\delta}\right)\right]^{1/\kappa}\right\} \tag{9.18}$$

$$= \exp\left\{-\left[n^{-\kappa}-\kappa\left(\frac{x-\lambda}{\delta n^{\kappa}}\right)\right]^{1/\kappa}\right\} \tag{9.19}$$

$$= \exp\left\{-\left[1-\kappa\left(\frac{x-(\lambda+\delta(1-n^{-\kappa})/\kappa}{\delta n^{-\kappa}}\right)\right]^{1/\kappa}\right\}.$$

Therefore, $[F_{GEVD_M}(x;\lambda,\delta,\kappa)]^n = F_{GEVD_M}(x;\lambda+\delta(1-n^{-\kappa})/\kappa, \delta n^{-\kappa}, \kappa)$. This proves the theorem for maxima. For minima, we have

$$1-[1-L_\kappa(x;\lambda,\delta)]^n = 1-\left[\exp\left\{-\left[1+\kappa\left(\frac{x-\lambda}{\delta}\right)\right]^{1/\kappa}\right\}\right]^n$$

$$= 1-\exp\left\{-n\left[1+\kappa\left(\frac{x-\lambda}{\delta}\right)\right]^{1/\kappa}\right\}$$

$$
\begin{aligned}
&= 1 - \exp\left\{ - \left[n^{\kappa} + \kappa \left(\frac{x - \lambda}{\delta n^{-\kappa}} \right) \right]^{1/\kappa} \right\} \\
&= 1 - \exp\left\{ - \left[1 + \kappa \left(\frac{x - (\lambda - \delta(1 - n^{-\kappa})/\kappa)}{\delta n^{-\kappa}} \right) \right]^{1/\kappa} \right\}.
\end{aligned}
$$

Therefore, $1-[1-F_{GEVD_m}(x;\lambda,\delta,\kappa)]^n = F_{GEVD_m}(x;\lambda-\delta(1-n^{-\kappa})/\kappa, \delta n^{-\kappa}, \kappa)$. This proves the theorem for minima. ∎

Example 9.2 (Maximum stability of the maximal Weibull family). The cdf of the maximum of an iid sample drawn from a maximal Weibull family $W_M(\lambda, \delta, \beta)$ is

$$
\begin{aligned}
F_{max}(x) &= [F_{W(\lambda,\delta,\beta)}]^n = \exp\left[- \left(\frac{\lambda - x}{\delta} \right)^{\beta} \right]^n = \exp\left[-n \left(\frac{\lambda - x}{\delta} \right)^{\beta} \right] \\
&= \exp\left[- \left(\frac{\lambda - x}{\delta n^{-1/\beta}} \right)^{\beta} \right] = F_{W(\lambda, \delta n^{-1/\beta}, \beta)}(x),
\end{aligned}
$$

which shows that it is a Weibull $W(\lambda, \delta n^{-1/\beta}, \beta)$ distribution. It is interesting to note that only the scale parameter is changed. ∎

Example 9.3 (Minimum stability of the minimal Gumbel family). The cdf of the minimum of an iid sample drawn from a Gumbel distribution for minima, $G_m(\lambda, \delta)$ is

$$
\begin{aligned}
F_{min}(x) &= 1 - [1 - F_{G(\lambda,\delta)}]^n = 1 - \exp\left[- \exp\left(\frac{x - \lambda}{\delta} \right) \right]^n \\
&= 1 - \exp\left[-n \exp\left(\frac{x - \lambda}{\delta} \right) \right] \\
&= 1 - \exp\left[- \exp\left(\frac{x - (\lambda - \delta \log n)}{\delta} \right) \right] = F_{G(\lambda - \delta \log n, \delta)}(x),
\end{aligned}
$$

which shows that it is a Gumbel $G_m(\lambda - \delta \log n, \delta)$ distribution. Note that only the location parameter is changed. ∎

9.1.4 Determining the Domain of Attraction of a CDF

An interesting problem from the point of view of extremes is about knowing the domain of attraction of a given cdf $F(x)$. To identify the domain of attraction of a given distribution $F(x)$ and the associated sequences a_n and b_n or c_n and d_n, we give two theorems that allow solving this problem (Castillo (1988), Castillo, Galambos, and Sarabia (1989), and Galambos (1987)).

Theorem 9.5 (Maximal domain of attraction of a given distribution). *A necessary and sufficient condition for the continuous cdf $F(x)$ to belong to* *the maximal domain of attraction, $H_\kappa(x)$, is that*

$$\lim_{\varepsilon \to 0} \frac{F^{-1}(1-\varepsilon) - F^{-1}(1-2\varepsilon)}{F^{-1}(1-2\varepsilon) - F^{-1}(1-4\varepsilon)} = 2^{-\kappa}, \tag{9.20}$$

where κ is the shape parameter of the associated limit distribution, GEVD_M.

This implies that

1. If $\kappa > 0$, $F(x)$ belongs to the Weibull maximal domain of attraction,
2. if $\kappa = 0$, $F(x)$ belongs to the Gumbel maximal domain of attraction, and
3. If $\kappa < 0$, $F(x)$ belongs to the Fréchet maximal domain of attraction.

The constants a_n and b_n can be chosen as follows:

1. Weibull:

$$a_n = w(F) \quad \text{and} \quad b_n = w(F) - F^{-1}\left(1 - \frac{1}{n}\right); \tag{9.21}$$

2. Gumbel:

$$a_n = F^{-1}\left(1 - \frac{1}{n}\right) \quad \text{and} \quad b_n = F^{-1}\left(1 - \frac{1}{ne}\right) - a_n; \tag{9.22}$$

3. Fréchet:

$$a_n = 0 \quad \text{and} \quad b_n = F^{-1}\left(1 - \frac{1}{n}\right); \tag{9.23}$$

where $w(F) = \sup\{x|F(x) < 1\}$ is the upper end of the cdf $F(x)$ and e is the base of the natural logarithm.

Theorem 9.6 (Minimal domain of attraction of a given distribution). *A necessary and sufficient condition for the continuous cdf $F(x)$ to belong to the domain of attraction for minima of $L_\kappa(x)$ is that*

$$\lim_{\varepsilon \to 0} \frac{F^{-1}(\varepsilon) - F^{-1}(2\varepsilon)}{F^{-1}(2\varepsilon) - F^{-1}(4\varepsilon)} = 2^{-\kappa}, \tag{9.24}$$

where κ is the shape parameter of the associated limit distribution, $GEVD_m$. This implies that

1. If $\kappa > 0$, $F(x)$ belongs to the Weibull minimal domain of attraction,
2. If $\kappa = 0$, $F(x)$ belongs to the Gumbel minimal domain of attraction, and
3. If $\kappa < 0$, $F(x)$ belongs to the Fréchet minimal domain of attraction.

The constants c_n and d_n can be chosen as follows:

1. Weibull:

$$c_n = v(F) \quad \text{and} \quad d_n = F^{-1}\left(\frac{1}{n}\right) - v(F); \tag{9.25}$$

2. Gumbel:

$$c_n = F^{-1}\left(\frac{1}{n}\right) \quad \text{and} \quad d_n = c_n - F^{-1}\left(\frac{1}{ne}\right); \tag{9.26}$$

3. Fréchet:

$$c_n = 0 \quad \text{and} \quad d_n = \left|F^{-1}\left(\frac{1}{n}\right)\right|; \tag{9.27}$$

where $v(F) = \inf(x|F(x) > 0)$ is the lower end of the cdf $F(x)$ and e is the base of the natural logarithm.

Note that Theorems 9.5 and 9.6 completely identify the limit distribution of a given cdf, that is, they not only give the sequences a_n, b_n, c_n and d_n but also the corresponding value of the shape parameter κ.

Example 9.4 (Exponential distribution: Maxima). The cdf of the exponential distribution is

$$F(x) = 1 - e^{-x/\lambda}, \quad x > 0,$$

and its inverse (quantile) function is

$$x_p = F^{-1}(p) = -\lambda \log(1 - p).$$

Then, the limit in (9.20) becomes

$$\lim_{\varepsilon \to 0} \frac{-\lambda \log[1 - (1 - \varepsilon)] + \lambda \log[1 - (1 - 2\varepsilon)]}{-\lambda \log[1 - (1 - 2\varepsilon)] + \lambda \log[1 - (1 - 4\varepsilon)]}$$

$$= \lim_{\varepsilon \to 0} \frac{-\log(\varepsilon) + \log 2 + \log(\varepsilon)}{-\log 2 - \log(\varepsilon) + \log 4 + \log(\varepsilon)} = 1 = 2^0 \Longrightarrow \kappa = 0,$$

which shows that the exponential distribution belongs to the Gumbel domain of attraction for maxima.

A possible selection of the constants, according to (9.22), is

$$a_n = F^{-1}\left(1 - \frac{1}{n}\right) = \lambda \log(n) \quad \text{and} \quad b_n = \lambda \log(ne) - \lambda \log(n) = \lambda.$$

■

Example 9.5 (Exponential distribution: Minima). The limit in (9.24) for the exponential distribution becomes

$$\lim_{\varepsilon \to 0} \frac{-\log(1 - \varepsilon) + \log(1 - 2\varepsilon)}{-\log(1 - 2\varepsilon) + \log(1 - 4\varepsilon)} = \lim_{\varepsilon \to 0} \frac{\varepsilon - 2\varepsilon}{2\varepsilon - 4\varepsilon} = 2^{-1} \Longrightarrow \kappa = 1,$$

which shows that the exponential distribution belongs to the Weibull domain of attraction for minima.

According to (9.25), a possible selection of the constants is

$$c_n = v(F) = 0 \quad \text{and} \quad d_n = F^{-1}\left(\frac{1}{n}\right) - v(F) = -\log\left(1 - \frac{1}{n}\right) \approx \frac{1}{n}.$$

■

Note that the limits in (9.20) and (9.24) can be calculated approximately by replacing a small number for ε, say $\varepsilon = 0.001$. For example, the limit in Example 9.5 can be approximated by

$$\begin{aligned}\lim_{\varepsilon\to 0}\frac{-\log(1-\varepsilon)+\log(1-2\varepsilon)}{-\log(1-2\varepsilon)+\log(1-4\varepsilon)} &\approx \frac{-\log(1-0.001)+\log(1-0.002)}{-\log(1-0.002)+\log(1-0.004)}\\ &= 0.499 \approx 2^{-1},\end{aligned}$$

as obtained in Example 9.5. Though not very rigorous, it is a practical procedure to determine κ when the limit is difficult to calculate exactly.

Example 9.6 (Cauchy distribution: Minima). The cdf of the Cauchy distribution is

$$F(x) = \frac{1}{2} + \frac{\tan^{-1}(x)}{\pi}, \quad -\infty < x < \infty,$$

and its inverse (quantile) function is

$$x_p = F^{-1}(p) = \tan[(p - 1/2)\pi].$$

Then, (9.24) gives

$$\lim_{\varepsilon\to 0}\frac{\tan\left[(\varepsilon-0.5)\,\pi\right]-\tan\left[(2\varepsilon-0.5)\,\pi\right]}{\tan\left[(2\varepsilon-0.5)\,\pi\right]-\tan\left[(4\,\varepsilon-0.5)\,\pi\right]} = \lim_{\varepsilon\to 0}\frac{-1/(2\varepsilon)}{-1/(4\varepsilon)} = 2^1 \Longrightarrow \kappa = -1,$$

which shows that the Cauchy distribution belongs to the Fréchet minimal domain of attraction. A possible selection of the constants, according to (9.27), is

$$c_n = 0 \quad \text{and} \quad d_n = \tan\left[\pi\left(\frac{1}{2} - \frac{1}{n}\right)\right].$$

Alternatively, approximating the above limit by replacing ε with 0.001, yields

$$\begin{aligned}&\lim_{\varepsilon\to 0}\frac{\tan\left[(\varepsilon-0.5)\,\pi\right]-\tan\left[(2\varepsilon-0.5)\,\pi\right]}{\tan\left[(2\varepsilon-0.5)\,\pi\right]-\tan\left[(4\,\varepsilon-0.5)\,\pi\right]}\\ &\qquad\approx \frac{\tan\left[(0.001-0.5)\,\pi\right]-\tan\left[(0.002-0.5)\,\pi\right]}{\tan\left[(0.002-0.5)\,\pi\right]-\tan\left[(0.004-0.5)\,\pi\right]} = 1.99996 \approx 2,\end{aligned}$$

as obtained using the exact limit. ■

Example 9.7 (Weibull distribution: Maxima). The cdf of the maximal Weibull distribution is

$$F(x) = \exp\left[-(-x)^{\beta}\right], \quad x \le 0, \beta > 0,$$

and its inverse (quantile) function is

$$x_p = F^{-1}(p) = -(-\log p)^{1/\beta}.$$

Then, (9.20) gives

$$\lim_{\varepsilon \to 0} \frac{-(-\log(1-\epsilon))^{1/\beta} + (-\log(1-2\epsilon))^{1/\beta}}{-(-\log(1-2\epsilon))^{1/\beta} + (-\log(1-4\epsilon))^{1/\beta}} = \lim_{\varepsilon \to 0} \frac{(2\epsilon)^{1/\beta} - \epsilon^{1/\beta}}{(4\epsilon)^{1/\beta} - (2\epsilon)^{1/\beta}} = 2^{-1/\beta},$$

which shows that the Weibull distribution for maxima belongs to the Weibull maximal domain of attraction, as expected. ■

Table 9.5 shows the maximal and minimal domains of attractions of some common distributions.

Table 9.5: Domains of Attraction of the Most Common Distributions.

Distribution	Domain of Attraction: Maximal	Domain of Attraction: Minimal
Normal	Gumbel	Gumbel
Exponential	Gumbel	Weibull
Lognormal	Gumbel	Gumbel
Gamma	Gumbel	Weibull
Gumbel_M	Gumbel	Gumbel
Gumbel_m	Gumbel	Gumbel
Rayleigh	Gumbel	Weibull
Uniform	Weibull	Weibull
Weibull_M	Weibull	Gumbel
Weibull_m	Gumbel	Weibull
Cauchy	Fréchet	Fréchet
Pareto	Fréchet	Weibull
Fréchet_M	Fréchet	Gumbel
Fréchet_m	Gumbel	Fréchet
M = maxima	m = minima	

Some important implications of the previous theorems are:

1. Only three distributions (Weibull, Gumbel, and Fréchet) can occur as limit distributions for maxima and minima.

2. Rules for determining if a given distribution $F(x)$ belongs to the domain of attraction of these three distributions and the corresponding κ values are available.

3. Rules for obtaining the corresponding sequences $\{a_n\}$ and $\{b_n\}$ or $\{c_n\}$ and $\{d_n\}$ are available.

4. A distribution with no finite end in the associated tail cannot belong to the Weibull domain of attraction (see Galambos (1987)).

5. A distribution with finite end in the associated tail cannot belong to the Fréchet domain of attraction (see Galambos (1987)).

6. The Gumbel distribution can be approximated as much as desired by the Weibull and Fréchet distributions. So, the choice between Gumbel and Weibull or Gumbel and Fréchet can be resolved by rejecting the Gumbel (see Fig. 9.2).

9.1.5 Asymptotic Distributions of Order Statistics

As in the case of maxima and minima, we wish to find the asymptotic nondegenerate distributions of the rth largest (smallest) order statistic, $H_r(x)$, such that

$$\lim_{n\to\infty} H_{n-r+1:n}(a_n + b_n x) = H_r(x), \quad \forall\ x, \tag{9.28}$$

$$\lim_{n\to\infty} L_{r:n}(c_n + d_n x) = L_r(x), \quad \forall\ x. \tag{9.29}$$

We shall distinguish between order statistics of central, low, and high order, because they show a completely different limit behavior.

Definition 9.3 (Central order statistics). *The order statistic $X_{r(n):n}$ is said to be a central order statistic iff*

$$\lim_{n\to\infty} \sqrt{n}\left(\frac{r(n)}{n} - p\right) = 0, \tag{9.30}$$

where $p \in (0,1)$.

For example, assume a sequence of samples of increasing size n, and the statistic of order $r(n) = 0.2n$ (also increasing), then for $n = 100$ we get $r(100) = 0.2 \times 100 = 20$, which implies that the 20th order statistic is central because $r(n) = 0.2n$ satisfies (9.30). Alternatively, if $r(n) = 0.2\sqrt{n}$, the second order statistic is not central because $r(100) = 0.2\sqrt{100} = 2$ does not satisfy (9.30). Note, however, that in practice this is somehow arbitrary, since we have a sample and not a sequence of samples, and then we can always find sequences, including our sample, that lead to central or low-order statistics.

Theorem 9.7 (Asymptotic distribution of central order statistics). *Let $F(x)$ be a continuous cdf with associated pdf, $f(x)$. Let $p_1, p_2, \ldots, p_k$ be a set of real numbers in the interval $(0,1)$ such that $f(F^{-1}(p_j)) \neq 0, 1 \leq j \leq k$. If $r_j(n)$ are such that*

$$\lim_{n\to\infty} \sqrt{n}\left(\frac{r_j(n)}{n} - p_j\right) = 0, \quad 1 \leq j \leq k,$$

then the vector

$$\sqrt{n}\left(X_{r_j:n} - F^{-1}(p_j)\right), \quad 1 \leq j \leq k,$$

is asymptotically a k-variate normal random vector with zero expectation and covariance matrix

$$\frac{p_i(1-p_j)}{f(F^{-1}(p_i))f(F^{-1}(p_j))}, \quad p_i \leq p_j.$$

This theorem states that central order statistics are jointly asymptotically normal.

Definition 9.4 (Order statistics of low and high order). *The order statistic $X_{r(n):n}$ is said to be of low order and the order statistic $X_{n-r(n)+1:n}$ is said to be of high order iff*

$$\lim_{n\to\infty} r(n) = k < \infty.$$

For these types of order statistics, the following results hold (Galambos (1987)).

Theorem 9.8 (Asymptotic distributions of high-order statistics). *If some sequences $\{a_n\}$ and $\{b_n\}$ exist such that*

$$\lim_{n\to\infty} [F(a_n + b_n x)]^n = H_\kappa(x), \quad \forall\ x. \tag{9.31}$$

Then

$$\lim_{n\to\infty} H_{n-r+1:n}(a_n + b_n x) = \begin{cases} H_\kappa(x) \displaystyle\sum_{i=0}^{r-1} \frac{[-\log H_\kappa(x)]^i}{i!}, & \textit{if } H_\kappa(x) > 0, \\ 0, & \textit{otherwise.} \end{cases} \tag{9.32}$$

This theorem justifies the following approximation for large x:

$$H_{n-r+1:n}(a_n + b_n x) \approx \exp\left\{-\left[1 - \kappa\left(\frac{x-\lambda}{\delta}\right)\right]^{1/\kappa}\right\} \sum_{i=0}^{r-1} \frac{\left[1 - \kappa\left(\dfrac{x-\lambda}{\delta}\right)\right]^{i/\kappa}}{i!}, \tag{9.33}$$

for $\kappa \neq 0$, and

$$H_{n-r+1:n}(x) \approx \exp\left\{-\exp\left(\frac{\lambda - x}{\delta}\right)\right\} \sum_{i=0}^{r-1} \frac{\exp\left(\dfrac{i(\lambda - x)}{\delta}\right)}{i!}, \tag{9.34}$$

for $\kappa = 0$.

Theorem 9.9 (Asymptotic distributions of low-order statistics). *If some sequences $\{c_n\}$ and $\{d_n\}$ exist such that*

$$\lim_{n\to\infty} 1-[1-F(c_n+d_n x)]^n = L_\kappa(x), \quad \forall\ x, \tag{9.35}$$

then

$$\lim_{n\to\infty} L_{r:n}(c_n+d_n x) = \begin{cases} 1-[1-L_\kappa(x)] \sum\limits_{i=0}^{r-1} \dfrac{-\log[1-L_\kappa(x)]^i}{i!}, & \text{if } L_\kappa(x)<1, \\ 1, & \text{otherwise.} \end{cases} \tag{9.36}$$

Similarly, this theorem justifies the approximations, for small x,

$$L_{r:n}(x) \approx 1-\exp\left\{-\left[1+\kappa\left(\frac{x-\lambda}{\delta}\right)\right]^{1/\kappa}\right\} \sum_{i=0}^{r-1} \frac{1}{i!}\left[1+\kappa\left(\frac{x-\lambda}{\delta}\right)\right]^{i/\kappa}, \tag{9.37}$$

for $\kappa \neq 0$, and

$$L_{r:n}(x) \approx 1-\exp\left\{-\exp\left(\frac{x-\lambda}{\delta}\right)\right\} \sum_{i=0}^{r-1} \frac{1}{i!} \exp\left(\frac{i(x-\lambda)}{\delta}\right), \tag{9.38}$$

for $\kappa = 0$.

The above two theorems have the following very important practical consequences:

1. If a limit distribution exists for the maximum (minimum), then a limit distribution also exists for high- (low-) order statistics.

2. The same normalizing sequences can be used for all of them (including the corresponding extremes).

3. They give the limit distributions for the high- (low-) order statistics as a function of the limit distributions for the corresponding extremes.

In addition, we have the following theorem for the joint limit distribution of a set of order statistics.

Theorem 9.10 (Asymptotic distribution of high-order statistics). *If some sequences $\{a_n\}$ and $\{b_n\}$ exist such that*

$$\lim_{n\to\infty} [F(a_n+b_n x)]^n = H(x), \quad \forall\ x, \tag{9.39}$$

with $H(x)$ nondegenerate, and letting $Z_j = a_n + b_n X_{n-j+1:n}$, then the limiting joint density of

$$(a_n+b_n X_{n:n}, a_n+b_n X_{n-1:n}, \ldots, a_n+b_n X_{n-r+1:n}),$$

where all are high-order statistics, is

$$\begin{aligned} f(z_1, z_2, \ldots, z_r) &= \exp\left\{-\left[1-\kappa\left(\frac{z_r-\lambda}{\delta}\right)\right]^{1/\kappa}\right\} \\ &\quad \times \prod_{j=1}^{r} \frac{1}{\delta}\left[1-\kappa\left(\frac{z_j-\lambda}{\delta}\right)\right]^{1/\kappa-1}, \end{aligned} \tag{9.40}$$

where $z_r \leq z_{r-1} \leq \ldots \leq z_1$ *and* $1-\kappa(z_j-\lambda)/\delta > 0$, *for* $\kappa \neq 0$, *and*

$$f(z_1, z_2, \ldots, z_r) = \exp\left\{-\exp\left(\frac{\lambda - z_r}{\delta}\right)\right\} \prod_{j=1}^{r} \frac{1}{\delta} \exp\left(\frac{\lambda - z_j}{\delta}\right), \tag{9.41}$$

for $\kappa = 0$.

Theorem 9.11 (Asymptotic distribution of low-order statistics). *If some sequences* $\{c_n\}$ *and* $\{d_n\}$ *exist such that*

$$\lim_{n\to\infty} 1 - [1 - F(c_n + d_n x)]^n = L(x), \quad \forall\ x, \tag{9.42}$$

with $L(x)$ *nondegenerate, and letting* $Z_j = c_n + d_n X_{j:n}$, *then the limiting joint density of*

$$(c_n + d_n X_{1:n}, c_n + d_n X_{2:n}, \ldots, c_n + d_n X_{r:n}),$$

where all are low-order statistics, is

$$\begin{aligned} f(z_1, z_2, \ldots, z_r) &= \exp\left\{-\left[1+\kappa\left(\frac{z_r-\lambda}{\delta}\right)\right]^{1/\kappa}\right\} \\ &\quad \times \prod_{j=1}^{r} \frac{1}{\delta}\left[1+\kappa\left(\frac{z_j-\lambda}{\delta}\right)\right]^{1/\kappa-1}, \end{aligned}$$

where $z_1 \leq z_2 \leq \ldots \leq z_r$ *and* $1+\kappa(z_j-\lambda)/\delta > 0$, *for* $\kappa \neq 0$, *and*

$$f(z_1, z_2, \ldots, z_r) = \exp\left\{-\exp\left(\frac{z_r - \lambda}{\delta}\right)\right\} \prod_{j=1}^{r} \frac{1}{\delta} \exp\left(\frac{z_j - \lambda}{\delta}\right), \tag{9.43}$$

for $\kappa = 0$.

9.2 Estimation for the Maximal GEVD

In this section, we apply the estimation methods discussed in Chapter 5 to estimate the parameters and quantiles of the maximal Generalized Extreme Value distribution ($GEVD_M$). The estimation for the minimal GEVD is discussed in Section 9.3.

To set the stage up, we need the pdf of the $GEVD_M$. The cdf of the maximal GEVD is given in (9.3) and (9.4), but repeated here for convenience:

$$H(x;\lambda,\delta,\kappa) = \begin{cases} \exp\left[-\left[1-\kappa\left(\dfrac{x-\lambda}{\delta}\right)\right]^{1/\kappa}\right], & 1-\kappa\left(\dfrac{x-\lambda}{\delta}\right) \geq 0, \kappa \neq 0, \\ \exp\left[-\exp\left(\dfrac{\lambda-x}{\delta}\right)\right], & -\infty < x < \infty, \kappa = 0. \end{cases}$$

The corresponding pth quantile can be derived by inverting $H(x;\lambda,\delta,\kappa)$ as

$$x_p = \begin{cases} \lambda + \delta\left[1-(-\log p)^{\kappa}\right]/\kappa, & \text{if } \kappa \neq 0, \\ \lambda - \delta\log(-\log p), & \text{if } \kappa = 0. \end{cases} \tag{9.44}$$

The pdf of the maximal GEVD is obtained by taking the derivative of the hazard function $H(x;\lambda,\delta,\kappa)$ with respect to x. For $\kappa \neq 0$, the pdf is

$$h(x;\lambda,\delta,\kappa) = \exp\left[-\left[1-\kappa\left(\frac{x-\lambda}{\delta}\right)\right]^{1/\kappa}\right]\left[1-\kappa\left(\frac{x-\lambda}{\delta}\right)\right]^{1/\kappa-1}\frac{1}{\delta}, \tag{9.45}$$

where the support is $x \leq \lambda + \delta/\kappa$, if $\kappa > 0$, or $x \geq \lambda + \delta/\kappa$, if $\kappa < 0$.

For $\kappa = 0$, the pdf is

$$h(x;\lambda,\delta) = \exp\left[-\exp\left(\frac{\lambda-x}{\delta}\right)\right]\exp\left(\frac{\lambda-x}{\delta}\right)\frac{1}{\delta}, \tag{9.46}$$

where the support is $-\infty < x < \infty$

9.2.1 The Maximum Likelihood Method

Point Estimates of the Parameters

For $\kappa \neq 0$, from (9.45), the loglikelihood function becomes

$$\ell = -n\log\delta - (1-\kappa)\sum_{i=1}^{n} z_i - \sum_{i=1}^{n}\exp\{-z_i\}, \tag{9.47}$$

where

$$z_i = -\frac{1}{\kappa}\log\{1-\kappa(x_i-\lambda)/\delta\}. \tag{9.48}$$

For $\kappa = 0$, the loglikelihood function is

$$\ell(\lambda,\delta) = -n\log\delta - \sum_{i=1}^{n}\exp\left[-\left(\frac{x_i-\lambda}{\delta}\right)\right] - \sum_{i=1}^{n}\left(\frac{x_i-\lambda}{\delta}\right). \tag{9.49}$$

One can use standard optimization packages to find the estimates of the parameters that maximize the loglikelihoods in (9.47) or (9.49). In order to avoid problems, one should do the following:

1. Use (9.49), if we believe that κ is close to 0.
2. Use (9.47), if we believe that κ is far enough from 0, and check to make sure that the conditions

$$1-\kappa\left(\frac{x_i-\lambda}{\delta}\right)\geq 0, \quad i=1,2,\ldots,n, \tag{9.50}$$

hold; otherwise, the loglikelihood will be infinite.

Alternative to the use of standard optimization packages, one can use the following iterative method.

The Case When $\kappa \neq 0$

The maximum likelihood estimates of $\theta = (\lambda, \delta, \kappa)$ can be obtained by the iterative formula:

$$\theta_{j+1} = \theta_j + \mathbf{I}_\theta^{-1}\nabla_\theta \ell, \quad j=0,1,\ldots, \tag{9.51}$$

where θ_0 is an initial estimate and $\mathbf{I}_\theta^{-1}$ and $\nabla_\theta \ell$ are evaluated at θ_j. Here, $\nabla_\theta \ell$ is the gradient vector of ℓ with respect to θ and

$$\mathbf{I}_\theta = \begin{bmatrix} m_{11} & m_{12} & m_{13} \\ m_{21} & m_{22} & m_{23} \\ m_{31} & m_{32} & m_{33} \end{bmatrix} \tag{9.52}$$

is the Fisher information matrix. The elements of $\mathbf{I}_\theta$ are given by (see Prescott and Walden (1980)):

$$\begin{aligned}
m_{11} &= E\left(-\frac{\partial^2 \ell}{\partial\lambda^2}\right) = \frac{n}{\delta^2}p, \\
m_{22} &= E\left(-\frac{\partial^2 \ell}{\partial\delta^2}\right) = \frac{n}{\delta^2\kappa^2}\left\{1-2\Gamma(2-\kappa)+p\right\}, \\
m_{33} &= E\left(-\frac{\partial^2 \ell}{\partial\kappa^2}\right) = \frac{n}{\kappa^2}\left\{\frac{\pi^2}{6}+\left(1-\gamma-\frac{1}{\kappa}\right)^2+\frac{2q}{\kappa}+\frac{p}{\kappa^2}\right\}, \\
m_{12} &= m_{21} = E\left(-\frac{\partial^2 \ell}{\partial\lambda\partial\delta}\right) = \frac{n}{\delta^2\kappa}\left\{p-\Gamma(2-\kappa)\right\}, \\
m_{13} &= m_{31} = E\left(-\frac{\partial^2 \ell}{\partial\lambda\partial\kappa}\right) = -\frac{n}{\delta\kappa}\left(q+\frac{p}{\kappa}\right), \\
m_{23} &= m_{32} = E\left(-\frac{\partial^2 \ell}{\partial\delta\partial\kappa}\right) = \frac{n}{\delta\kappa^2}\left[1-\gamma-\frac{\{1-\Gamma(2-\kappa)\}}{\kappa}-q-\frac{p}{\kappa}\right],
\end{aligned}$$

where

$$\Gamma(u) = \int_0^\infty y^{u-1}e^{-y}dy$$

is the Gamma function,

$$\psi(u) = \frac{d \log \Gamma(u)}{du} \tag{9.53}$$

is the Psi function,

$$p = (1-\kappa)^2 \Gamma(1-2\kappa), \tag{9.54}$$

$$q = \Gamma(2-\kappa)\left[\psi(1-\kappa) - (1-\kappa)/\kappa\right], \tag{9.55}$$

and $\gamma = 0.5772157$ is the Euler's constant. The regularity conditions are satisfied only for $\kappa < 1/2$, and, in this case, the asymptotic variances and covariances are given by

$$\boldsymbol{\Sigma}_{\hat{\theta}} = \mathbf{I}_{\theta}^{-1}. \tag{9.56}$$

When $\hat{\theta}$ is substituted for θ in (9.56), we obtain an estimate of the covariance matrix

$$\hat{\boldsymbol{\Sigma}}_{\hat{\theta}} = \mathbf{I}_{\theta}^{-1}\Big|_{\theta=\hat{\theta}}. \tag{9.57}$$

Remark 9.1 *The above numerical solutions used to obtain the MLE require an initial estimate $\theta_0 = \{\lambda_0, \delta_0, \kappa_0\}$ (see (9.51)). As possible initial values for the parameter estimates, one can use Gumbel estimates for λ_0 and δ_0 and $\kappa_0 = 0$. A simple version of the EPM (see Section 9.2.3) or the QLS (see Section 9.2.4) can also be used as an initial starting point.*

The Case When $\kappa = 0$

The maximum likelihood estimates of $\theta = (\lambda, \delta)$ can be obtained by the iterative formula:

$$\theta_{j+1} = \theta_j + \mathbf{I}_{\theta}^{-1} \nabla_{\theta} \ell, \quad j = 0, 1, \ldots, \tag{9.58}$$

where θ_0 is an initial estimate and $\mathbf{I}_{\theta}^{-1}$ and $\nabla_{\theta}\ell$ are evaluated at θ_j. Here, $\nabla_{\theta}\ell$ is the gradient vector and

$$\mathbf{I}_{\theta} = \begin{bmatrix} \dfrac{n(1-\kappa)^2\Gamma(1-2\kappa)}{\delta^2} & \dfrac{n(\gamma-1)}{\delta^2} \\ \dfrac{n(\gamma-1)}{\delta^2} & \dfrac{n(6-12\gamma+6\gamma^2+n^2)}{6\delta^2} \end{bmatrix} \tag{9.59}$$

is the Fisher information matrix.

In this case, the regularity conditions hold, and the asymptotic variances and covariances are given by

$$\boldsymbol{\Sigma}_{\hat{\theta}} = \mathbf{I}_{\theta}^{-1}. \tag{9.60}$$

When $\hat{\theta}$ is substituted for θ in (9.60), we obtain an estimate of the covariance matrix

$$\hat{\boldsymbol{\Sigma}}_{\hat{\theta}} = \mathbf{I}_{\theta}^{-1}\Big|_{\theta=\hat{\theta}}. \tag{9.61}$$

Confidence Intervals for the Parameters ($\kappa \neq 0$)

For $\kappa \neq 0$, confidence intervals for the parameters can be obtained using (5.43). Let $\theta = \{\lambda, \delta, \kappa\}$. Obtain the MLE of θ by maximizing the loglikelihood function in (9.47). The Fisher information matrix is the matrix $\mathbf{I}_\theta$ in (9.52), the inverse of which, evaluated at $(\hat{\lambda}, \hat{\delta}, \hat{\kappa})$, is the estimated covariance matrix of $(\hat{\lambda}, \hat{\delta}, \hat{\kappa})$. The square root of the diagonal elements of this matrix are the standard errors, $(\hat{\sigma}_{\hat{\lambda}}, \hat{\sigma}_{\hat{\delta}}, \hat{\sigma}_{\hat{\kappa}})$, of the estimates $(\hat{\lambda}, \hat{\delta}, \hat{\kappa})$, respectively. Accordingly, the $(1-\alpha)100\%$ confidence intervals for the parameters are

$$\begin{aligned} \lambda &\in \left(\hat{\lambda} \pm z_{\alpha/2}\, \hat{\sigma}_{\hat{\lambda}}\right), \\ \delta &\in \left(\hat{\delta} \pm z_{\alpha/2}\, \hat{\sigma}_{\hat{\delta}}\right), \\ \kappa &\in \left(\hat{\kappa} \pm z_{\alpha/2}\, \hat{\sigma}_{\hat{\kappa}}\right). \end{aligned} \tag{9.62}$$

Confidence Intervals for the Parameters ($\kappa = 0$)

For $\kappa = 0$, confidence intervals for the parameters $\theta = \{\lambda, \delta\}$ can be obtained using (5.43). Obtain the MLE of θ by maximizing the loglikelihood function in (9.49). The Fisher information matrix is $\mathbf{I}_\theta$ in (9.59), the inverse of which, evaluated at $(\hat{\lambda}, \hat{\delta})$, is the estimated covariance matrix of $(\hat{\lambda}, \hat{\delta})$. The square root of the diagonal elements of this matrix are the standard errors, $(\hat{\sigma}_{\hat{\lambda}}, \hat{\sigma}_{\hat{\delta}})$, of the estimates $(\hat{\lambda}, \hat{\delta})$, respectively. Accordingly, the $(1-\alpha)100\%$ confidence intervals for the parameters are given by

$$\lambda \in \left(\hat{\lambda} \pm z_{\alpha/2}\, \hat{\sigma}_{\hat{\lambda}}\right) \quad \text{and} \quad \delta \in \left(\hat{\delta} \pm z_{\alpha/2}\, \hat{\sigma}_{\hat{\delta}}\right). \tag{9.63}$$

Example 9.8 (Data sets). Table 9.6 shows the maximum likelihood estimates of the parameters of the $GEVD_M$ distribution for some of the data sets in Chapter 1, for which the right tail is of interest. To judge the overall goodness-of-fit, we use the *average scaled absolute error*,

$$ASAE = \frac{1}{n} \sum_{i=1}^{n} \frac{|x_{i:n} - \hat{x}_{i:n}|}{x_{n:n} - x_{1:n}}. \tag{9.64}$$

The ASAE are given in Table 9.6. To construct confidence intervals for the parameters, we also need the standard errors of the estimates. For example, for the wind data, the estimated variances of $\hat{\lambda}, \hat{\delta}$, and $\hat{\kappa}$ are 0.704, 0.5811, and 0.0225, respectively. Thus, a 95% confidence interval for κ, for example, can be obtained using (9.62) as follows:

$$\kappa \in \left(-0.42 \pm 1.96\, \sqrt{0.0225}\right) = (-0.714,\ -0.126), \tag{9.65}$$

which indicates that κ is likely to be negative, hence supporting the hypothesis that the distribution of the wind data follow a Fréchet domain of attraction.

Table 9.6: Parameter Estimates Obtained from Fitting the Maximal GEVD to Some of the Data Sets in Chapter 1 for Which the Right Tail Is of Interest Using the Maximum Likelihood Method.

Data Set	$\hat{\lambda}$	$\hat{\delta}$	$\hat{\kappa}$	ASAE
Wind	28.09	5.06	-0.42	0.015
Bilbao	8.02	0.67	0.13	0.031
Men	102.7	1.54	0.24	0.029
Women	104.0	1.38	0.22	0.018
Flood	38.86	7.85	0.06	0.017
Wave	11.37	5.65	0.02	0.014

As another example, for the Bilbao wave heights data, the estimated variances of $\hat{\lambda}, \hat{\delta}$, and $\hat{\kappa}$ are 0.0036, 0.0021, and 0.0066, respectively, and a 95% confidence interval for κ is

$$\kappa \in \left(0.134 \pm 1.96\ \sqrt{0.0066}\right) = (-0.025,\ 0.293)\,, \tag{9.66}$$

which includes zero. Accordingly, a null hypothesis that states that the Gumbel model is the domain of attraction for the distribution of the Bilbao wave heights data cannot be rejected at 5% level of significance. ■

Point Estimates of the Quantiles

The quantiles can be estimated by substituting the parameter estimates in (9.44) and obtaining

$$\hat{x}_p = \begin{cases} \hat{\lambda} + \hat{\delta}\left[1 - (-\log p)^{\hat{\kappa}}\right]/\hat{\kappa}, & \text{if } \kappa \neq 0, \\ \hat{\lambda} - \hat{\delta}\log(-\log p), & \text{if } \kappa = 0. \end{cases} \tag{9.67}$$

Confidence Intervals for the Quantiles ($\kappa \neq 0$)

For $\kappa \neq 0$, confidence intervals for the quantiles can be obtained using (5.45) and the delta method (Section 5.1.3). The point estimates of the quantiles are given in (9.67). The asymptotic variance of $\hat{x}_p$ is

$$\sigma^2_{\hat{x}_p} \approx \nabla^T_{\theta} x_p \Sigma_{\hat{\theta}} \nabla_{\theta} x_p, \tag{9.68}$$

where $\Sigma_{\hat{\theta}}$ is the asymptotic covariance matrix of $\hat{\theta} = (\hat{\lambda}, \hat{\delta}, \hat{\kappa})$ in (9.56) and

$$\nabla_{\theta} x_p = \begin{bmatrix} \dfrac{\partial x_p}{\partial \lambda} \\ \dfrac{\partial x_p}{\partial \delta} \\ \dfrac{\partial x_p}{\partial \kappa} \end{bmatrix} = \begin{bmatrix} 1 \\ \dfrac{1 - (-\log p)^{\kappa}}{\kappa} \\ -\dfrac{\delta[1 - (-\log p)^{\kappa}]}{\kappa^2} - \dfrac{\delta}{\kappa}(-\log p)^{\kappa} \log(-\log p) \end{bmatrix}. \quad (9.69)$$

The estimated asymptotic variance of $\hat{x}_p$ is computed by evaluating $\Sigma_{\hat{\theta}}$ and $\nabla_{\theta} x_p$ at $\hat{\theta} = (\hat{\lambda}, \hat{\delta}, \hat{\kappa})$ and obtaining

$$\hat{\sigma}^2_{\hat{x}_p} \approx \nabla^T_{\hat{\theta}} x_p \hat{\Sigma}_{\hat{\theta}} \nabla_{\hat{\theta}} x_p. \quad (9.70)$$

Accordingly, a $(1 - \alpha)100\%$ confidence interval for x_p is given by

$$x_p \in \left(\hat{x}_p \pm z_{\alpha/2} \hat{\sigma}_{\hat{x}_p}\right). \quad (9.71)$$

Confidence Intervals for the Quantiles ($\kappa = 0$)

For $\kappa = 0$, confidence intervals for the parameters and quantiles can be obtained in a similar way. Here, we have $\theta = \{\lambda, \delta\}$. The point estimates of the quantiles are given in (9.67). The estimated asymptotic variance of $\hat{x}_p$ is

$$\hat{\sigma}^2_{\hat{x}_p} \approx \nabla^T_{\hat{\theta}} x_p \hat{\Sigma}_{\hat{\theta}} \nabla_{\hat{\theta}} x_p, \quad (9.72)$$

where $\hat{\Sigma}_{\hat{\theta}}$ is the estimated asymptotic covariance matrix of $\hat{\theta} = (\hat{\lambda}, \hat{\delta})$ in (9.61), and $\nabla_{\theta} x_p$ is given by

$$\nabla_{\theta} x_p = \begin{bmatrix} \dfrac{\partial x_p}{\partial \lambda} \\ \dfrac{\partial x_p}{\partial \delta} \end{bmatrix} = \begin{bmatrix} 1 \\ -\log(-\log p) \end{bmatrix}. \quad (9.73)$$

Accordingly, a $(1 - \alpha)100\%$ confidence interval for x_p is given by

$$x_p \in \left(\hat{x}_p \pm z_{\alpha/2} \hat{\sigma}_{\hat{x}_p}\right). \quad (9.74)$$

Example 9.9 (Point estimates and confidence intervals for quantiles). Table 9.7 shows the maximum likelihood estimates of the 0.95 and 0.99 quantiles of the $GEVD_M$ distribution for some of the data sets in Chapter 1, for which the right tail is of interest. The corresponding 95% confidence intervals (CI) are also given. Since sample sizes are small, these confidence intervals are imprecise for the wind, flood, and wave data but narrow for other data sets. As expected, the confidence intervals for $x_{0.99}$ are wider than those for $x_{0.95}$. ■

Table 9.7: The 95% Maximum Likelihood Confidence Intervals (CI) for the 0.95 and 0.99 Quantiles of the $GEVD_M$ for Some of the Data Sets in Chapter 1, for Which the Right Tail Is of Interest.

Data Set	$\hat{x}_{0.95}$	$CI(x_{0.95})$	$\hat{x}_{0.99}$	$CI(x_{0.99})$
Wind	58.02	(39.96, 76.08)	99.28	(33.63, 164.93)
Bilbao	9.67	(9.40, 9.94)	10.33	(9.75, 10.91)
Men	105.91	(105.16, 106.65)	106.91	(105.56, 108.26)
Women	106.98	(106.33, 107.63)	107.96	(106.89, 109.03)
Flood	60.13	(54.61, 65.65)	70.22	(60.08, 80.36)
Wave	27.56	(22.35, 32.77)	35.97	(25.2, 46.75)

9.2.2 The Probability Weighted Moments Method

In this section, we apply the Probability Weighted Method (PWM), discussed in Section 5.3, to estimate the parameters and quantiles of the maximal GEVD. It is convenient to use the β_s moments of the GEVD for $\beta \neq 0$. For $\kappa \leq -1$, $\beta_0 = E(X)$, but the rest of the β_s moments do not exist. For $\kappa > -1$, the β_s moments are

$$\beta_s = M(1, s, 0) = \frac{1}{s+1}\left[\lambda + \frac{\delta}{\kappa}\left(1 - \frac{\Gamma(1+\kappa)}{(s+1)^{\kappa}}\right)\right]. \tag{9.75}$$

The corresponding sample moments $b_s = m(1, s, 0)$ are

$$b_s = \frac{1}{n}\sum_{i=1}^{n} x_{i:n}\; p_{i:n}^{s}. \tag{9.76}$$

Now, assuming $\kappa > -1$ and using β_0, β_1 and β_2, after some algebraic manipulations, we obtain

$$\begin{aligned} \beta_0 &= \lambda + \frac{\delta}{\kappa}\{1 - \Gamma(1+\kappa)\}, \\ 2\beta_1 - \beta_0 &= \frac{\delta}{\kappa}\Gamma(1+\kappa)(1 - 2^{-\kappa}), \\ \frac{3\beta_2 - \beta_0}{2\beta_1 - \beta_0} &= \frac{1 - 3^{-\kappa}}{1 - 2^{-\kappa}}. \end{aligned} \tag{9.77}$$

Replacing the moments by the corresponding sample moments, we obtain

$$b_0 = \lambda + \frac{\delta}{\kappa}\{1 - \Gamma(1+\kappa)\}, \tag{9.78}$$

$$2b_1 - b_0 = \frac{\delta}{\kappa}\Gamma(1+\kappa)(1 - 2^{-\kappa}), \tag{9.79}$$

$$\frac{3b_2 - b_0}{2b_1 - b_0} = \frac{1 - 3^{-\kappa}}{1 - 2^{-\kappa}}. \tag{9.80}$$

The PWM estimators are the solutions of (9.78)–(9.80) for λ, δ, and κ. Note that (9.80) is a function of only one parameter κ, so it can be solved by an appropriate numerical method to obtain the PWM estimate $\hat{\kappa}_{PWM}$ of κ. This estimate can then be substituted in (9.79) and an estimate of δ emerges as

$$\hat{\delta}_{PWM} = \frac{\hat{\kappa}(2b_1 - b_0)}{\Gamma(1+\hat{\kappa})(1-2^{-\hat{\kappa}})}. \tag{9.81}$$

Finally, $\hat{\kappa}$ and $\hat{\delta}$ are substituted in (9.78), and an estimate of λ is obtained as

$$\hat{\lambda}_{PWM} = b_0 - \frac{\hat{\delta}}{\hat{\kappa}}\{1 - \Gamma(1+\hat{\kappa})\}. \tag{9.82}$$

Alternatively, Hosking, Wallis, and Wood (1985) propose the approximate estimators:

$$\begin{aligned} \hat{\kappa} &= 7.859q + 2.9554q^2, \\ \hat{\delta} &= \frac{(2m_1 - \bar{x})\hat{\kappa}}{\Gamma(1+\hat{\kappa})(1-2^{\hat{\kappa}})}, \\ \hat{\lambda} &= \bar{x} + \hat{\delta}\{\Gamma(1+\hat{\kappa}) - 1\}/\hat{\kappa}, \end{aligned} \tag{9.83}$$

where $\bar{x}$ is the sample mean,

$$q = \frac{2m_1 - \bar{x}}{3m_2 - \bar{x}} - \frac{\log 2}{\log 3},$$

and

$$m_j = \frac{1}{n}\sum_{i=1}^{n} \frac{(i-1)(i-2)\dots(i-j)}{(n-1)(n-2)\dots(n-j)}\, x_{i:n}, \quad j = 1, 2.$$

The variance-covariance matrix has the form

$$\hat{\Sigma}(\hat{\lambda}, \hat{\delta}, \hat{\kappa}) = \frac{1}{n}\begin{bmatrix} \delta^2 w_{11} & \delta^2 w_{12} & \delta^2 w_{13} \\ \delta^2 w_{12} & \delta^2 w_{22} & \delta^2 w_{23} \\ \delta^2 w_{13} & \delta^2 w_{23} & \delta^2 w_{33} \end{bmatrix}. \tag{9.84}$$

The w_{ij} are functions of κ and have complicated formulas. However, they can be evaluated numerically. Hosking, Wallis, and Wood (1985) provide a table for some important values of κ. These are given here in Table 9.8.

Example 9.10 (Data sets). Table 9.9 shows the PWM method's estimates of the parameters of the GEVD_M distribution and the ASAE in (9.64) for some of the data sets in Chapter 1, for which the right tail is of interest. Unfortunately, the variances of these estimates are too complicated (see (9.84)), hence the PWM method's confidence intervals are difficult to obtain. By comparison with the MLE in Tables 9.6 one can see that, with the exception of $\hat{\delta}$ estimates for the men and women data sets, which are less reliable due to the high ASAE values, the PWM method's estimates are close to the corresponding MLEs. This is especially the case for the parameter κ, which is arguably the most important of the three parameters. ∎

Table 9.8: Elements of the Asymptotic Covariance Matrix of the PWM Estimators of the Parameters in the GEVD According to Hosking et al. (1985).

κ	w_{11}	w_{12}	w_{13}	w_{22}	w_{23}	w_{33}
-0.4	1.6637	1.3355	1.1405	1.8461	1.1628	2.9092
-0.3	1.4153	0.8912	0.5640	1.2574	0.4442	1.4090
-0.2	1.3322	0.6727	0.3926	1.0013	0.2697	0.9139
-0.1	1.2915	0.5104	0.3245	0.8440	0.2240	0.6815
0.0	1.2686	0.3704	0.2992	0.7390	0.2247	0.5633
0.1	1.2551	0.2411	0.2966	0.6708	0.2447	0.5103
0.2	1.2474	0.1177	0.3081	0.6330	0.2728	0.5021
0.3	1.2438	-0.0023	0.3297	0.6223	0.3033	0.5294
0.4	1.2433	-0.1205	0.3592	0.6368	0.3329	0.5880

Table 9.9: Parameter Estimates Obtained from Fitting the Maximal GEVD to Some of the Data Sets in Chapter 1, for Which the Right Tail Is of Interest Using the PWM Method.

Data Set	$\hat{\lambda}$	$\hat{\delta}$	$\hat{\kappa}$	ASAE
Wind	28.14	5.61	-0.33	0.016
Bilbao	8.00	0.71	0.12	0.025
Men	102.3	2.54	0.22	0.115
Women	103.6	2.41	0.25	0.117
Flood	38.56	7.68	0.01	0.017
Wave	11.25	5.68	0.00	0.013

9.2.3 The Elemental Percentile Method

In the case of the maximal GEVD, the maximum likelihood method have problems because

1. The range of the distribution depends on the parameters, hence, the MLE do not have the usual asymptotic properties.

2. The MLE requires numerical solutions.

3. For some samples, the likelihood may not have a local maximum.

Here we use the elemental percentile method (EPM), discussed in general in Section 5.4, for estimating the parameters and quantiles of the maximal GEVD, as proposed by Castillo and Hadi (1995c). The estimates are obtained in two stages as explained next.

First Stage: Initial Estimates

The initial point estimates of the parameters are obtained as follows. For the case $\kappa \neq 0$, the maximal GEVD$_M$ distribution, $H_\kappa(x; \lambda, \delta)$ in (9.3), has three parameters, so each elemental set contains three distinct order statistics. Let $I = \{i, j, r\} : i < j < r \in \{1, 2, \ldots, n\}$ be the indices of three distinct order statistics. Equating sample and theoretical GEVD$_M$ quantiles in (9.44), we obtain

$$\begin{aligned} x_{i:n} &= \lambda + \delta \left[1 - (-\log\, p_{i:n})^\kappa\right] / \kappa, \\ x_{j:n} &= \lambda + \delta \left[1 - (-\log\, p_{j:n})^\kappa\right] / \kappa, \\ x_{r:n} &= \lambda + \delta \left[1 - (-\log\, p_{r:n})^\kappa\right] / \kappa, \end{aligned} \tag{9.85}$$

where $p_{i:n} = (i - 0.35)/n$. Subtracting the third from the second of the above equations, and also the third from the first, then taking the ratio, we obtain

$$\frac{x_{j:n} - x_{r:n}}{x_{i:n} - x_{r:n}} = \frac{C_r^\kappa - C_j^\kappa}{C_r^\kappa - C_i^\kappa} = \frac{1 - A_{jr}^\kappa}{1 - A_{ir}^\kappa}, \tag{9.86}$$

where $C_i = -\log(p_{i:n})$ and $A_{ir} = C_i / C_r$. This is an equation in one unknown κ. Solving (9.86) for κ using the bisection method (see Algorithm 9.1), we obtain $\hat{\kappa}_{ijr}$, which indicates that this estimate is a function the three observations $x_{i:n}$, $x_{j:n}$, and $x_{r:n}$. Substituting $\hat{\kappa}_{ijr}$ in two of the equations in (9.85) and solving for δ and λ, we obtain

$$\hat{\delta}_{ijr} = \frac{\hat{\kappa}_{ijr}(x_{i:n} - x_{r:n})}{C_r^{\hat{\kappa}} - C_i^{\hat{\kappa}}} \tag{9.87}$$

and

$$\hat{\lambda}_{ijr} = x_{i:n} - \frac{\hat{\delta}_{ijr}(1 - C_i^{\hat{\kappa}_{ijr}})}{\hat{\kappa}_{ijr}}. \tag{9.88}$$

The estimates $\hat{\kappa}_{ijr}$, $\hat{\delta}_{ijr}$, and $\hat{\lambda}_{ijr}$ must satisfy that $x_{n:n} \leq \hat{\lambda} + \hat{\delta}/\hat{\kappa}$, when $\hat{\kappa} > 0$ and $x_{1:n} \geq \hat{\lambda} + \hat{\delta}/\hat{\kappa}$, when $\hat{\kappa} < 0$. In this way, we force parameter estimates to be consistent with the observed data.

Algorithm 9.1 (Solving Equation (9.86))

- **Input:** An elemental subset of three distinct order statistics, $x_{i:n}$, $x_{j:n}$, and $x_{r:n}$.

- **Output:** Initial estimates of κ.

Let $A_{ir} = \log(p_{i:n}) / \log(p_{r:n})$ and compute $D_{ijr} = (x_{j:n} - x_{r:n})/(x_{i:n} - x_{r:n})$, then solve the equation

$$D_{ijr} = \frac{1 - A_{jr}^\kappa}{1 - A_{ir}^\kappa}$$

in κ and obtain $\hat{\kappa}$ as follows:

1. If $D_{ijr} < \log(A_{jr}) / \log(A_{ir})$, then use the bisection method to obtain the zero of the function

$$h(\kappa) = \frac{1 - A_{jr}^\kappa}{1 - A_{ir}^\kappa} - D_{ijr}$$

on the interval

$$\left(0, \frac{\log D_{ijr}}{\log A_{ji}}\right).$$

2. Otherwise, use the bisection method on the interval

$$\left(\frac{\log(1 - D_{ijr})}{\log A_{jr}}, 0\right).$$

In the case of Gumbel distribution ($\kappa = 0$), we have two parameters, λ and δ. In this case, each elemental set contains two observations. Let $I = \{i, j\}$ be a set of two distinct indices, where $i < j \in \{1, 2, \ldots, n\}$. Equating sample and theoretical GEVD_M quantile in (9.44), we obtain

$$\begin{aligned} x_{i:n} &= \lambda - \delta \ \log(-\log \ p_{i:n}), \\ x_{j:n} &= \lambda - \delta \ \log(-\log \ p_{j:n}). \end{aligned} \tag{9.89}$$

The solution of this system gives

$$\hat{\delta}_{ij} = \frac{x_{i:n} - x_{j:n}}{\log C_j - \log C_i} \tag{9.90}$$

and

$$\hat{\lambda}_{ij} = x_{i:n} + \hat{\delta}_{ij} \log C_i, \tag{9.91}$$

where $C_i = -\log(p_{i:n})$. Thus, we have a closed form solution.

Second Stage: Final Estimates

The above initial estimates are based on only three distinct order statistics (two for the case of the Gumbel distribution). More statistically efficient and robust estimates are obtained using other order statistics as follows. Select a prespecified number, N, of elemental subsets each of size 3, either at random or using all possible subsets. For each of these elemental subsets, an elemental estimate of the parameters $\theta = \{\kappa, \delta, \lambda\}$ is computed. Let us denote these elemental estimates by $\hat{\theta}_1, \hat{\theta}_2, \ldots, \hat{\theta}_N$. The elemental estimates that are inconsistent with the data are discarded. These elemental estimates can then be combined, using some suitable (preferably robust) functions, to obtain an overall final estimate of θ. Examples of robust functions include the *median* (MED) and the *α-trimmed mean* (TM_α), where α indicates the percentage of trimming. Thus, a final estimate of $\theta = \{\kappa, \delta, \lambda\}$, can be defined as

$$\begin{aligned} \hat{\kappa}_{\text{MED}} &= \text{Median}(\hat{\kappa}_1, \hat{\kappa}_2, \ldots, \hat{\kappa}_N), \\ \hat{\delta}_{\text{MED}} &= \text{Median}(\hat{\delta}_1, \hat{\delta}_2, \ldots, \hat{\delta}_N), \\ \hat{\lambda}_{\text{MED}} &= \text{Median}(\hat{\lambda}_1, \hat{\lambda}_2, \ldots, \hat{\lambda}_N), \end{aligned} \tag{9.92}$$

or

$$\begin{aligned} \hat{\kappa}_{\text{TM}_\alpha} &= \text{TM}_\alpha(\hat{\kappa}_1, \hat{\kappa}_2, \ldots, \hat{\kappa}_N), \\ \hat{\delta}_{\text{TM}_\alpha} &= \text{TM}_\alpha(\hat{\delta}_1, \hat{\delta}_2, \ldots, \hat{\delta}_N), \\ \hat{\lambda}_{\text{TM}_\alpha} &= \text{TM}_\alpha(\hat{\lambda}_1, \hat{\lambda}_2, \ldots, \hat{\lambda}_N), \end{aligned} \tag{9.93}$$

where Median$(y_1, \ldots, y_N)$ is the median of the set of numbers $\{y_1, \ldots, y_N\}$ and $\text{TM}_\alpha(y_1, \ldots, y_N)$ is the mean obtained after trimming the $(\alpha/2)100\%$ largest and the $(\alpha/2)100\%$ smallest order statistics of $y_1, \ldots, y_N$.

The MED estimators are very robust but inefficient. The TM_α estimators are less robust but more efficient than the MED estimators. The larger the trimming, the more robust and less efficient are the TM_α estimators. Experience indicates that the properties of the MED and TM_α estimators are similar, but the root mean quare errors of TM_α estimators may be slightly smaller than those of the MED.

To avoid estimates that are inconsistent with the data, it is better to estimate the upper end $\lambda + \delta/\kappa$ instead of λ and, based on it, recover λ, that is, replace the last estimate in (9.92) by

$$(\hat{\lambda} + \hat{\delta}/\hat{\kappa})_{\text{MED}} = \text{Median}\left((\hat{\lambda} + \hat{\delta}/\hat{\kappa})_{1,2,n}, (\hat{\lambda} + \hat{\delta}/\hat{\kappa})_{1,3,n}, \ldots, (\hat{\lambda} + \hat{\delta}/\hat{\kappa})_{1,n-1,n}\right)$$

and recover $\hat{\lambda}$ by means of

$$\hat{\lambda}_{\text{MED}} = (\hat{\lambda} + \hat{\delta}/\hat{\kappa})_{\text{MED}} - \hat{\delta}_{\text{MED}}/\hat{\kappa}_{\text{MED}}.$$

Similarly, we replace the last estimates in (9.93) by

$$(\hat{\lambda} + \hat{\delta}/\hat{\kappa})_{\text{TM}\alpha} = \text{TM}_\alpha\left((\hat{\lambda} + \hat{\delta}/\hat{\kappa})_{1,2,n}, (\hat{\lambda} + \hat{\delta}/\hat{\kappa})_{1,3,n}, \ldots, (\hat{\lambda} + \hat{\delta}/\hat{\kappa})_{1,n-1,n}\right)$$

and recover $\hat{\lambda}$ by means of

$$\hat{\lambda}_{\text{TM}\alpha} = (\hat{\lambda} + \hat{\delta}/\hat{\kappa})_{\text{TM}\alpha} - \hat{\delta}_{\text{TM}\alpha}/\hat{\kappa}_{\text{TM}\alpha}.$$

The quantile estimates for any desired p are then obtained by substituting the above parameter estimates in (9.44).

Example 9.11 (Data sets). Table 9.10 shows the EPM-MED and EPM-TM estimates of the parameters and the corresponding ASAE in (9.64) for the maximal GEVD for some of the data sets in Chapter 1, for which the right tail is of interest. As can be seen from the two tables, both versions of the EPM are similar to each other and, in many cases, to those of the MLE. ∎

A Computationally Efficient Version of EPM

As per Remark 9.1, a computationally simple version of the EPM can be used as an initial starting point for the numerical solution of the MLE. Here, the final estimates in (9.92) or (9.93) can be computed using only a small number of elemental subsets. Let $i = 1$ and $r = n$ and compute $\hat{\kappa}_{1jn}, \hat{\delta}_{1jn}$ and $\hat{\lambda}_{1jn}, j = 2, 3, \ldots, n-1$. These are only $n - 2$ estimates, which is much smaller than the number N of elemental subsets. They can be combined to produce final estimates using (9.92) or (9.93). These estimates can then be used as a starting value in the MLE algorithm.

Table 9.10: Parameter Estimates Obtained from Fitting the Maximal GEVD to Some of the Data Sets in Chapter 1 for Which the Right Tail Is of Interest Using the EPM.

EPM-TM				
Data Set	$\hat{\lambda}$	$\hat{\delta}$	$\hat{\kappa}$	ASAE
Wind	27.96	5.44	-0.38	0.015
Bilbao	7.95	0.73	0.03	0.030
Men	102.6	1.43	0.15	0.023
Women	104.0	1.37	0.26	0.019
Flood	39.05	6.88	0.08	0.022
Wave	11.31	5.55	0.04	0.018
EPM-MED				
Data Set	$\hat{\lambda}$	$\hat{\delta}$	$\hat{\kappa}$	ASAE
Wind	27.91	5.39	-0.36	0.016
Bilbao	7.95	0.71	0.02	0.030
Men	102.6	1.48	0.16	0.028
Women	104.0	1.38	0.25	0.018
Flood	39.04	6.89	0.05	0.020
Wave	11.30	5.56	0.04	0.017

Confidence Intervals

Confidence intervals for the parameters or the quantiles can be obtained by simulation. Note that since the parameter and quantile estimates are well defined for all possible combinations of parameter and sample values, the variances of these estimates (hence, confidence intervals for the corresponding parameter or quantile values) can be obtained using sampling-based methods such as the jackknife or the bootstrap methods (Efron (1979) and Diaconis and Efron (1974)). A comparison of confidence intervals for the GEVD is given by Dupuis and Field (1998).

9.2.4 The Quantile Least Squares Method

The *quantile least squares* (QLS) method for the GEVD estimates the parameters by minimizing the sum of squares of the differences between the theoretical and the observed quantiles. More precisely, for $\kappa \neq 0$, the QLS method parameter estimates are the solution of the minimization problem

$$\underset{\lambda,\delta,\kappa}{\text{Minimize}} \quad \sum_{i=1}^{n} \left[x_{i:n} - \lambda - \delta\left[1 - (-\log p_{i:n})^{\kappa}\right]/\kappa\right]^2, \tag{9.94}$$

Table 9.11: Parameter Estimates Obtained from Fitting the Maximal GEVD to Some of the Data Sets in Chapter 1 for Which the Right Tail is of Interest Using the QLS Method.

Data Set	$\hat{\lambda}$	$\hat{\delta}$	$\hat{\kappa}$	ASAE
Wind	28.02	4.26	-0.60	0.014
Bilbao	8.04	0.72	0.19	0.028
Men	102.6	1.62	0.22	0.024
Women	103.9	1.45	0.21	0.016
Flood	38.36	7.34	-0.07	0.016
Wave	11.15	5.78	-0.04	0.011

and, for $\kappa = 0$, they are the solution of the minimization problem

$$\underset{\lambda, \delta}{\text{Minimize}} \quad \sum_{i=1}^{n} \left[x_{i:n} - \lambda + \delta \log(-\log p_{i:n})\right]^2. \tag{9.95}$$

These problems can be solved using standard optimization packages.

Example 9.12 (Data sets). Table 9.11 shows the QLS estimates of the parameters and the ASAE in (9.64) for the maximal GEVD for some of the data sets in Chapter 1, for which the right tail is of interest. One can see that this method gives similar estimates of the parameters to those of the previous methods. ∎

9.2.5 The Truncation Method

Since the tail of the distribution defines the domain of attraction, it could be convenient to use the tail of interest only to fit the limit GEVD. To this end, we can fix a threshold value u and then consider only the sample values exceeding the threshold u.

For $\kappa \neq 0$, it is clear from Section 3.3 and (9.47) that in the limit the loglikelihood of the truncated sample becomes

$$\begin{aligned} \ell(\lambda, \delta, \kappa) &= -n \log \delta - (1-\kappa) \sum_{i=1}^{n} z_i - \sum_{i=1}^{n} \exp\{-z_i\} \\ &\quad -n \log \left\{1 - \exp\left[-\left(1 - \kappa \frac{u-\lambda}{\delta}\right)^{1/\kappa}\right]\right\}, \end{aligned} \tag{9.96}$$

where

$$z_i = -\frac{1}{\kappa} \log\{1 - \kappa(x_i - \lambda)/\delta\}, \tag{9.97}$$

and n is now the truncated sample size.

For $\kappa = 0$, we get

$$\begin{aligned}\ell(\lambda, \delta) &= -n\log\delta - \sum_{i=1}^{n} \exp\left[-\left(\frac{x_i-\lambda}{\delta}\right)\right] - \sum_{i=1}^{n}\left(\frac{x_i-\lambda}{\delta}\right) \\ &\quad -n\log\left\{1-\exp\left[-\exp\left(\frac{\lambda-u}{\delta}\right)\right]\right\}. \end{aligned} \tag{9.98}$$

The estimates can be obtained by maximizing (9.96) or (9.98) using standard optimization packages.

9.3 Estimation for the Minimal GEVD

By Theorem 9.3, parameter estimates for the $GEVD_m$ can be obtained using the estimation methods for the $GEVD_M$. Thus, one can use the following algorithm:

1. Change the signs of the data: $x_i \to -x_i, \quad i = 1, 2, \ldots, n$.

2. Estimate the parameters of the $GEVD_M$, say $\hat{\lambda}, \hat{\delta}$ and $\hat{\kappa}$.

3. The $GEVD_m$ parameter estimates are then $\{-\hat{\lambda}, \hat{\delta}, \hat{\kappa}\}$, that is, we need only to change the sign of $\hat{\lambda}$.

Example 9.13 (Data sets). Table 9.12 gives estimates of the parameters and the ASAE in (9.64) for the minimal GEVD for some of the data sets in Chapter 1, for which the left tail is of interest for various estimation methods. The parameter estimates of λ and δ are similar to each other, but there is a wide variation in the estimates of the parameter κ.

Table 9.13 shows the 95% maximum likelihood confidence intervals (CI) for the 0.01 and 0.05 quantiles of the $GEVD_m$ distribution for some of the data sets in Chapter 1, for which the left tail is of interest. Due to the small sizes of the sample, these intervals are quite wide. ■

9.4 Graphical Methods for Model Selection

In Section 6.1, we discussed the use of the Probability Paper Plots (PPP) in general. In this section, we discuss their use in the problem of extremes. More precisely, we indicate how probability paper plots can be used to determine the domain of attraction of a given sample.[1]

---

[1]Some of the material in this section is reprinted from the book *Extreme Value Theory in Engineering*, by E. Castillo, Copyright © Academic Press (1988), with permission from Elsevier.

Table 9.12: Parameter Estimates Obtained from Fitting the Minimal GEVD to Some of the Data Sets in Chapter 1 for Which the Left Tail Is of Interest.

		Parameter Estimates			
Method	Data Set	$\hat{\lambda}$	$\hat{\delta}$	$\hat{\kappa}$	ASAE
MLE	Epicenter	158.3	58.64	0.49	0.160
	Chain	92.37	16.40	0.23	0.188
	Insulation	1135	177.8	0.39	0.146
	Precipitation	43.35	5.37	0.21	0.180
PWM	Epicenter	164.8	52.48	0.23	0.224
	Chain	92.70	12.78	0.05	0.193
	Insulation	1128	155.2	0.39	0.132
	Precipitation	43.67	4.69	0.10	0.197
EPM-TM	Epicenter	172.6	59.69	0.26	0.276
	Chain	94.64	16.79	0.11	0.240
	Insulation	1115	175.4	0.55	0.113
	Precipitation	44.28	5.86	0.13	0.243
EPM-MED	Epicenter	174.1	58.79	0.28	0.280
	Chain	94.88	16.85	0.15	0.238
	Insulation	1116	178.1	0.53	0.114
	Precipitation	44.29	5.68	0.14	0.237
QLS	Epicenter	163.9	57.31	0.31	0.214
	Chain	93.53	18.16	0.16	0.225
	Insulation	1136	196.1	0.45	0.147
	Precipitation	43.81	5.81	0.18	0.213

Table 9.13: The 95% Maximum Likelihood Confidence Intervals (CI) for the 0.01 and 0.05 Quantiles of the $GEVD_m$.

Data Set	$\hat{x}_{0.01}$	$CI(x_{0.01})$	$\hat{x}_{0.05}$	$CI(x_{0.05})$
Epicenter	50.46	(23.35,77.58)	66.08	(52.03,80.14)
Chain	45.75	(21.22,70.29)	57.04	(43.51,70.57)
Insulation	751.56	(666.94,836.18)	820.06	(755.18,884.95)
Precipitation	27.56	(20.92,34.19)	31.51	(28.03,34.99)

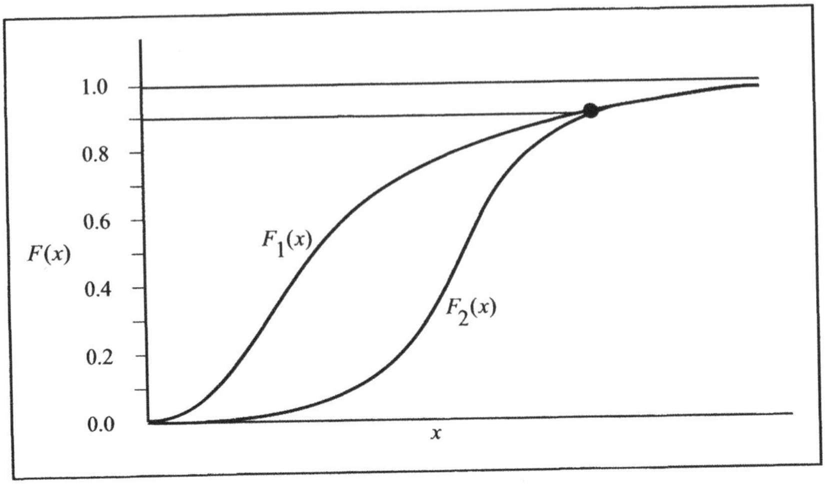

Figure 9.3: Two different distributions with the same right tail and, hence, the same domain of attraction.

9.4.1 Probability Paper Plots for Extremes

The problem of checking whether a sample comes from a maximal Gumbel family of distributions, and the problem of determining whether the domain of attraction of a given sample is maximal Gumbel, are different. In the former case, the whole sample is expected to follow a linear trend when plotted on a maximal Gumbel probability paper, while only the upper tail is expected to exhibit that property for the latter.

Consequently, it is incorrect to reject the assumption that a maximal Gumbel domain of attraction does not fit the data only because the data do not show a linear trend when plotted on a maximal Gumbel probability paper. This is because two distributions with the same tail, even though the rest be completely different, lead to exactly the same domain of attraction, and should be approximated by the same model. For example, Figure 9.3 shows two cdfs that are identical on the interval (0.9,1) but are different in the rest (see Smith (1987)).

Since only the right (left) tail governs the domain of attraction for maxima (minima), then one should focus only on the behavior of the high- (low-) order statistics. The remaining data values are not needed. In fact, they can be more of a hinderance than of help in solving the problem. The problem is then to determine how many data points should be considered in the analysis or to decide which weights must be assigned to each data point.

When selecting the weights for different data points, the following considerations should be taken into account:

1. The estimates of tail quantiles have larger variance than those not in the tail.

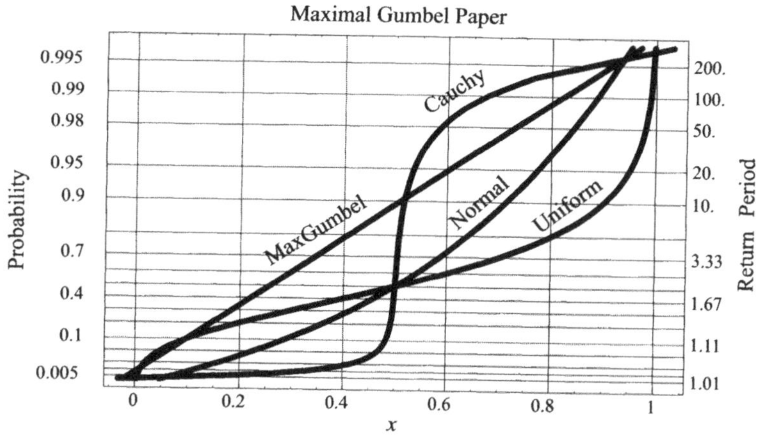

Figure 9.4: The uniform, normal, Cauchy, and maximal Gumbel distributions plotted on a maximal Gumbel probability paper.

2. Data on the tail of interest have more information on the limit distribution than those not in the tail.

These are contradicting considerations and should be balanced. However, on the other tail, the weights must be very small if not zero.

With the aim of illustrating graphically the role of the tails in the extreme behavior, Figures 9.4 and 9.5 show well-known distributions on maximal and minimal Gumbel probability papers, respectively. Note that since none of them is Gumbel, they exhibit nonlinear trends. Figure 9.4 shows an almost linear trend for the right tail of the normal distribution, confirming that the normal belongs to the maximal Gumbel domain of attraction. Note also the curvatures in the right tail of the uniform (positive curvature and vertical asymptote) and Cauchy distributions (negative curvature and horizontal trend) showing Weibull and Fréchet domains of attractions for maxima, respectively.

Figure 9.5 shows an almost linear trend for the left tail of the normal distribution, confirming that it belongs to the minimal Gumbel domain of attraction. Note also the curvatures in the left tail of the uniform (negative curvature and vertical asymptote) and the reversed Cauchy distributions (positive curvature and horizontal trend) showing minimal Weibull and Fréchet domains of attractions, respectively.

Finally, Figure 9.6 shows the minimal Weibull, uniform, and normal distributions plotted on a minimal Weibull probability paper. The Weibull distribution is a straight line, as expected. The uniform distribution is almost straight line. This is also expected because the uniform distribution belongs to the Weibull domain of attraction (see Table 9.5). The normal distribution is a curve, which is also expected because it belongs to the Gumbel, not to the Weibull, domain of attraction.

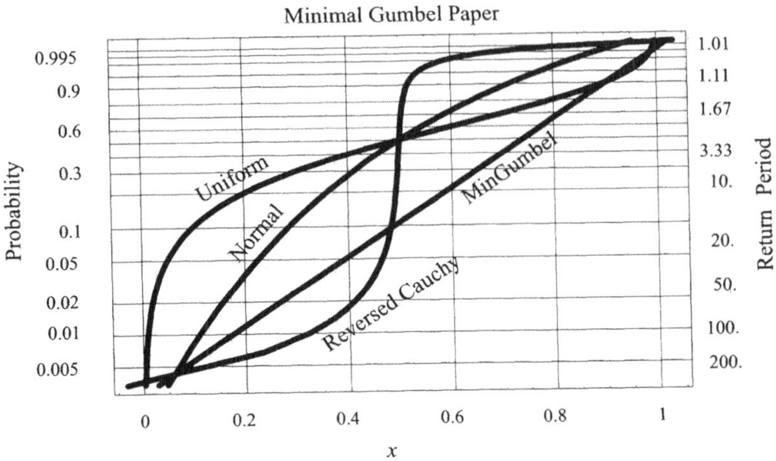

Figure 9.5: The uniform, normal, reversed Cauchy, and minimal Gumbel distributions plotted on a minimal Gumbel probability paper.

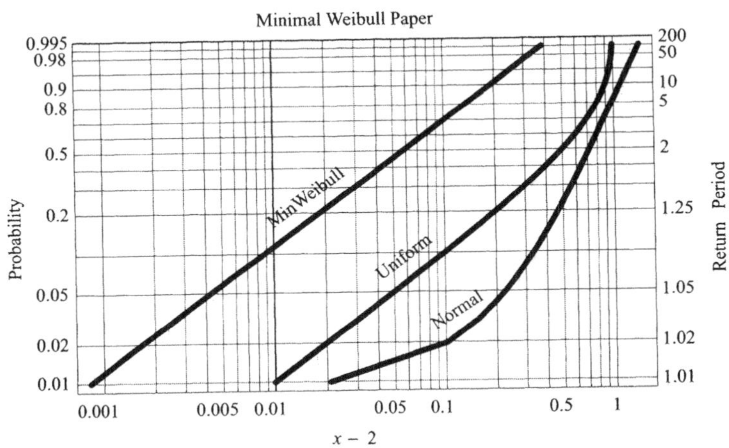

Figure 9.6: The uniform, normal, and minimal Weibull distributions plotted on a minimal Weibull probability paper.

To explain what happens when representing a distribution on a Gumbel probability paper, we mention the following. If a GEVD family in (9.3) is represented on a maximal Gumbel probability paper, then using (6.17) we obtain

$$v \quad = \quad -\log[-\log(p)] = -\log\{-\log[H_\kappa(u;\lambda,\delta)]\}$$

$$
\begin{aligned}
&= -\log\left\{-\log\left[\exp\left(-\left[1-\kappa\left(\frac{u-\lambda}{\delta}\right)\right]^{1/\kappa}\right)\right]\right\} \\
&= \frac{-1}{\kappa}\log\left[1-\kappa\left(\frac{u-\lambda}{\delta}\right)\right], \quad 1-\kappa\left(\frac{u-\lambda}{\delta}\right)\geq 0,
\end{aligned}
$$

where v and u are the ordinate and the abscissa, respectively. The domain is $u \leq \lambda + \delta/\kappa$, if $\kappa > 0$, or $u \geq \lambda + \delta/\kappa$, if $\kappa < 0$. Taking derivatives with respect to u twice, we get

$$
v' = \frac{1}{\delta\left[1-\kappa\left(\frac{u-\lambda}{\delta}\right)\right]} \quad \text{and} \quad v'' = \frac{\kappa}{\delta^2\left[1-\kappa\left(\frac{u-\lambda}{\delta}\right)\right]^2}. \tag{9.99}
$$

Note that when we approach the upper end of the distribution, v' tends to infinity, if $\kappa > 0$, and to zero, if $\kappa < 0$. This fact is very useful in the process of identifying maximal domain of attraction of Weibull and Fréchet, because they show vertical and horizontal slopes, in the tail of interest, respectively. Note also that v' can tend to zero or infinity for Gumbel type distribution if $\delta \to 0$ or $\delta \to \infty$. In addition, we have

$$
\begin{aligned}
v'' &> 0, \text{ iff } \kappa > 0, \\
v'' &= 0, \text{ iff } \kappa = 0, \\
v'' &< 0, \text{ iff } \kappa < 0.
\end{aligned}
$$

If we deal with the left tail instead of the right tail, the symbols "<" and ">" must be interchanged.

Consequently, on maximal (minimal) Gumbel probability papers, the distributions in the Weibull domain of attraction appear as concave (convex), the distributions in the Fréchet domain of attraction appear as convex (concave), and the distributions in the Gumbel domain of attraction are almost straight lines.

Summarizing, a practical method for determining the domain of attraction of a sample is as follows:

1. Determine whether we have a maximum or a minimum problem.

2. Plot the empirical cumulative distribution function (ecdf) on a maximal (minimal) Gumbel probability paper, depending on whether we have a maximum or a minimum problem.

3. Observe the curvature (concavity or convexity) and the slopes on the tail of interest. If the convexity is negligible and the slopes do not go to zero or infinity, accept the assumption of a Gumbel domain of attraction. Otherwise, and depending on the curvature accept a Weibull or Fréchet domain of attraction. More precisely, proceed according to the following rules:

   (a) If we deal with a maximum problem (maximal Gumbel probability paper), then

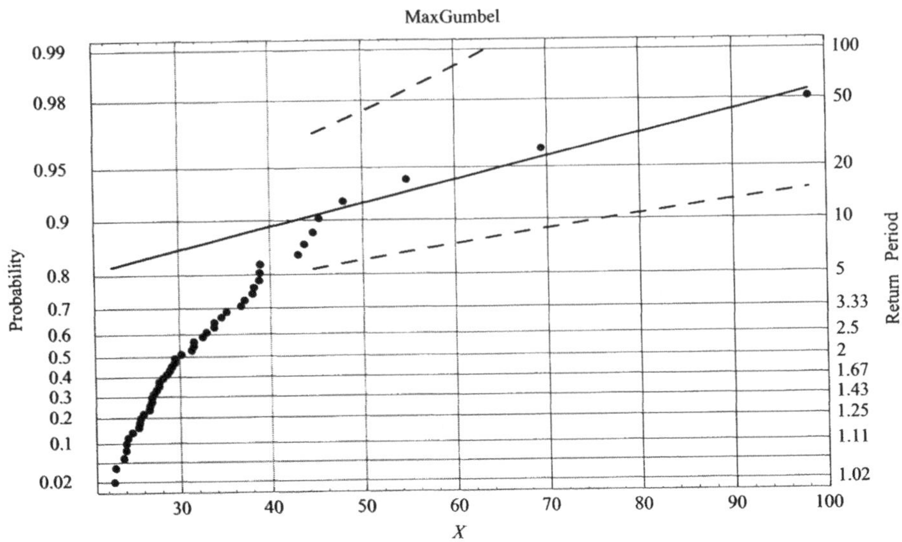

Figure 9.7: Yearly maxima wind data on maximal Gumbel probability paper.

- If $v'' > 0$, accept a maximal Weibull domain of attraction.
- If $v'' = 0$, accept a maximal Gumbel domain of attraction.
- If $v'' < 0$, accept a maximal Fréchet domain of attraction.

(b) If we deal with a minimum problem (minimal Gumbel probability paper), then

- If $v'' > 0$, accept a minimal Fréchet domain of attraction.
- If $v'' = 0$, accept a minimal Gumbel domain of attraction.
- If $v'' < 0$, accept a minimal Weibull domain of attraction.

Example 9.14 (Maximum wind speed data). The yearly maximum wind speed (in miles/hour) registered at a given location during a period of 50 years are given in Table 1.1. The maximal Gumbel PPP for these data is given in Figure 9.7. The pattern of points has a convex shape. However, the right tail shows an almost linear trend, suggesting a maximal Gumbel domain of attraction. A straight line has been adjusted visually leading to the estimates: $\lambda = -15$ and $\delta = 23.5$ miles/hour. However, due to the convexity of that tail, a maximal Fréchet domain of attraction could also be assumed. This is a more conservative decision, because Fréchet has a heavier tail.

The same data are plotted on a maximal Fréchet PPP in Figure 9.8 for three different values $\lambda = 0, 8$ and 18 miles/hour. The value of $\lambda = 18$ miles/hour provides the best fit. The scatter of points are clearly linear, which supports the assumption of a maximal Fréchet distribution. ∎

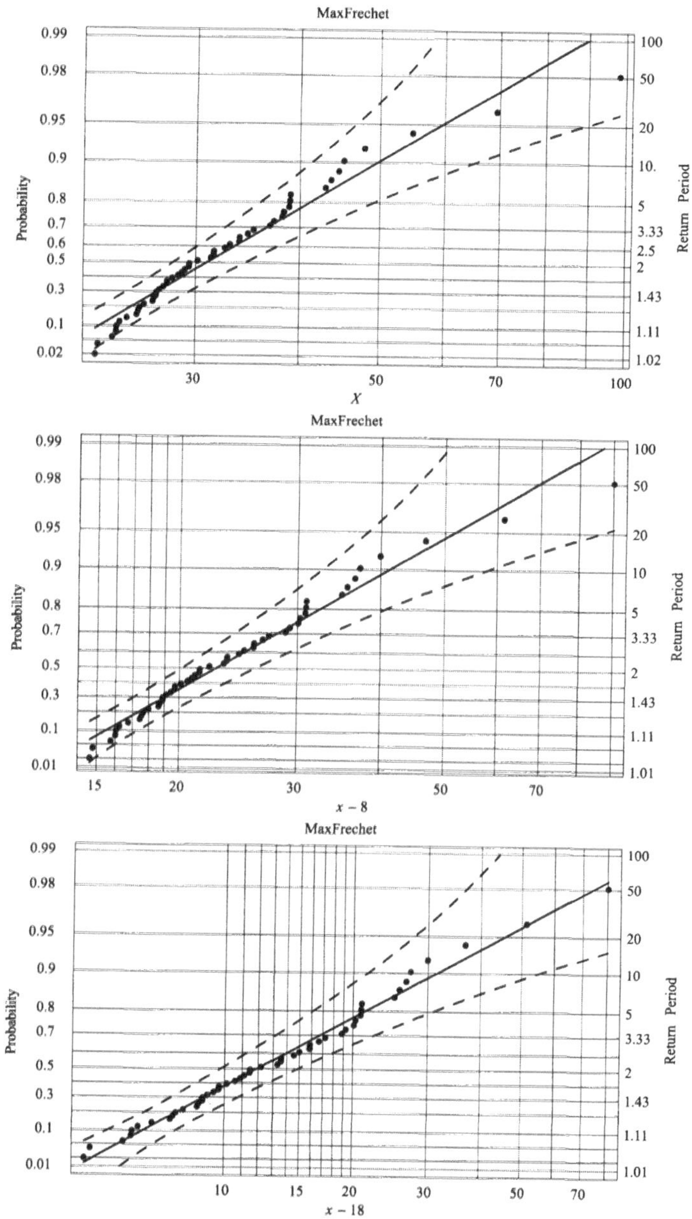

Figure 9.8: The wind data plotted on maximal Fréchet PPP for three different values of $\lambda = 0, 8, 18$.

Example 9.15 (Telephone Calls Data). The times (in minutes) between 48 consecutive telephone calls are given in Table 1.8. The upper panel in Figure

9.9 shows these data on minimal Gumbel PPP. The pattern of points exhibits a convex trend and a slope going to infinity in the left tail, thus suggesting a minimal Weibull domain of attraction and a lower end value $\lambda = 0$. The data are then plotted on a minimal Weibull PPP in the lower panel in Figure 9.9. Observe that a linear trend is seen not just for the right tail but also for the entire data. This is expected because the data are neither minima nor maxima; they are the times between consecutive telephone calls. Such interarrival times are usually modeled in practice using the exponential distribution, which is a special case of the Weibull distribution. ■

9.4.2 Selecting a Domain of Attraction from Data

To summarize the use of PPP in extremes, we give some guidelines for selecting the appropriate domain of attraction from data. By Theorems 9.1 and 9.2, the only nondegenerate family of distributions for maxima and minima of iid samples are the maximal and minimal GEVDs given in (9.3) and (9.6), respectively. The maximal GEVD family includes the maximal Weibull, Gumbel, and Fréchet domain of attractions given by (9.10), (9.11), and (9.12), respectively. The minimal GEVD family includes the minimal Weibull, Gumbel, and Fréchet domain of attractions given by (9.14), (9.15), and (9.16), respectively.

Now, given a set of maxima or minima data, which one of these families provides a good fit for the data. We offer the following guidelines:

1. Use first physical considerations to eliminate some of the possible domains of attractions. If the random variable is limited in the tail of interest, eliminate the Fréchet domain of attraction; otherwise, eliminate the Weibull domain of attraction.

2. If the data is maxima (minima), draw the data on the maximal (minimal) Gumbel PPP.

3. If the tail of interest (the right tail for maxima and left tail for minima) shows a linear trend, then the domain of attraction is the maximal (minimal) Gumbel family.

4. If the tail of interest has a vertical asymptote, then the domain of attraction is the maximal (minimal) Weibull. The value of λ required to plot the data on a Weibull PPP is the value of X associated with the asymptote.

5. If the tail of interest has a horizontal asymptote, then the domain of attraction is the maximal (minimal) Fréchet. The value of λ required to plot the data on a maximal (minimal) Fréchet PPP is chosen by iterations until the corresponding tail shows a linear trend.

6. In case of doubt between Gumbel and Weibull, a conservative choice is Gumbel because it has a heavier tail than Weibull. In case of doubt between Gumbel and Fréchet, a conservative choice is the Fréchet model. In other words, to be on the safe side, use the model with the heavier tail.

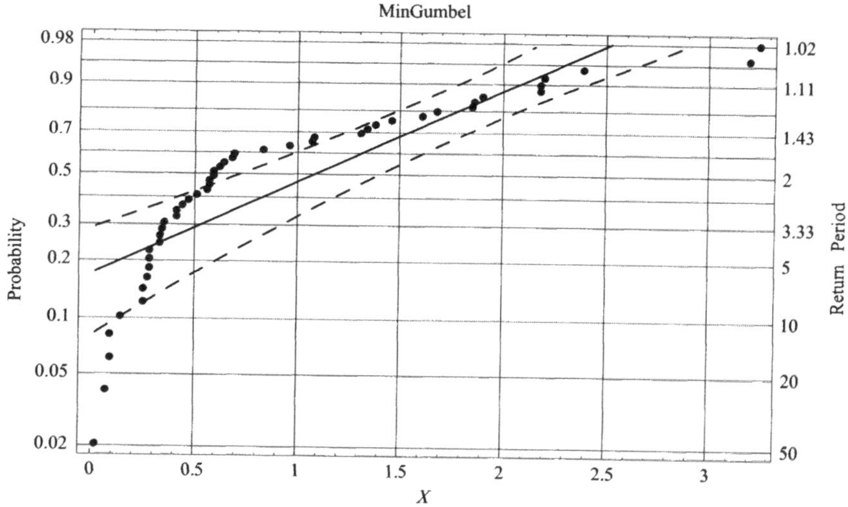

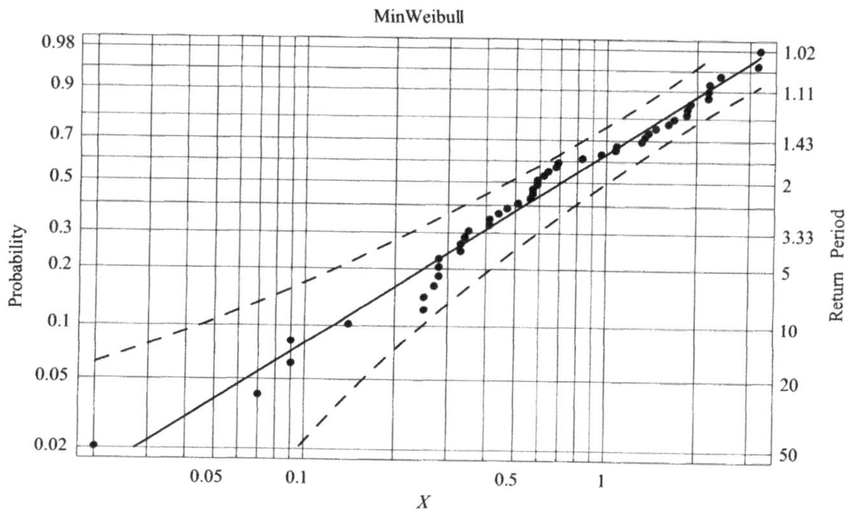

Figure 9.9: Times between 48 consecutive telephone calls (in seconds) on a minimal Gumbel and minimal Weibull probability papers.

Recall that a Gumbel-type cdf can be approximated as accurately as we desire by Weibull and Fréchet type cdfs. Thus, from a practical point of view, the wrong rejection of a Gumbel type distribution can be corrected in the estimation process that usually follows this decision. That is, if the true model is Gumbel and we used Fréchet or Weibull instead, we should find the estimate of κ to be close to zero.

The main drawback of this method is its subjectivity. Note that no precise criteria is given in order to know what negligible convexity means or what is exactly meant by tail. However, it has been proven very useful in practical applications because of its simplicity and accuracy.

9.5 Model Validation

In Section 6.3, we discussed the P-P and Q-Q plots as tools for model validation. The P-P plot is a scatter plot of the estimated versus the actual percentiles, that is, the scatter plot of

$$\hat{F}(x_{i:n}) \quad \text{versus} \quad p_{i:n}, \quad i = 1, 2, \ldots, n.$$

The Q-Q plot is a scatter plot of the estimated versus the observed quantiles:

$$\hat{F}^{-1}(p_{i:n}) \quad \text{versus} \quad x_{i:n}, \quad i = 1, 2, \ldots, n.$$

If the model fits the data well, the pattern of points in both plots would exhibit a 45-degree linear trend. Let us now examine the P-P and Q-Q plots for some of the data in Chapter 1.

Example 9.16 (Plots for the GEVD$_M$ distribution). The P-P and Q-Q plots for some of the data in Chapter 1 for which the right tail is of interest are given in Figures 9.10–9.13 for the MLE, PWM, EPM-TM, and the QLS. The plots for the EPM-MED is similar to those of the EPM-TM. As would be expected, the graphs that exhibit the most linear trend are those for the QLS method because the method minimizes the difference between the theoretical and observed quantiles. The P-P and Q-Q plots for all methods are similar except for the Men data where the fit of the PWM is not as good as those obtained by the other methods. ■

Example 9.17 (Plots for the GEVD$_m$ distribution). The P-P and Q-Q plots for some of the data in Chapter 1 for which the left tail is of interest are given in Figures 9.14 and 9.15. For space-saving purposes, we give the plots only for the MLE and the QLS. As can be seen from the graphs, the trend is linear for the chain and precipitation data, but deviation from linearity can be seen in the graphs for the epicenter and insulation data. ■

9.6 Hypothesis Tests for Domains of Attraction

The above graphical displays may be complemented by formal hypothesis testing procedures, as explained in the next section. As we have seen, the GEVD is the only limiting distribution for extrema in the case of iid samples and in many cases of dependent samples. But the GEVD has three special cases: Weibull, Gumbel, and Fréchet, that have very different physical meanings. In fact, we saw that a limited distribution in the tail of interest cannot be in the

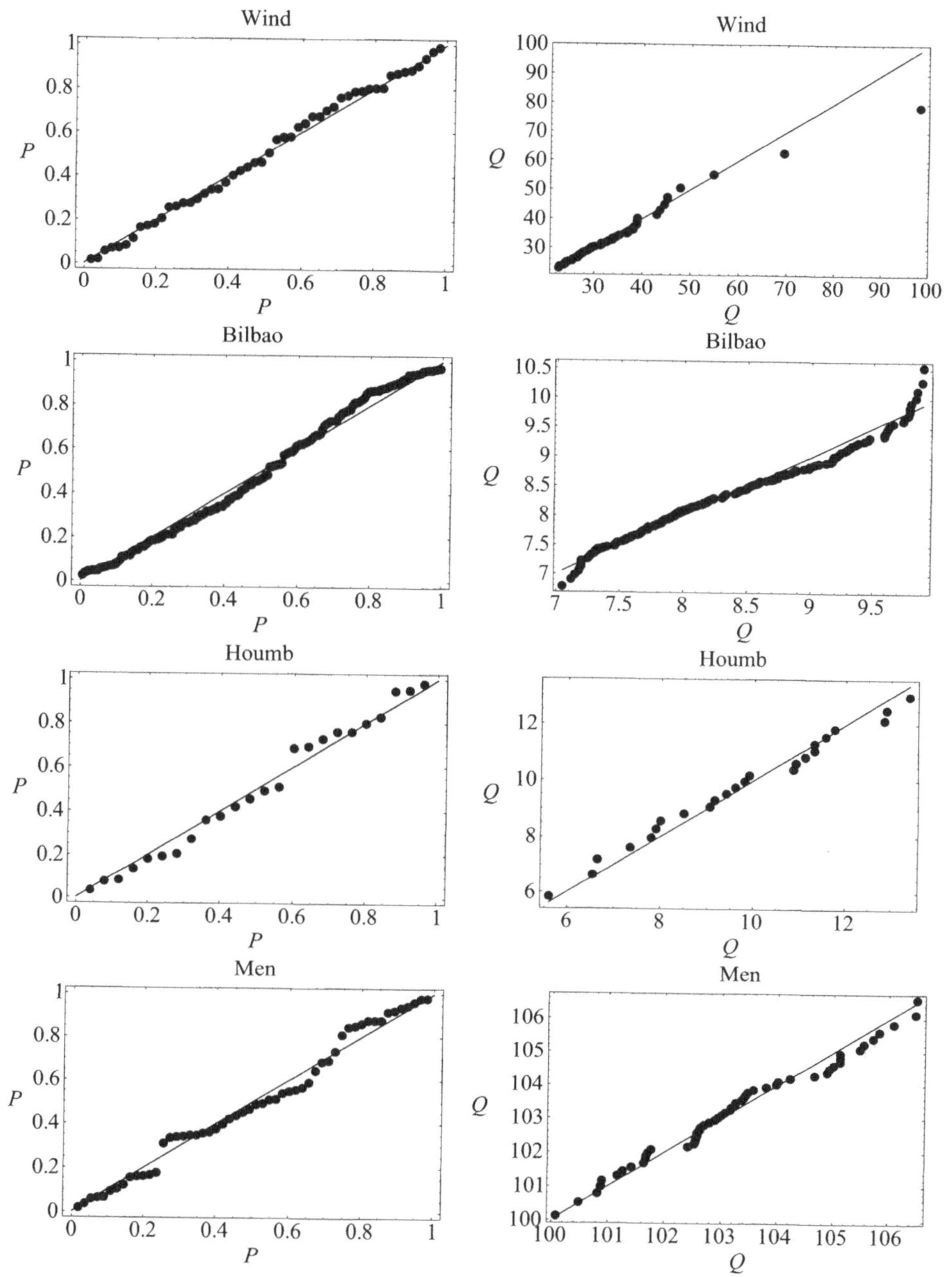

Figure 9.10: P-P and Q-Q plots obtained from fitting the maximal GEVD to four data sets using the maximum likelihood method.

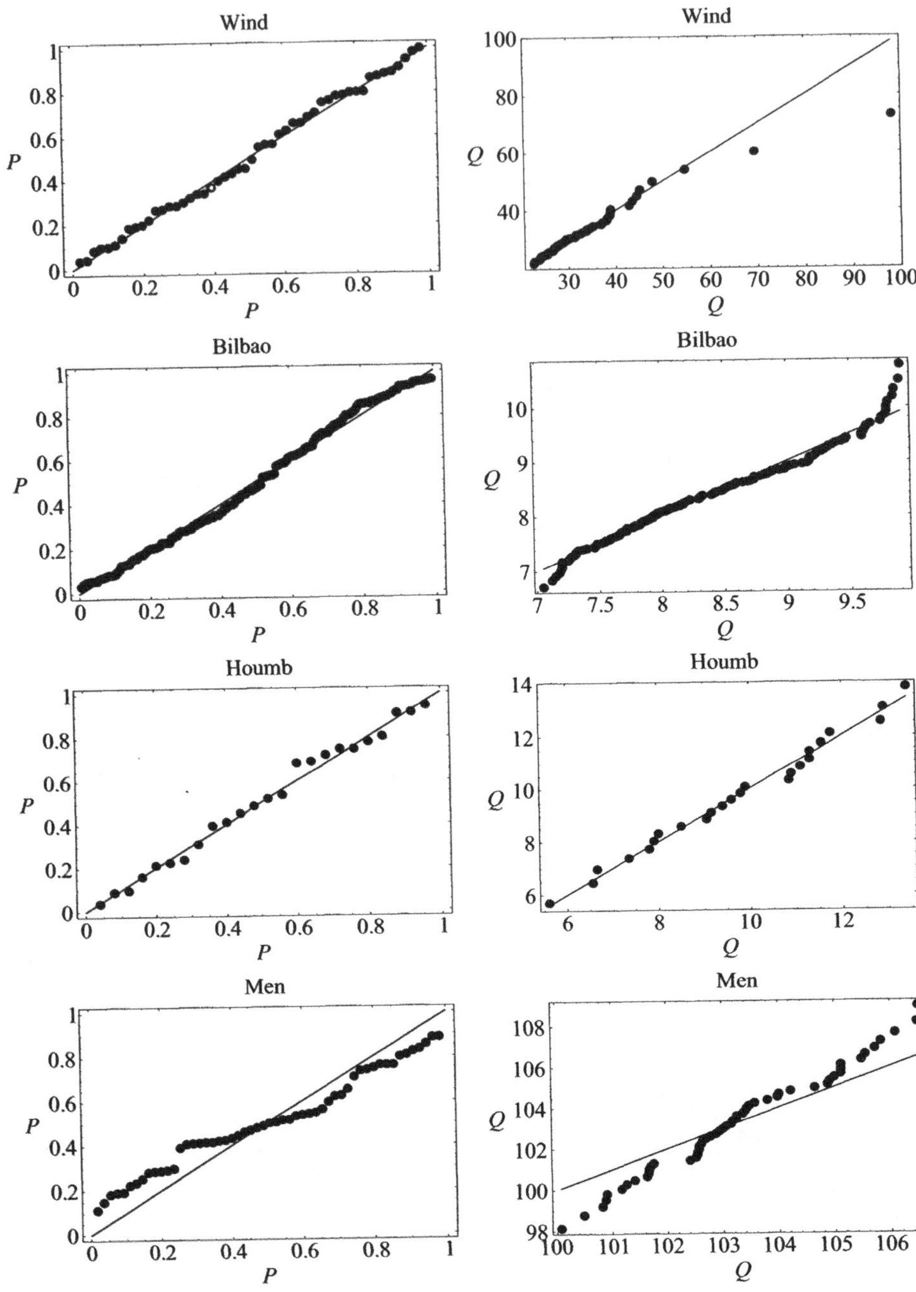

Figure 9.11: P-P and Q-Q plots obtained from fitting the maximal GEVD to four data sets using the PWM method.

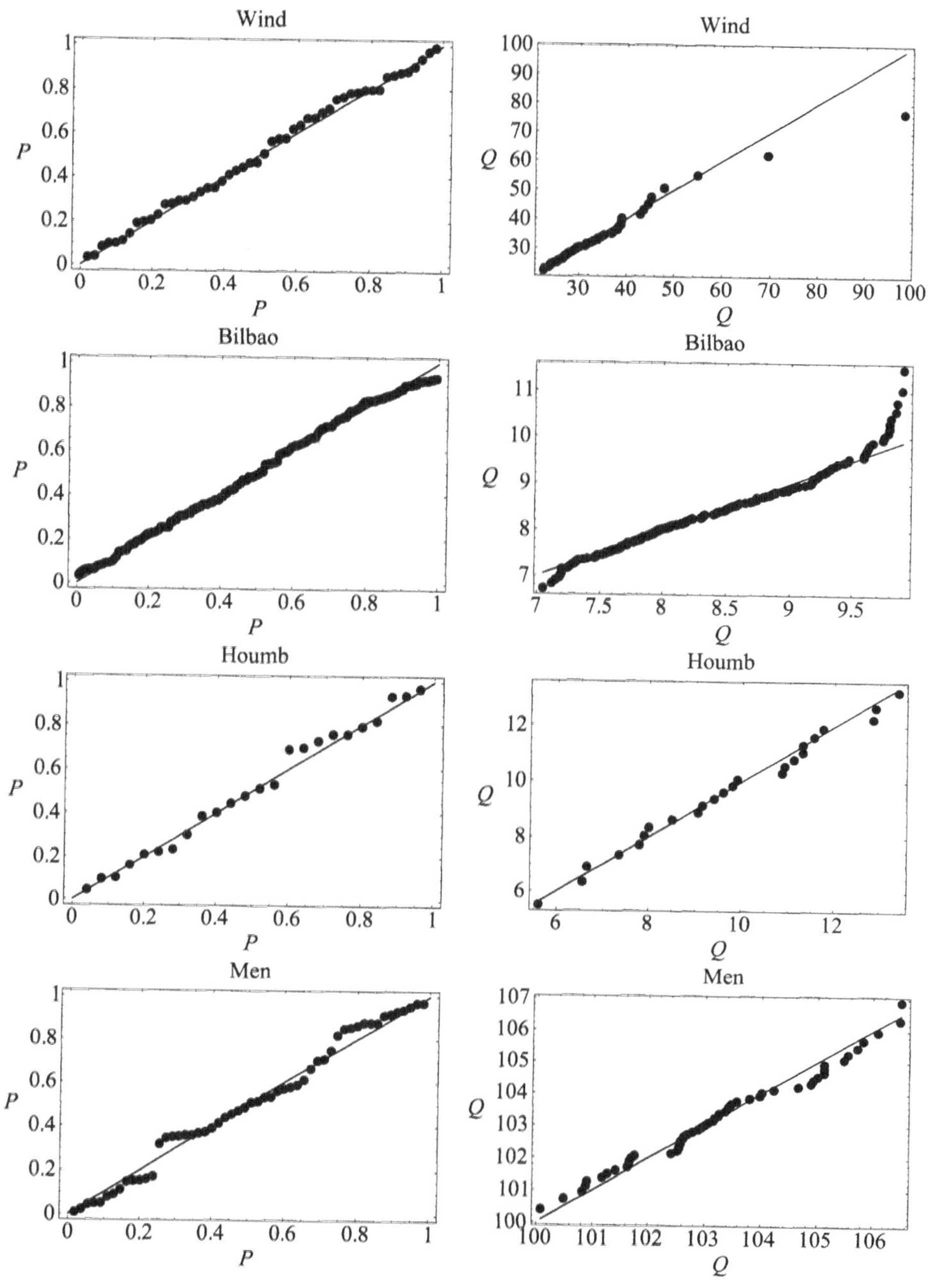

Figure 9.12: P-P and Q-Q plots obtained from fitting the maximal GEVD to four data sets using the EPM-TM method.

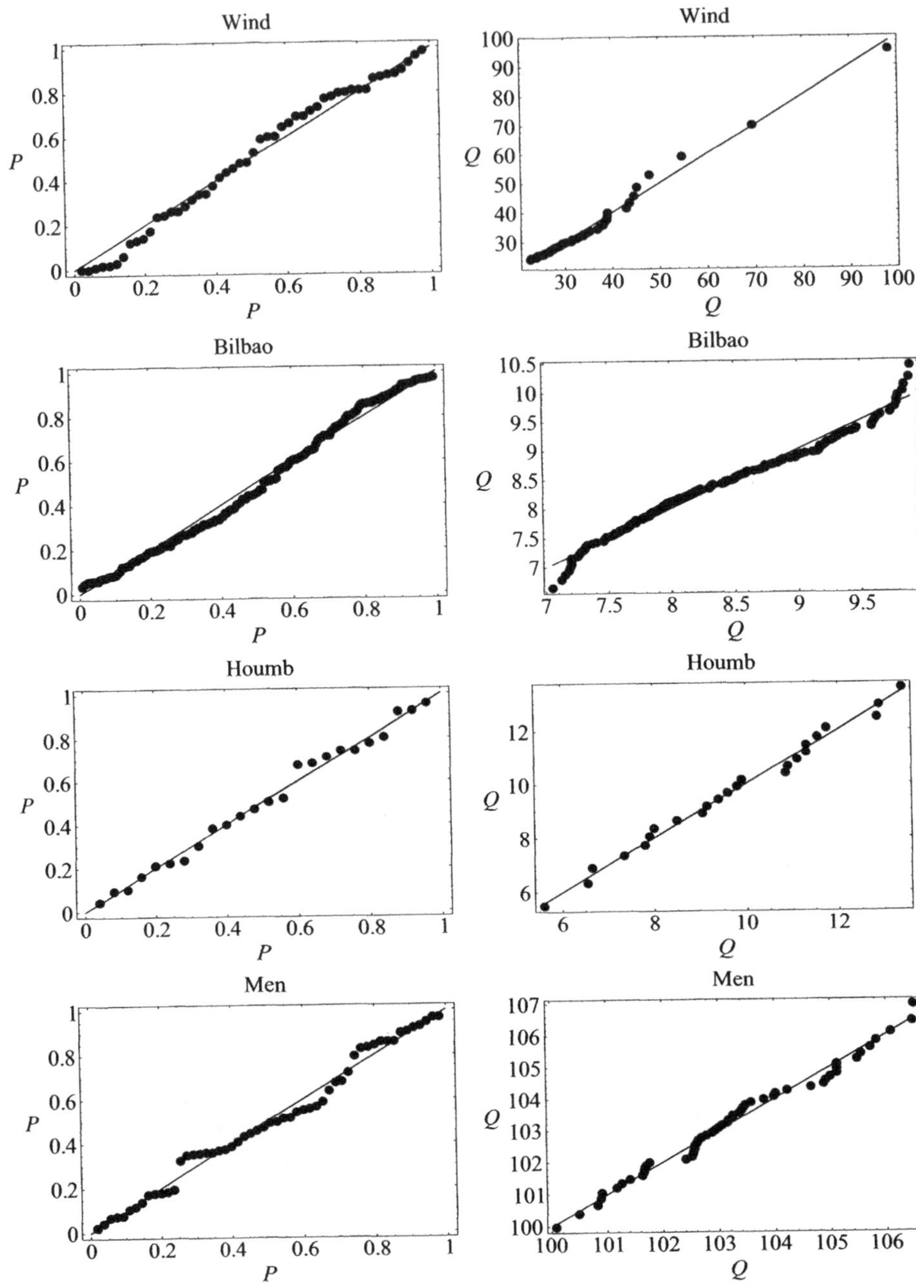

Figure 9.13: P-P and Q-Q plots obtained from fitting the maximal GEVD to four data sets using the QLS method.

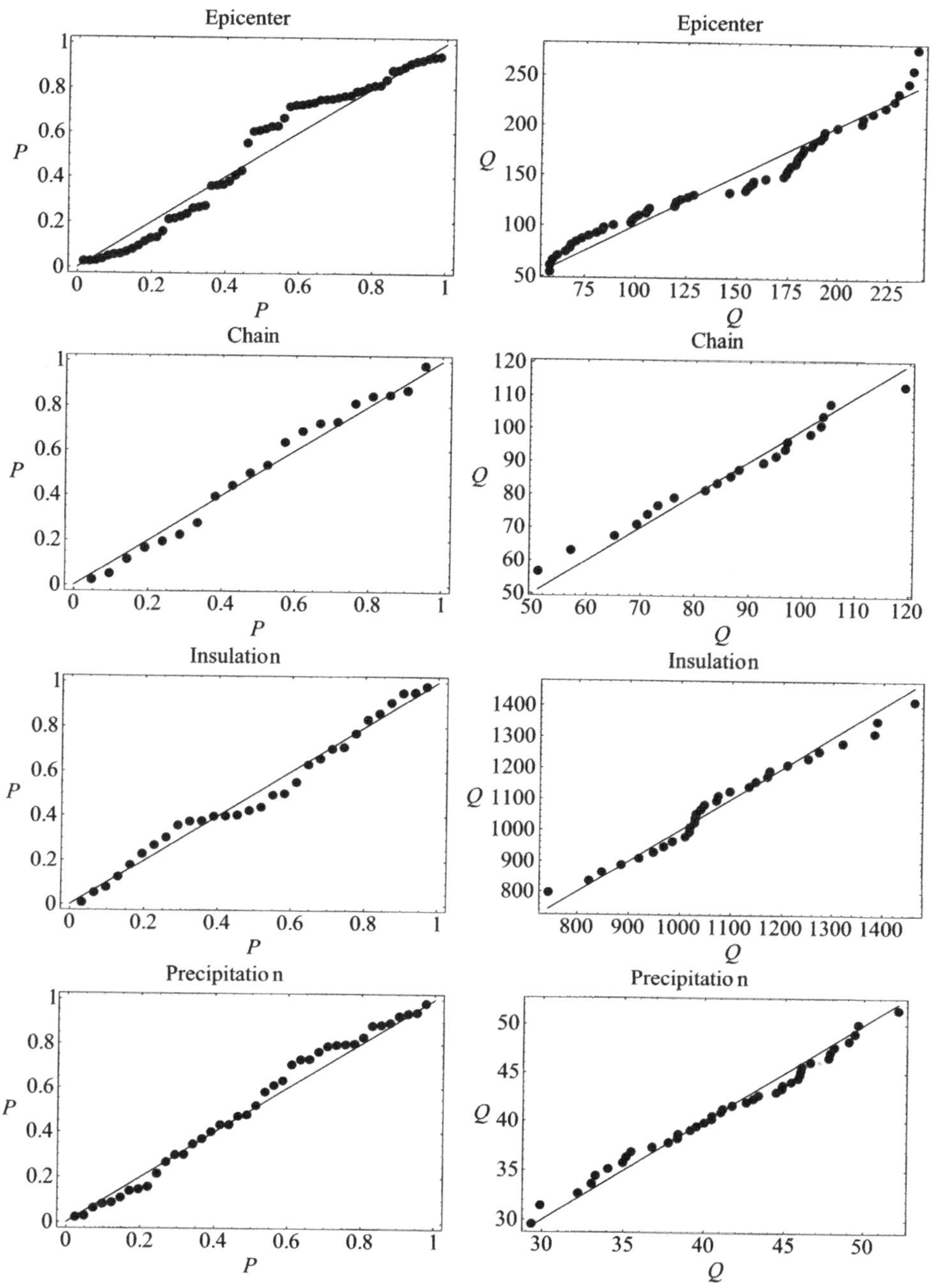

Figure 9.14: P-P and Q-Q plots obtained from fitting the minimal GEVD to four data sets using the maximum likelihood method.

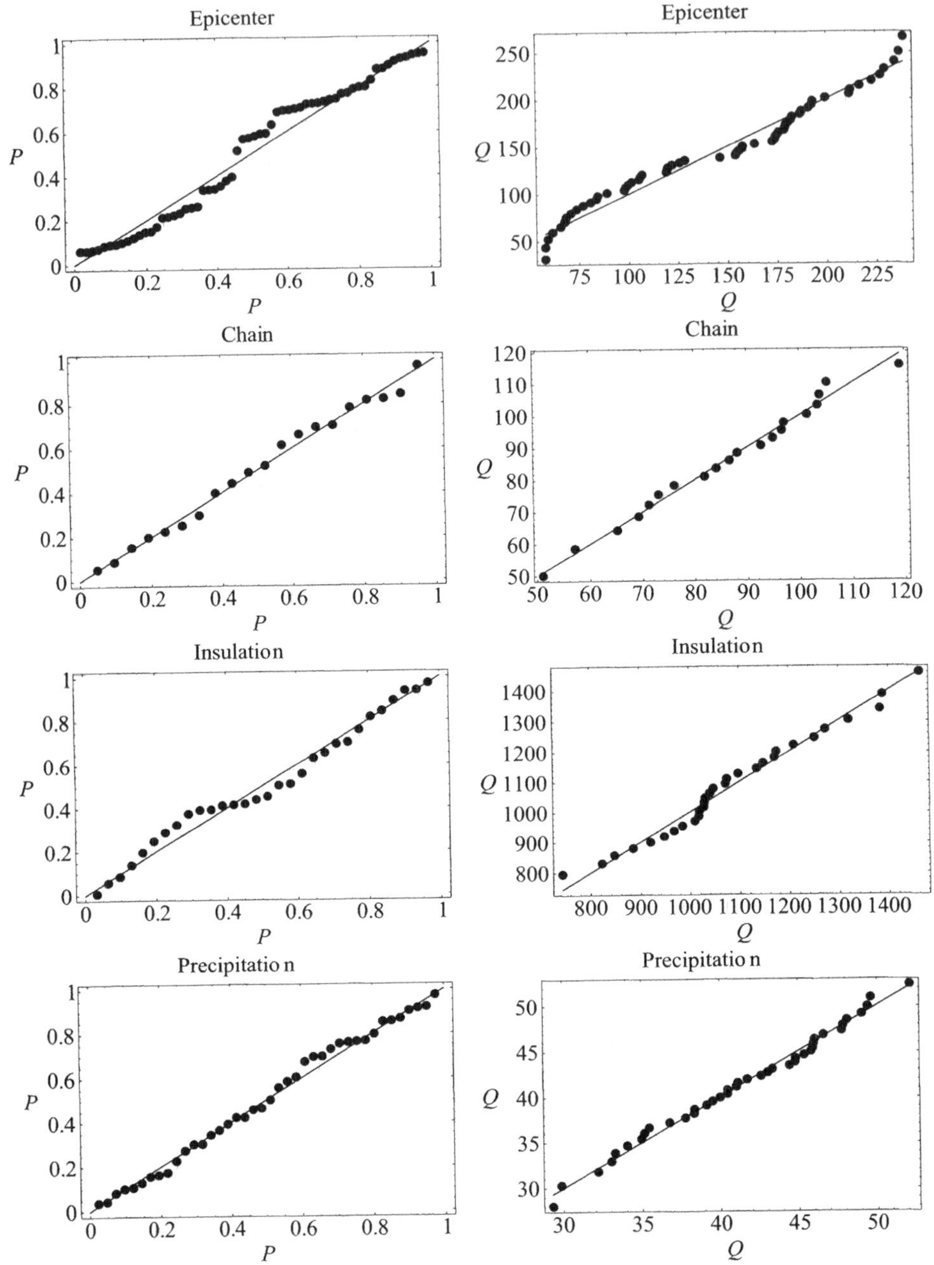

Figure 9.15: P-P and Q-Q plots obtained from fitting the minimal GEVD to four data sets using the QLS method.

Fréchet domain of attraction, and that an unlimited distribution cannot be in the Weibull domain of attraction. In addition, Fréchet models have heavier tails than the Gumbel models and the latter have heavier tails than the Weibull models. On the other hand, the Gumbel models are simpler because they have one less parameter than the Fréchet and Weibull models. Selection of the wrong model can be costly in practice because it leads to the selection of erroneous design parameters. Thus, the problem of identification of one of these three subfamilies is of practical as well as theoretical interest. This problem is known as model selection, that is, given the data, we need to decide which one of these models best fits the data.

In Section 9.4, we discussed the use of the probability plot papers for model selection. In this section, we use the hypothesis testing approach to test for domains of attraction. In particular, we discuss methods that can be used for the GEVD family to test whether the inclusion of the shape parameter κ improves the quality of the fit. Two approaches are presented: (a) methods based on likelihood and (b) methods based on curvature.

9.6.1 Methods Based on Likelihood

The classical large-sample likelihood theory provides asymptotically optimal tests of parametric hypothesis, and some variants can also be considered. In the GEVD, we wish to test

$$H_0 : \kappa = 0 \text{ (Gumbel)} \quad \textit{versus} \quad H_1 : \kappa \neq 0 \text{ (Fréchet or Weibull)}$$

based on a data set $\mathbf{x} = \{x_1, \ldots, x_n\}$. The log likelihood, $\ell(\mathbf{x}; \boldsymbol{\theta})$, is a function of $\boldsymbol{\theta} = (\lambda, \delta, \kappa)$. Let $\hat{\boldsymbol{\theta}}_0 = (\hat{\lambda}_0, \hat{\delta}_0, 0)$ and $\hat{\boldsymbol{\theta}}_1 = (\hat{\lambda}_1, \hat{\delta}_1, \hat{\kappa}_1)$ be the maximum likelihood estimates of $\boldsymbol{\theta}$ under H_0 and H_1, respectively (see Section 9.2.1). Note that under H_0, the estimate of κ is constrained to zero. From this, two asymptotically equivalent tests emerge:

1. Likelihood ratio test, and
2. Wald tests.

The Likelihood Ratio Test

This test compares the loglikelihood evaluated at $\hat{\boldsymbol{\theta}}_0$ with the loglikelihood evaluated at $\hat{\boldsymbol{\theta}}_1$, that is, it compares $\ell(\mathbf{x}; \hat{\boldsymbol{\theta}}_\mathbf{0})$ with $\ell(\mathbf{x}; \hat{\boldsymbol{\theta}}_\mathbf{1})$. Specifically, the likelihood ratio test is given by

$$LR = 2\{\ell(\mathbf{x}, \hat{\boldsymbol{\theta}}_\mathbf{1}) - \ell(\mathbf{x}, \hat{\boldsymbol{\theta}}_\mathbf{0})\}. \tag{9.100}$$

Under H_0, LR is a χ^2 with 1 degree of freedom.

Hosking (1984) suggests the following modification of the likelihood ratio test:

$$LR^* = \left(1 - \frac{2.8}{n}\right) LR. \tag{9.101}$$

This modification gives a more accurate approximation to the asymptotic distribution of LR. Thus, H_0 is rejected at the significance level α if

$$LR^* > \chi_1^2(1-\alpha), \tag{9.102}$$

where $\chi_1^2(1-\alpha)$ is the $(1-\alpha)$ quantile of the χ^2 distribution with 1 degree of freedom. The critical values $\chi_1^2(1-\alpha)$ can be obtained from Table A.3 in the Appendix.

Wald Tests

Wald tests compare $\hat{\kappa}_1$ with its estimated standard error, $\hat{\sigma}_{\hat{\kappa}_1}$. The standard error, $\hat{\sigma}_{\hat{\kappa}_1}$, is the square-root of the third diagonal element of the inverse of the information matrix. The *expected* version of the information matrix is

$$\mathbf{I}_E = -\frac{1}{n}E\left(\frac{\partial^2 \ell(\mathbf{x};\theta)}{\partial\theta\partial\theta^T}\right), \tag{9.103}$$

whose elements are given in (9.52). Its *observed* version is

$$\mathbf{I}_O = -\frac{1}{n}\left(\frac{\partial^2 \ell(\mathbf{x};\theta)}{\partial\theta\partial\theta^T}\right). \tag{9.104}$$

The Wald test statistic is then given by

$$W = \frac{1}{n}\frac{\hat{\kappa}_1^2}{\hat{\sigma}_{\hat{\kappa}_1}^2}. \tag{9.105}$$

Two versions of the Wald test in (9.105) emerge depending on whether one uses the expected information matrix in (9.103) or the observed information matrix in (9.104) for the standard error in (9.105).

As mentioned above, the likelihood ratio and Wald tests are asymptotically equivalent and asymptotically optimal for the test $H_0 : \kappa = 0$ versus $H_1 : \kappa \neq 0$. Under H_0, all statistics have a χ^2 distribution with 1 degree of freedom as the limit distribution. Thus, one can use the rejection rule in (9.102) with any of the above tests in place of LR^*.

For a one-sided hypothesis test, that is, if we wish to test

$$H_0 : \kappa = 0 \text{ (Gumbel)} \quad \text{versus} \quad H_1 : \kappa > 0 \text{ (Weibull)}, \tag{9.106}$$

or

$$H_0 : \kappa = 0 \text{ (Gumbel)} \quad \text{versus} \quad H_1 : \kappa < 0 \text{ (Fréchet)}, \tag{9.107}$$

the square roots of the above statistics may be used, with the sign of the square root being that of $\hat{\kappa}_1$. These statistics are also asymptotically optimal and have the standard normal distribution as the limit under H_0. Positive deviations indicate $\kappa > 0$ and negative deviations $\kappa < 0$, and the standard normal one-tailed critical values may be used. For example, H_0 in (9.106) is rejected if

$$\sqrt{LR^*} > \Phi(1-\alpha), \tag{9.108}$$

Table 9.14: Some Hypothesis Tests and Their P-Values (PV) for the Maximal GEVD for Some of the Data Sets in Chapter 1.

Right Tail								
Data Set	n	$\hat{\kappa}$	LR	PV	LR*	PV	Wald[1]	PV
Wind	50	-0.42	14.78	0.00	13.95	0.00	10.32	0.00
Bilbao	179	0.13	2.55	0.01	2.51	0.01	8.58	0.00
Men	54	0.24	3.62	0.00	3.43	0.00	10.76	0.00
Women	54	0.22	4.62	0.00	4.38	0.00	8.34	0.00
Flood	60	0.06	0.63	0.53	0.60	0.55	0.55	0.58
Wave	50	0.02	0.05	0.96	0.05	0.96	0.06	0.95
Left Tail								
Data Set	n	$\hat{\kappa}$	LR	PV	LR*	PV	Wald[1]	PV
Epicenter	60	0.49	6.48	0.00	6.18	0.00	88.67	0.00
Chain	20	0.23	1.29	0.20	1.11	0.27	3.38	0.00
Insulation	30	0.39	7.57	0.00	6.86	0.00	20.8	0.00
Precipitation	40	0.21	1.32	0.19	1.23	0.22	5.53	0.00

[1] Based on the observed information matrix in (9.104).

where $\Phi(1-\alpha)$ is the $(1-\alpha)$ quantile of the standard normal distribution. The critical values $\Phi(1-\alpha)$ can be obtained from Table A.1 in the Appendix.

Example 9.18 (Hypothesis tests). The results of the above hypothesis tests and their p-values (PV) are shown in Table 9.14 for some of the data sets in Chapter 1 for which the right (left) tail is of interest, respectively. The tests suggest Fréchet domain of attraction for the wind data, Gumbel domain of attraction for the flood and wave data sets, some doubts between Weibull and Gumbel for the chain data, and Gumbel for the remaining data sets. In general, the different tests agree with the exception of the chain and precipitation data sets.

■

9.6.2 The Curvature Method

The method to be described below has the same appealing geometrical property of the basic idea that is used for the probability paper method, that is, the statistics upon which a decision will be made is based on the tail curvature (see Castillo, Galambos, and Sarabia (1989)). This curvature can be measured in different ways. For example, by the difference or the quotient of slopes at two points. In addition, any of these two slopes can be measured by utilizing two or more data points. The latter option seems to be better in order to reduce variances.

Here we propose to fit two straight lines, by least-squares, to two tail intervals and to use the quotient of their slopes to measure the curvature. More precisely, we use the statistic

$$S = \frac{S_{n_1,n_2}}{S_{n_3,n_4}}, \tag{9.109}$$

where $S_{i,j}$ is the slope of the least-squares straight line fitted on Gumbel probability paper, to the rth order statistics with $i \leq r \leq j$. Thus, we can write

$$S_{ij} = \frac{m\Sigma_{11} - \Sigma_{10}\Sigma_{01}}{m\Sigma_{20} - \Sigma_{10}\Sigma_{01}}, \tag{9.110}$$

where

$$m = n_j - n_i + 1, \tag{9.111}$$

$$\Sigma_{10} = \sum_{k=n_i}^{n_j} \left\{ -\log\left[-\log\left(\frac{k-0.5}{n}\right)\right]\right\}, \tag{9.112}$$

$$\Sigma_{01} = \sum_{k=n_i}^{n_j} x_k, \tag{9.113}$$

$$\Sigma_{11} = \sum_{k=n_i}^{n_j} \left\{ -x_k \log\left[-\log\left(\frac{k-0.5}{n}\right)\right]\right\}, \tag{9.114}$$

$$\Sigma_{20} = \sum_{k=n_i}^{n_j} \left\{ -\log\left[-\log\left(\frac{k-0.5}{n}\right)\right]\right\}^2, \tag{9.115}$$

and n is the sample size.

An important property of the least squares slope S_{ij} is that it is a linear combination of order statistics with coefficients which add up to zero. This property makes the statistic S location and scale invariant.

The selection of n_1, n_2, n_3, and n_4 must be based on the sample size and the speed of convergence to the asymptotic distribution, which sometimes can be inferred from the sample. Apart from speed of convergence considerations, we have selected the following values when the right tail is of interest:

$$\begin{aligned} n_1 &= n + 1 - [2\sqrt{n}], \\ n_2 &= n_3 = n + 1 - \frac{[2\sqrt{n}]}{2}, \\ n_4 &= n, \end{aligned} \tag{9.116}$$

where $[x]$ means the integer part of x. The $\sqrt{n}$ is selected to ensure using only high-order statistics.

According to the above theory and with the values in (9.116), if the statistic S is well above 1, we can decide that the domain of attraction is Weibull type.

And, if it is well below 1, the decision is in favor of a Fréchet type. However, in order to be able to give significance levels of the test, we need to know the cdf of S. Due to the analytical difficulties associated with this problem, this distribution has been approximated by Monte Carlo simulation techniques assuming a maximal Gumbel parent. After 5,000 repetitions for samples of size $n = 10, 20, 40, 60, 80, 100$, and 200, the cdfs for S in Table 9.15 were obtained. From this table, critical values associated with given significance levels can be obtained. However, in selecting these values it must be taken into account that a wrong decision in rejecting a Gumbel type domain of attraction can, in many applications, be corrected, if estimation follows this decision. The asymptotic properties of this method have been studied by Castillo, Galambos, and Sarabia (1989).

Table 9.15: Simulated CDF of S (Maximal Gumbel Parent).

	Sample size						
CDF	10	20	40	60	80	100	200
0.01	0.119	0.157	0.223	0.260	0.279	0.302	0.363
0.02	0.157	0.201	0.271	0.307	0.331	0.353	0.412
0.05	0.239	0.286	0.357	0.398	0.419	0.442	0.498
0.1	0.340	0.390	0.457	0.497	0.515	0.537	0.588
0.2	0.520	0.563	0.615	0.652	0.665	0.678	0.716
0.3	0.702	0.730	0.763	0.790	0.798	0.803	0.825
0.5	1.157	1.128	1.087	1.088	1.074	1.056	1.045
0.7	1.951	1.758	1.554	1.503	1.459	1.393	1.327
0.8	2.716	2.314	1.935	1.833	1.760	1.650	1.531
0.9	4.438	3.447	2.628	2.421	2.282	2.089	1.870
0.95	6.817	4.833	3.403	3.047	2.814	2.533	2.214
0.98	11.554	7.219	4.560	3.975	3.601	3.154	2.661
0.99	17.125	9.512	5.649	4.815	4.299	3.648	3.024

Example 9.19 (Test for Gumbel versus GEVD). Table 9.16 gives the values of the S statistic, and the corresponding cdf, p-values, and domains of attractions resulting from the curvature method when applied to some data sets in Chapter 1. The corresponding domains of attractions have been determined using the p-values (see Table 9.15). Note that the S statistic is a unilateral test so that the p-value is the $\min(p, 1-p)$. For large S, we test Weibull versus Gumbel and for small S, we test Fréchet versus Gumbel.we test Weibull versus Gumbel. Note also that the p-values for the left tail data are obtained by computing S after changing sign of the data and treating them as right tail data. ∎

Table 9.16: The Values of the S Statistic and the Corresponding CDF, p-values (PV), and Domains of Attractions Resulting from the Curvature Method when Applied to some Data Sets in Chapter 1.

Tail	Data Set	n	S	CDF	PV	Domain of Attraction
Right	Wind	50	0.355	0.04	0.04	Fréchet
	Bilbao	179	6.115	1.00	0.00	Weibull
	Men	54	1.876	0.78	0.22	Gumbel
	Women	54	0.629	0.19	0.19	Gumbel
	Flood	60	0.562	0.14	0.14	Gumbel
	Wave	50	0.750	0.28	0.28	Gumbel
Left	Epicenter	60	7.075	1.00	0.00	Weibull
	Chain	20	1.490	0.63	0.37	Gumbel
	Insulation	30	1.587	0.69	0.31	Gumbel
	Precipitation	40	2.567	0.91	0.09	Gumbel

9.7 The Case of Dependent Observations

In the previous sections, we discussed the limit distributions of order statistics in the case of an iid sample. However, as we have stated in Section 7.6, one can argue that dependent observations are more likely in real practice than independent ones. Consequently, we need to know the limit distributions of the order statistics in the case of dependent observations.

Questions similar to those formulated for the case of independent observations also arise in the case of dependent observations. In particular, we address the following questions:

1. Is the GEVD family of distributions the only limit family?

2. Under what conditions do the limit distributions for the independent observations case remain valid for the case of dependent observations?

As would be expected, the dependent observations case is more complicated than the independent case. One of the main reasons for this is that, while the independent case can be formulated in terms of the marginal distributions $F_i(x_i)$, the dependence case requires more information about the joint distribution of the random variables involved. However, only partial knowledge is required. This implies that different joint distributions (different dependence conditions) can lead to the same limit distributions (this happens when the partial information required coincide for both cases).

We should also point out here that, unlike in the independent observations case, where only a limited family can arise as limit distributions, in the dependent observations case, any distribution can arise as the limit. Consider, for example, the esoteric case of a sequence of random variables $\{X_1, X_2, \ldots, X_n\}$, such that $X_i = X$ for all i. Then, we have $X_{i:n} = X$ for $i = 1, \ldots, n$, which implies that the limit distribution for any order statistic is $F(x)$, the cdf of X.

In the following sections, we discuss some important dependent sequences and the limit distributions of their order statistics under some dependence structures. These include the following sequences:

- Stationary sequences
- Exchangeable variables
- Markov sequences
- The m-dependent sequences
- Moving average sequences
- Normal sequences

9.7.1 Stationary Sequences

Some *stationary sequences* are important examples of dependent observations. To define the stationary sequences, we first need the definition of a condition known as the *$D(u_n)$ dependence condition*, which plays an important role, because the limit distributions for the maxima can be identified.

Definition 9.5 (The $D(u_n)$ dependence condition). *Let $\{u_n\}$ be a real sequence. The condition $D(u_n)$ is said to hold if for any set of integers $i_1 < i_2 < \ldots < i_p$ and $j_1 < j_2 < \ldots < j_q$ such that $j_1 - i_p \geq s \geq 1$, we have*

$$|F_{i_1,i_2,\ldots,i_p,j_1,j_2,\ldots,j_q}(u_n) - F_{i_1,i_2,\ldots,i_p}(u_n)F_{j_1,j_2,\ldots,j_q}(u_n)| \leq \alpha_{n,s},$$

where $\alpha_{n,s}$ is nondecreasing in s and

$$\lim_{n\to\infty} \alpha_{n,[n\delta]} = 0, \qquad \forall\ \delta > 0.$$

Note that for independent sequences, the dependence condition, $D(u_n)$, holds trivially with $\alpha_{n,s} = 0$.

Definition 9.6 (Stationary sequence). *A sequence $X_1,\ X_2,\ \ldots$ of random variables is called stationary if*

$$F_{i_1,i_2,\ldots,i_k}(x_1, x_2, \ldots, x_k) = F_{i_1+s,i_2+s,\ldots,i_k+s}(x_1, x_2, \ldots, x_k), \tag{9.117}$$

for every pair of integers k and s.

The following theorem gives the limit distributions for maxima of stationary sequences satisfying the $D(u_n)$ condition (see Leadbetter, Lindgren, and Rootzén (1983)).

Theorem 9.12 (Limit distributions of maxima: The $D(u_n)$ condition). *Let $\{X_n\}$ be a stationary sequence and let $\{a_n\}$ and $\{b_n\}$ be two sequences of real numbers such that*

$$\lim_{n\to\infty} \Pr(a_n + b_n X_{n:n} \le x) = F(x),$$

where $F(x)$ is a cdf. If the $D(u_n)$ dependence condition holds for the sequence $\{U_n = a_n + b_n X_{n:n}\}$ for each x, then $F(x)$ is one of the limit distributions for the independence case.

Example 9.20 (Marshall-Olkin model). Consider a longitudinal element (electric cord, railway conductor rail, wire, chain, etc.), hypothetically or really subdivided into n pieces of unit length. Assume that independent Poisson processes govern the occurrence of shocks destroying k consecutive pieces starting at the jth piece ($j = 1, 2, \ldots, n-k+1$, for $k = 1, 2, \ldots, n$). Assume further that the intensity of the processes are λ^k. This means that we have n Poisson processes of intensity λ destroying one piece, $(n-1)$ processes of intensity λ^2 destroying two pieces, and so on. Since the intensity of the Poisson processes must decrease as the damaged length increases, we must have $(\lambda < 1)$. Note also that, due to boundary effects, the extreme pieces are the strongest and the central are the weakest, because extreme pieces are affected by fewer processes than the central pieces.

The survival function of the n pieces is given by the Marshall-Olkin model

$$\begin{aligned} S(x_1, x_2, \ldots, x_n) &= \Pr(X_1 > x_1, X_2 > x_2, \ldots, X_n > x_n) \\ &= \exp\left[-\left(\lambda \sum_{i=1}^{n} x_i + \lambda^2 \sum_{i=1}^{n-1} \max(x_i, x_{i+1}) + \ldots \right.\right. \\ &\quad \left.\left. + \lambda^n \sum_{i=1}^{1} \max(x_i, x_{i+1}, \ldots, x_n)\right)\right], \end{aligned}$$

where $X_1, X_2, \ldots, X_n$ are the lifetime of the pieces. Hence, we have

$$\begin{aligned} S(x, x, \ldots, x) &= \exp\left\{-[\lambda n + \lambda^2(n-1) + \ldots + \lambda^n]x\right\} \\ &= \exp\left[\left(\frac{-\lambda(n(1-\lambda) - \lambda + \lambda^{n+1})}{(1-\lambda)^2}\right) x\right]. \end{aligned}$$

If the element is assumed to fail as soon as one of the pieces fails (series system), the lifetime of the element is that of its weakest piece. Thus, the cdf, $F(x)$, of its lifetime is given by

$$F(x) = 1 - S(x, x, \ldots, x) = 1 - \exp\left[\left(\frac{-\lambda(n(1-\lambda) - \lambda + \lambda^{n+1})}{(1-\lambda)^2}\right) x\right].$$

Choosing the constants

$$c_n = 0 \quad \text{and} \quad d_n = \frac{1-\lambda}{\lambda n},$$

we obtain

$$\lim_{n\to\infty} F(c_n + d_n x) = 1 - \exp(-x),$$

which proves that the limit distribution for the minimum is the exponential distribution, which is a special case of the Weibull distribution. ∎

Example 9.21 (Marshall-Olkin model). Assume now that Example 9.20 is modified in such a way that all pieces are affected by the same number of Poisson processes. Assuming that there are $(n+k-1)$ Poisson processes of intensity λ^k, the jth process destroying k consecutive pieces (real or hypothetical) starting at the piece number $j-k+1$, for $k = 1, 2, \ldots, n$. This means that any given piece is affected by one Poisson process destroying it alone, two Poisson processes destroying that piece and a contiguous one, three Poisson processes destroying that piece and two contiguous pieces, and so on.

In this case, the sequence $X_1, X_2, \ldots, X_n$ is stationary and we have

$$\begin{aligned} S(x_1, x_2, \ldots, x_n) &= \Pr(X_1 > x_1, X_2 > x_2, \ldots, X_n > x_n) \\ &= \exp\left[-\left(\lambda \sum_{i=1}^{n} x_i + \lambda^2 \sum_{i=0}^{n} \max(x_i, x_{i+1}) + \ldots \right.\right. \\ &\quad \left.\left. + \lambda^n \sum_{i=2-n}^{n} \max(x_i, x_{i+1}, \ldots, x_{i+n-1})\right)\right], \end{aligned}$$

where $x_i = 0$ if $i < 1$ or $i > n$. Condition $D(u_n)$ holds because we have

$$|S_{i_1,i_2,\ldots,i_p,j_1,j_2,\ldots,j_q}(u_n) - S_{i_1,i_2,\ldots,i_p}(u_n) S_{j_1,j_2,\ldots,j_q}(u_n)| \le 1 - \exp[-\lambda^{s+1} p(\lambda) u_n],$$

where $p(\lambda)$ is a polynomial of order less than $p+q$ and s is as defined in Definition 9.5. If now we choose $u_n = c_n + d_n x$ with $c_n = 0$ and $d_n = 1/n$, we get

$$\alpha_{n,s} = 1 - \exp\left(-\lambda^{s+1} p(\lambda) \frac{x}{n}\right),$$

and

$$\lim_{n\to\infty} \alpha_{n,[n\delta]} = 1 - \exp\left(-\lambda^{n\delta+1} p(\lambda) \frac{x}{n}\right) = 0, \qquad \forall\, \delta > 0.$$

This shows that condition $D(u_n)$ is satisfied. Thus, according to Theorem 9.12, the limit distribution coincides with that of the independent observations case.

The same conclusion could have been obtained by noting that the cdf of the lifetime of the element is

$$\begin{aligned} F(x) &= 1 - S(x, x, \ldots, x) \\ &= 1 - \exp\left\{-[\lambda + \lambda^2(n+1) + \ldots + \lambda^n(2n-1)]x\right\} \\ &= 1 - \exp\left[-\lambda\left(\frac{n(1-2\lambda^n)}{1-\lambda} + \frac{\lambda(1-\lambda^n)}{(1-\lambda)^2}\right)x\right], \end{aligned}$$

from which one gets

$$\lim_{n\to\infty} F\left(\frac{(1-\lambda)x}{\lambda n}\right) = 1 - \exp(-x),$$

which is the exponential distribution (a special case of the Weibull distribution). ■

9.7.2 Exchangeable Variables

Another important dependent sequence is the exchangeable variables sequence.

Definition 9.7 (Exchangeable variables). *The random variables X_1, X_2, ..., X_n are said to be exchangeable if the distribution of the vector X_{i_1}, X_{i_2}, ..., X_{i_n} is the same for all possible permutation of the indices $\{i_1, i_2, \ldots, i_n\}$.*

It is clear that exchangeable random variables are stationary.

Consider, for example, a system of n identical elements such that all of them are under the same working conditions. Then their lifetimes $X_1, X_2, \ldots, X_n$ are exchangeable random variables.

Definition 9.8 (Exchangeable events). *The events $C_1, C_2, \ldots, C_n$ are said to be exchangeable if the probability $\Pr(C_{i_1}, C_{i_2}, \ldots, C_{i_k})$ is the same for all possible permutations of the indices $\{1 \le i_1 < i_2 < \ldots < i_k \le n\}$.*

Galambos (1987) shows that for any given set of events $A = \{A_1, A_2, \ldots, A_n\}$, a set of exchangeable events $C = \{C_1, C_2, \ldots, C_n\}$ can be found such that

$$\Pr(m_n(A) = t) = \Pr(m_n(C) = t),$$

where $m_n(A)$ and $m_n(C)$ represent the number of events of the sets A and C that occur.

The main practical implication of this is that a set of events can be replaced by an exchangeable set for which the calculations of probabilities become easier. In fact, for exchangeable events one has

$$S_{k,n}(C) = \binom{n}{k} \Pr(C_1, C_2, \ldots, C_k) = \binom{n}{k} \alpha_k,$$

where $S_{k,n}(C)$ is as defined in (7.30) and $\alpha_k = \Pr(C_1, C_2, \ldots, C_k)$.

If the set of random variables can be extended to a larger set of N exchangeable variables, the following theorem gives the limit distribution for the order statistics.

Theorem 9.13 (Limit distributions for exchangeable variables). *Let $X_1, \ldots, X_n$ be such that there exist additional random variables X_{n+1}, X_{n+2}, ..., X_N, with distributions which satisfy*

$$S_{k,N}(x) = \frac{\binom{N}{k} S_{k,n}(x)}{\binom{n}{k}}$$

and

$$\lim_{n\to\infty} \frac{N}{n} = \infty.$$

Then, there exist constants a_n and $b_n > 0$ such that

$$\lim_{n\to\infty} \Pr(a_n + b_n X_{n-k+1:n} < x) = A_k(x)$$

iff $\Pr\left(m_N(a_n + b_n X) < \frac{Ny}{n}\right)$ *converges in distribution to $U(y) = U(y; x)$, and*

$$A_k(x) = \sum_{t=0}^{k-1} \int_0^\infty \frac{y^t \exp(-y) dU(y; x)}{t!}.$$

Example 9.22 (Mardia's multivariate distribution). Assume a sequence $\{X_n\}$ of random variables such that the joint distribution of the first n terms in the sequence follows a Mardia's distribution with survival function

$$S(x_1, x_2, \ldots, x_n) = \left[\sum_{j=1}^{n} \exp\left(\frac{x_j}{w_n}\right) - n + 1\right]^{-w_n}$$

and such that

$$\lim_{n\to\infty} w_n = \infty.$$

Note that this distribution has unit exponential marginals and that it is a sequence of exchangeable variables.

The minimum order statistic $X_{1:n}$ has cdf

$$L_n(x) = 1 - \left[n \exp\left(\frac{x}{w_n}\right) - n + 1\right]^{-w_n} = 1 - \left\{1 - \exp\left(\frac{x}{w_n}\right)\right\}^{-w_n}.$$

Thus, we can write

$$\lim_{n\to\infty} L_n(c_n + d_n x) = \lim_{n\to\infty} 1 - \left\{1 - \exp\left(\frac{c_n + d_n x}{w_n}\right)\right\}^{-w_n}.$$

From this, it is clear that we get a nondegenerate limit distribution if

$$\lim_{n\to\infty} \frac{n(c_n + d_n x)}{w_n} = 0.$$

Hence, we have

$$\begin{aligned} \lim_{n\to\infty} L_n(c_n + d_n x) &= \lim_{n\to\infty} \left[1 - \left(1 + \frac{n(c_n + d_n x)}{w_n}\right)^{-w_n}\right] \\ &= \lim_{n\to\infty} \{1 - \exp[-n(c_n + d_n x)]\}. \end{aligned}$$

We can choose $c_n = 0$ and $d_n = 1/n$, so that

$$\lim_{n\to\infty} L_n(c_n + d_n x) = 1 - \exp(-x),$$

which is the exponential distribution with parameter $\lambda = 1$. ∎

9.7.3 Markov Sequences of Order p

Definition 9.9 (Markov sequence of order p). *A sequence $\{X_n\}$ of random variables is said to be a Markov sequence of order p iff $(\ldots, X_{m-1}, X_m)$ is independent of $(X_{m+r}, X_{m+r+1}, \ldots)$, given $(X_{m+1}, X_{m+2}, \ldots, X_{m+p})$ for any $r > p$ and $m > 0$.*

In practical terms, this definition means that the past $(\ldots, X_{m-1}, X_m)$ and the future $(X_{m+r}, X_{m+r+1}, \ldots)$ are independent, given the present $(X_{m+1}, X_{m+2}, \ldots, X_{m+p})$. In other words, that knowledge of the past does not add new information on the future if the present is known, which is the well-known Markov property.

Remark 9.2 (Condition $D(u_n)$ for Markov sequences of order 1). *The $D(u_n)$ condition holds for any stationary Markov sequence of order 1 with cdf $F(x)$ such that*

$$\lim_{n\to\infty} F(u_n) = 1.$$

This remark allows applying the results in Section 9.7.1 to the case of Markov sequences of order 1.

9.7.4 The m-Dependent Sequences

Definition 9.10 (The m-dependent sequences). *A sequence $\{X_n\}$ of random variables is said to be an m-dependent sequence iff the random variable $(X_{i_1}, X_{i_2}, \ldots, X_{i_k})$ and $(X_{j_1}, X_{j_2}, \ldots, X_{j_k})$ are independent for*

$$\min(j_1, j_2, \ldots, j_k) - \max(i_1, i_2, \ldots, i_k) \geq m.$$

Note that X_i and X_j can be dependent if they are close ($|i - j| < m$) but they are independent if they are far apart.

Theorem 9.14 (Limit distributions for m-dependent sequences). *Let $X_1, X_2, \ldots, X_n$ be a m-dependent stationary sequence with common $F(x)$ such that*

$$\lim_{n\to\infty} n[1 - F(a_n + b_n x)] = w(x), \quad 0 < w(x) < \infty.$$

Then,

$$\lim_{n\to\infty} \Pr(a_n + b_n Z_n < x) = \exp[-w(x)]$$

iff

$$\lim_{w\to u(F)} \frac{\Pr(X_1 \geq w, X_i \geq w)}{1 - F(w)} = 0, \quad 1 \leq i < m, \tag{9.118}$$

where $u(F)$ is the upper end of $F(\cdot)$.

Example 9.23 (Marshall-Olkin model). Assume an m-dependent sequence $\{X_n\}$ of random variables such that the joint survival function of any m consecutive elements $\{X_i, X_{i+1}, \ldots, X_{i+m-1}\}$ in the sequence is the Marshall-Olkin survival function

$$S(x_1, x_2, \ldots, x_n) = \exp\left(-\lambda_1 \sum_{i=1}^{n} x_i - \lambda_2 \sum_{i<j} \max(x_i, x_j)\right).$$

Then, condition (9.118) becomes

$$\begin{aligned}\lim_{w\to\infty} \frac{\Pr(X_1 \geq w, X_i \geq w)}{1 - F(w)} &= \lim_{w\to\infty} \frac{\exp[-(2\lambda_1 + \lambda_2)w]}{\exp[-(\lambda_1 + \lambda_2)w]} \\ &= \lim_{w\to\infty} \exp[-\lambda_1 w] = \begin{cases} 0, & \text{if } \lambda_1 > 0, \\ 1, & \text{if } \lambda_1 = 0. \end{cases}\end{aligned}$$

Consequently, the limit distribution of the maximum of $\{X_n\}$ coincides with the independent case (Gumbel type) only if $\lambda_1 > 0$.

■

9.7.5 Moving Average Models

Definition 9.11 (Finite moving average stationary models). *A moving average model, denoted by MA(q), is a model of the form*

$$X_t = \epsilon_t + c_1\epsilon_{t-1} + c_2\epsilon_{t-2} + \ldots + c_q\epsilon_{t-q},$$

where $\{\epsilon_t\}$ is a sequence of iid random variables.

It is clear that a moving average model is m-dependent. The next theorem gives the asymptotic distribution of sequences of moving average models (see Leadbetter, Lindgren, and Rootzén (1983)).

Theorem 9.15 (Limit distributions of MA models). *Consider the following moving average model*

$$Y_t = \sum_{i=-\infty}^{\infty} C_i X_{t-i}, \tag{9.119}$$

where the X_t for $t > 1$ are independent and stable random variables, that is, with characteristic function (see (3.4.3)) of the form

$$\psi(t) = \exp\left[-\gamma^\alpha |t|^\alpha \left(1 - \frac{i\beta h(t,\alpha)t}{|t|}\right)\right],$$

where

$$h(t,\alpha) = \begin{cases} \dfrac{2\log|t|}{\pi}, & \text{if } \alpha = 1, \\ \tan\left(\dfrac{\pi\alpha}{2}\right), & \text{otherwise,} \end{cases}$$

and

$$\gamma \geq 0, \quad 0 < \alpha \leq 2, \quad |\beta| \leq 1.$$

Assume also that the constants C_i $(-\infty < i < \infty)$ satisfy

$$\sum_{i=-\infty}^{\infty} |C_i|^\alpha < \infty$$

and

$$\sum_{i=-\infty}^{\infty} C_1 \log |C_i| < \infty, \quad \textit{if } \alpha = 1, \ \beta \neq 0.$$

Then, we have

$$\lim_{n\to\infty} \Pr(X_{n:n} \leq n^{1/\alpha} x) = \begin{cases} \exp\left\{-K_\alpha[c_+^\alpha(1+\beta) + c_-^\alpha(1-\beta)]x^{-\alpha}\right\}, & \textit{if } x > 0, \\ 0, & \textit{if } x \leq 0, \end{cases}$$

where

$$c_+ = \max_{-\infty<i<\infty} \max(0, C_i),$$

$$c_- = \max_{-\infty<i<\infty} \max(0, -C_i),$$

and

$$K_\alpha = \frac{1}{\pi}\Gamma(\alpha)\sin\left(\frac{\alpha\pi}{2}\right).$$

Note that the normal distribution is a particular case of a stable distribution ($\alpha = 2$). Note also that the *autoregressive moving average model* (ARMA) model is a moving average model of the form (9.119), where the coefficients C_i can be obtained by inverting the original ARMA model.

9.7.6 Normal Sequences

An important case of stationary sequences is the Gaussian sequences for which we have the following results (see Galambos (1978)).

Theorem 9.16 (Asymptotic distribution of normal sequences). *If $\{X_n\}$ is a stationary sequence of standard normal random variables with correlation function*

$$r_m = \rho[X_j, X_{j+m}],$$

we have that:

1. *If $\lim_{m\to\infty} r_m \log m = 0$, then*

$$\lim_{n\to\infty} \Pr(Z_n < a_n + b_n x) = H_\infty(x).$$

2. *If* $\lim_{m\to\infty} r_m \log m = \tau$, $0 < \tau < \infty$, *then*

$$\lim_{n\to\infty} \Pr(Z_n < a_n + b_n x) = H(x),$$

where $H(x)$ *is the convolution of* $H_0(x+\tau)$ *and* $\Phi(x(2\tau)^{-1/2})$, *where the convolution of two functions f(x) and g(x) is given by*

$$f * g = \int f(\tau)g(t-\tau)d\tau.$$

3. *If* $\lim_{m\to\infty} r_m \log m = \infty$, $\lim_{m\to\infty} r_m(\log m)^{1/3} = 0$, *and* $\{r_m\}$ *is decreasing, then*

$$\lim_{n\to\infty} \Pr(Z_n < (1-r_n)^{1/2} a_n + b_n x r_n^{1/2}) = \Phi(x),$$

where

$$a_n = \frac{1}{b_n} - \frac{b_n[\log\log n + \log(4\pi)]}{2}, \quad (9.120)$$

$$b_n = (2\log n)^{-1/2}. \quad (9.121)$$

Example 9.24 (ARMA(p, q) Box-Jenkins model). Consider a sequence $\{X_n\}$, which follows an $ARMA(p, q)$ Box-Jenkins autoregressive moving average model of order p and q of the form

$$X_t = \phi_1 X_{t-1} + \ldots + \phi_p X_{t-p} + \epsilon_t - \theta_1 \epsilon_{t-1} - \ldots - \theta_q \epsilon_{t-q},$$

where $\phi_1, \phi_2, \ldots \phi_p$ and $\theta_1, \theta_2, \ldots, \theta_q$ are constants and $\{\epsilon_t\}$ is a sequence of iid $N(0, \sigma^2)$ random variables. Assume that the above model is stationary and invertible, that is, the roots of the polynomials

$$p(x) = x^p - \phi_1 x^{p-1} - \ldots - \phi_p$$

and

$$q(x) = x^q - \theta_1 x^{q-1} - \ldots - \theta_q$$

lie inside the unit circle. Then the *autocorrelation* function $r_k = E(X_j X_{j+k})$ satisfies the difference equation

$$r_k = \phi_1 r_{k-1} + \phi_2 r_{k-2} + \ldots + \phi_p r_{k-p}, \quad k > q+1,$$

which has a solution

$$r_k = \sum_{s=1}^{r} \left((m_s)^k \left[\left(\sum_{j=0}^{p_s-1} c_{ij} k^j \right) \cos(k\alpha_s) + \left(\sum_{j=0}^{p_s-1} d_{ij} k^j \right) \sin(k\alpha_s) \right] \right),$$

where c_{ij} and d_{ij} are constants, m_s and α_s are the modulus and argument, respectively, of the r roots of the polynomial $p(x)$, and p_s are their degrees of multiplicity, which satisfy the equation

$$\sum_{s=1}^{r} p_s = p.$$

It is clear that

$$\lim_{m \to \infty} r_m \log m = 0.$$

Then, if σ^2 is chosen such that X_t is $N(0,1)$, Theorem 9.16 guarantees that the limit distribution of the maximum order statistic $X_{n:n}$ is

$$\lim_{n \to \infty} \Pr(Z_n < a_n + b_n x) = \exp[-\exp(-x)],$$

with a_n and b_n given by (9.120) and (9.121), respectively. ■

Leadbetter, Lindgren, and Rootzén (1983) show that if

$$\lim_{m \to \infty} r_m \log m = 0$$

and $n[1 - \Phi(u_n)]$ is bounded, then the condition $D(u_n)$ holds for normal sequences.

Exercises

9.1 Derive the quantiles in (9.5) and (9.8) from their corresponding distributions.

9.2 Prove the converse result in Theorem 9.3.

9.3 Obtain the Weibull, Gumbel, and Fréchet distributions for minima from the Weibull, Gumbel, and Fréchet distributions for maxima, and vice versa using Theorem 9.3.

9.4 Prove the maximum stability of the maximal Fréchet family, that is, that the maximum of two Fréchet independent distributions is a Fréchet distribution. Show that the minimum Fréchet does not have such a property.

9.5 Prove the minimum stability of the minimal Weibull family, that is, that the minimum of two Weibull independent distributions is a Weibull distribution. Show that the maximal Weibull does not have such a property.

9.6 Is there a family of distributions stable with respect to max and min operations? If it exists, obtain such a family.

9.7 Determine the limit distributions for maxima and minima of a distribution with pdf $f(x) = 2x, if\ 0 \le x \le 1$, and obtain sequences of constants such that (9.1) and (9.2) hold.

9.8 Obtain the limit distribution for maxima and minima of the uniform random variable and one of the corresponding sequences of normalizing constants. Can any of the three candidates be eliminated for each of the cases (maxima and minima)?

9.9 In Examples 9.4–9.7, we derived the domain of attraction for some of the distributions in Table 9.5 and their corresponding normalizing constants. Show that the domains of attraction for the other distributions are as presented in Table 9.5 and find their corresponding normalizing constants.

9.10 Obtain the asymptotic joint distribution of the order statistics $r_1(n)$ and $r_2(n)$ from a uniform distribution if $r_1(n) = 0.2n$ and $r_2(n) = 0.3n$. Do the same if $r_1(n) = 3\sqrt{n}$ and $r_2(n) = r_1(n) = 5\sqrt{n}$.

9.11 The lifetimes of 40 test items subjected to a fatigue test are shown in Table 1.12.

(a) Determine the lifetime associated with a probability of failure of 10^{-4}.

(b) Compute the above probability if the real structure has a length 1000 times more than the one of the test items.

(c) Discuss the independence assumption and try to improve it by suggesting an alternative.

9.12 Give real examples (e.g., in engineering and science) where the independence assumption does not hold for maxima and minima. What is the alternative for the limit distribution in this case?

9.13 Give real examples (e.g., in engineering and science) of exchangeable and not exchangeable random variables.

9.14 Derive the formulas for confidence intervals of the percentiles of the Gumbel distribution.

9.15 Derive formulas for the QLS estimates for the GEVD from (9.94) and (9.95) by differentiating with respect to the parameters. Are they closed or implicit formulas?

9.16 Plot two of the data sets in Chapter 1 on a Gumbel probability paper and decide about its domain of attraction based on the tail curvature. If it is not Gumbel type, suggest a location parameter value, plot again the data on the adequate paper and check its domain of attraction type. If the result is not satisfactory, change the location parameter value until you obtain a satisfactory graph.

9.17 Use the wind data to test the assumption that it comes from a Gumbel parent versus a GEVD parent. Use different tests of hypothesis, compare the results, and make conclusions accordingly.

9.18 Use the curvature method to test the domain of attraction of two data sets in Chapter 1. Use different subsets of data, compare the results and state your conclusions.

9.19 Give examples where we have interest in order statistics of low and high order. Discuss whether or not this definition is arbitrary in real examples where a single sample is considered.

Chapter 10

Limit Distributions of Exceedances and Shortfalls

In Chapter 9, we studied the limit distributions of order statistics. Sometimes, it is more useful to analyze the values of random variables that exceed or fall below a given threshold value. For example, in modeling rain falls or wave heights, it would be better to have data on yearly exceedances over a threshold (all rain falls that exceed a certain amount or all wave heights that exceed a certain height) than to have data that consist of only yearly maximum rain fall over a period of n years. This is because maxima in some years can be much below several high-order statistics in other years. Thus, in many practical applications an important part of the information (large (small) values other than the maxima (minima) occurring during the same year) would be lost if we use only extremes.

Observations that exceed a given threshold u are called *exceedances* over a threshold. Lacking better terminology, we refer to observations that fall below a given threshold u as *shortfalls.*

As has been indicated, exceedances and shortfalls play an important role in extremes. Thus, their extremal behavior is of a great practical importance. This chapter is devoted to the limit distributions of exceedances and shortfalls.

This chapter is organized as follows. Sections 10.1 and 10.2 show how exceedances and shortfalls can be modeled as a Poisson process. Sections 10.3 and 10.5 discuss the maximal and minimal GPD as the domains of attractions of exceedances and shortfalls, respectively. Sections 10.4 and 10.6 obtain some approximations based on the maximal and minimal GPD, respectively. Section 10.7 shows how to obtain the minimal from the maximal GPD. Section 10.8 discusses several methods for the estimation of the parameters and quantiles of the GPD families. Section 10.9 uses the P-P and Q-Q plots for model validation. Finally, Section 10.10 provides some tests of hypothesis associated with the GPD families.

10.1 Exceedances as a Poisson Process

In Example 2.10, we computed the probability of r exceedances of u in n iid repetitions of an experiment as

$$\Pr(M_{u_n} = r) = \binom{n}{r} (p_u)^r (1 - p_u)^{n-r}, \quad r \in \{0, 1, \ldots, n\}, \tag{10.1}$$

where M_{u_n} is the number of exceedances over a threshold u in n iid repetitions of an experiment and p_u is the probability that an experiment will result in a value exceeding u. If we now assume that u depends on n in such a way that

$$\lim_{n\to\infty} n\,[1 - F(u_n)] = \lambda, \quad 0 \le \lambda < \infty, \tag{10.2}$$

then, the probabilities in (10.1) can be approximated by a Poisson random variable and we have the following theorem (see de Haan and Sinha (1999) and Leadbetter, Lindgren, and Rootzén (1983)).

Theorem 10.1 (Exceedances as a Poisson process). *Let $\{X_n\}$ be a sequence of iid random variables with cdf $F(x)$. If a sequence of real numbers $\{u_n\}$ satisfies (10.2), then, we have*

$$\lim_{n\to\infty} \Pr(M_{u_n} = r) = \frac{\exp(-\lambda)\lambda^r}{r!}, \quad r \ge 0, \tag{10.3}$$

where M_{u_n} is the number of X_i $(i = 1, 2, \ldots, n)$ that exceed u_n and the right-hand side term of (10.3) must be taken as 1 or 0 when $\lambda = 0$ or $\lambda = \infty$, respectively.

10.2 Shortfalls as a Poisson Process

The number of shortfalls under a threshold u can also be modeled in the same way as the number of exceedances. Letting m_{u_n} denote the number of shortfalls under a threshold u in n iid repetitions of an experiment, then the pmf of m_{u_n} is

$$\Pr(m_{u_n} = r) = \binom{n}{r} (p_u)^r (1 - p_u)^{n-r}, \quad r \in \{0, 1, \ldots, n\}, \tag{10.4}$$

where now p_u is the probability that an experiment will result in a value below u. Assuming that u depends on n in such a way that

$$\lim_{n\to\infty} n\,[F(u_n)] = \lambda, \quad 0 \le \lambda < \infty, \tag{10.5}$$

then, the probabilities in (10.4) can be approximated by a Poisson random variable as given in the following theorem:

Theorem 10.2 (Shortfalls as a Poisson process). *Let $\{X_n\}$ be a sequence of iid random variables with cdf $F(x)$. If a sequence of real numbers $\{u_n\}$ satisfies (10.5), then, we have*

$$\lim_{n\to\infty} \Pr(m_{u_n} = r) = \frac{\exp(-\lambda)\lambda^r}{r!}, \quad r \geq 0, \tag{10.6}$$

where m_{u_n} is the number of X_i $(i = 1, 2, \ldots, n)$ that fall below u_n and the right-hand side term of (10.3) must be taken as 1 or 0 when $\lambda = 0$ or $\lambda = \infty$, respectively.

The practical importance of Theorems 10.1 and 10.2 is that exceedances or shortfalls of rare events follow a Poisson law.

Note that (10.3) and (10.6) give the limit pdf of the number of exceedances and the number of shortfalls, respectively, but we are also interested in the distribution of the exceedances and shortfalls themselves, and not just their numbers. These limit distributions are discussed in the following two sections.

10.3 The Maximal GPD

Pickands (1975) demonstrates that when the threshold tends to the upper end of the random variable, the exceedances follow a maximal generalized Pareto distribution, $GPD_M(\lambda, \kappa)$, with cdf

$$F(x;\lambda,\kappa) = \begin{cases} 1 - \left(1 - \dfrac{\kappa x}{\lambda}\right)^{1/\kappa}, & \left(1 - \dfrac{\kappa x}{\lambda}\right) \geq 0, \\ & \kappa \neq 0, \lambda > 0, \\ 1 - e^{-x/\lambda}, & x \geq 0, \kappa = 0, \lambda > 0, \end{cases} \tag{10.7}$$

where λ and κ are scale and shape parameters, respectively. For $\kappa \neq 0$, the range of x is $0 \leq x \leq \lambda/\kappa$, if $\kappa > 0$, and $x \geq 0$, if $\kappa \leq 0$. The case $\kappa = 0$, that is, the exponential distribution, is obtained by taking the limit of (10.7) as $\kappa \to 0$. The distributions in (10.7) are called the *maximal generalized Pareto family of distributions* and are denoted by $GPD_M(\lambda, \kappa)$.

The speed of convergence to the generalized Pareto distribution is discussed in Worms (2001), and the selection of the threshold values is discussed in Guillou and Hall (2001).

One of the most important and interesting properties of $GPD_M(\lambda, \kappa)$ is that it is stable with respect to truncations from the left. In fact, we have

$$\begin{aligned} P(X - u \leq x | X > u) &= \frac{P(u < X \leq x + u)}{P(X > u)} \\ &= \frac{F(x+u;\lambda,\kappa) - F(u;\lambda,\kappa)}{1 - F(u;\lambda,\kappa)} \end{aligned}$$

$$= \frac{\left[1 - \frac{\kappa u}{\lambda}\right]^{1/\kappa} - \left[1 - \frac{\kappa(x+u)}{\lambda}\right]^{1/\kappa}}{\left[1 - \frac{\kappa u}{\lambda}\right]^{1/\kappa}}$$

$$= 1 - \left[1 - \frac{\kappa x}{\lambda - \kappa u}\right]^{1/\kappa}, \tag{10.8}$$

which means that if X is a $GPD(\lambda, \kappa)$, then $X - u$, given $X > u$, is also a $GPD(\lambda - \kappa u, \kappa)$, and this holds for any value of the threshold u. This property implies that if a given data set and a threshold u_0 is consistent for this law, it will be for any other threshold value $u_1 > u_0$. Note that $X - u$ is the difference between the actual value and the threshold value.

A justification of the GPD_M as a limit distribution for exceedances is as follows. Let $F(x)$ be the cdf of X; then the limit distribution theory for maxima (see Section 9.1.1) allows us to say that for large enough n,

$$[F(x)]^n \approx \exp\left\{-\left[1 - \kappa\left(\frac{x-\lambda}{\delta}\right)\right]^{1/\kappa}\right\}.$$

Taking the log on both sides, we obtain

$$n \log F(x) \approx -\left[1 - \kappa\left(\frac{x-\lambda}{\delta}\right)\right]^{1/\kappa}. \tag{10.9}$$

For large x, $F(x) \to 1$, that is, $F(x) = 1 - \epsilon(x)$ with $\epsilon(x) \to 0$, then

$$\log F(x) = \log(1 - \epsilon(x)) \approx -\epsilon(x) = F(x) - 1,$$

and from (10.9) we have

$$1 - F(x) \approx \frac{1}{n}\left[1 - \kappa\left(\frac{x-\lambda}{\delta}\right)\right]^{1/\kappa}. \tag{10.10}$$

Then, for $x > 0$,

$$\begin{aligned}
F_{X-u|X>u}(x) &= 1 - \Pr(X > u + x | X > u) \\
&= 1 - \frac{\Pr(X > u + x)}{\Pr(X > u)} \\
&= 1 - \frac{1 - F(u+x)}{1 - F(u)} \\
&= 1 - \left[\frac{1 - \kappa(u + x - \lambda)/\delta}{1 - \kappa(u - \lambda)/\delta}\right]^{1/\kappa} \\
&= 1 - \left[1 - \frac{\kappa x}{\delta - \kappa(u - \lambda)}\right]^{1/\kappa} \\
&= 1 - [1 - \kappa x / \lambda^*]^{1/\kappa}
\end{aligned}$$

where $\lambda^* = \delta - \kappa(u - \lambda)$. This is the cdf of the GPD$_M$ as given in (10.7). Note that for $\kappa > 0$ we have a limited distribution at the right end.

Some special cases of the $GPD_M(\lambda, \kappa)$ are:

- When $\kappa = 0$, the GPD_M reduces to the exponential distribution with mean λ.
- When $\kappa = 1$, the GPD_M reduces to the Uniform $U[0, \lambda]$.
- When $\kappa \leq -1/2$, $Var(X) = \infty$.
- When $\kappa < 0$, the GPD_M becomes the Pareto distribution, hence the name generalized Pareto distribution.

The pth quantile of the GPD_M is

$$x_p = \begin{cases} \lambda\left[1 - (1-p)^{\kappa}\right]/\kappa, & \text{if } \kappa \neq 0, \\ -\lambda \log(1-p), & \text{if } \kappa = 0. \end{cases} \tag{10.11}$$

10.4 Approximations Based on the Maximal GPD

The original cdf, $F(x)$, where the sample data $X_1, X_2, \ldots, X_n$, were drawn from, can be approximated using only exceedances, that is, the values of $X_i > u$, as follows. Since

$$\begin{aligned} F_{X-u|X>u}(x) &= \frac{\Pr(X - u \leq x, X > u)}{\Pr(X > u)} \\ &= \frac{F(x+u) - F(u)}{1 - F(u)} = \frac{F(x+u) - 1 + \theta}{\theta}, \end{aligned}$$

where $\theta = 1 - F(u)$, we see that the original distribution can be approximated by

$$F(x) = (1 - \theta) + \theta F_{X-u|X>u}(x - u), \quad x > u,$$

that is, using the GPD, the approximation of the distribution of the original random variable X is

$$F(x) = \begin{cases} (1-\theta) + \theta\left[1 - \left(1 - \dfrac{\kappa(x-u)}{\lambda}\right)^{1/\kappa}\right] & \text{if } \left(1 - \dfrac{\kappa(x-u)}{\lambda}\right) \geq 0, \\ & \lambda > 0, \kappa \neq 0, \\ (1-\theta) + \theta\left[1 - \exp\left(-(x-u)/\lambda\right]\right)\right], & \text{if } x \geq u, \lambda > 0, \kappa = 0. \end{cases}$$

The corresponding initial sample log-likelihood function, $\ell(\boldsymbol{\theta}|\mathbf{x}) = \log L(\boldsymbol{\theta}|\mathbf{x})$, becomes

$$\begin{cases} (n-m_u)\log(1-\theta) + m_u \log\left(\frac{\theta}{\lambda}\right) + \sum\limits_{x_i>u}\left(\frac{1}{\kappa}-1\right)\log\left(1-\frac{\kappa(x_i-u)}{\lambda}\right), \\ \qquad \text{if } \left(1-\frac{\kappa(x_i-u)}{\lambda}\right) \ge 0,\ i=1,2,\ldots,m_u, \lambda>0, \kappa\neq 0, \\ (n-m_u)\log(1-\theta) + m_u \log\left(\frac{\theta}{\lambda}\right) - \sum\limits_{x_i>u}\frac{x_i-u}{\lambda} \\ \qquad \text{if } x_i \ge u,\ i=1,2,\ldots,m_u, \lambda>0, \kappa=0, \end{cases}$$

where m_u is the number of x_i values above the threshold value u.

10.5 The Minimal GPD

If one is dealing with a minimum problem, instead of the exceedances above the threshold, one is interested in the *shortfalls* below the threshold. In this case, the limit distribution of $X-u|X<u$ is

$$F(x;\lambda,\kappa) = \begin{cases} \left(1+\frac{\kappa x}{\lambda}\right)^{1/\kappa}, & \left(1+\frac{\kappa x}{\lambda}\right)\ge 0, x\le 0, \\ & \kappa\neq 0, \lambda>0, \\ e^{x/\lambda}, & x\le 0, \kappa=0, \lambda>0, \end{cases} \tag{10.12}$$

where λ and κ are scale and shape parameters, respectively. For $\kappa \neq 0$, the range of x is $-\lambda/\kappa \le x \le 0$, if $\kappa > 0$, and $x < 0$, if $\kappa \le 0$.

The distributions in (10.12) are called the *minimal generalized Pareto family of distributions* and are denoted by $GPD_m(\lambda,\kappa)$. Note that for $\kappa > 0$, the GPD_m is limited on the left. The $GPD_m(\lambda,\kappa)$ is also stable with respect to truncations from the right. The proof of this is similar to that for the $GPD_M(\lambda,\kappa)$.

Some special cases of the $GPD_m(\lambda,\kappa)$ are:

- When $\kappa = 0$, the GPD_m reduces to the reversed exponential distribution (see Section 3.2.2) with mean λ.
- When $\kappa = 1$, the GPD_m reduces to the Uniform $U[-\lambda, 0]$.
- When $\kappa \le -1/2$, $Var(X) = \infty$.
- When $\kappa < 0$, the GPD_m becomes the reversed Pareto distribution.

The pth quantile of the GPD_m is

$$x_p = \begin{cases} \lambda\left(p^\kappa - 1\right)/\kappa, & \text{if } \kappa\neq 0, \\ \lambda\log(p), & \text{if } \kappa = 0. \end{cases} \tag{10.13}$$

10.6 Approximations Based on the Minimal GPD

The original cdf, $F(x)$, where the sample data $X_1, X_2, \ldots, X_n$ were drawn from, can be approximated using only the shortfalls, that is, the values of $X_i < u$, as follows. Since

$$\begin{aligned} F_{X-u|X\leq u}(x) &= \frac{\Pr(X-u \leq x, X \leq u)}{\Pr(X \leq u)} \\ &= \frac{\Pr(X \leq x+u)}{\Pr(X \leq u)} \\ &= \frac{F(x+u)}{F(u)} = \frac{F(x+u)}{1-\theta}, \end{aligned} \tag{10.14}$$

we have

$$F(x) = (1-\theta)F_{X-u|X\leq u}(x-u).$$

Thus, the approximation of the original cdf, $F(x)$, of the random variable X is

$$\begin{cases} (1-\theta)\left(1+\dfrac{\kappa(x-u)}{\lambda}\right)^{1/\kappa}, & \text{if } \left(1+\dfrac{\kappa(x-u)}{\lambda}\right) \geq 0,\ x \leq u, \lambda > 0, \kappa \neq 0, \\ (1-\theta)e^{((x-u)/\lambda)}, & \text{if } x \leq u, \lambda > 0, \kappa = 0. \end{cases}$$

The corresponding initial sample log-likelihood function, $\ell(\boldsymbol{\theta}|\mathbf{x}) = \log L(\boldsymbol{\theta}|\mathbf{x})$, becomes

$$\begin{cases} m_u \log\theta + (n-m_u)\log\left(\dfrac{1-\theta}{\lambda}\right) + \left(\dfrac{1}{\kappa}-1\right)\sum\limits_{x_i\leq u}\log\left(1+\dfrac{\kappa(x_i-u)}{\lambda}\right), \\ \qquad \text{if } \left(1+\dfrac{\kappa(x_i-u)}{\lambda}\right) \geq 0,\ i=1,2,\ldots,n-m_u, \lambda > 0, \kappa \neq 0, \\ m_u \log(\theta) + (n-m_u)\log\left(\dfrac{1-\theta}{\lambda}\right) + \sum\limits_{x_i\leq u}\dfrac{x_i-u}{\lambda}, \\ \qquad \text{if } x_i \leq u,\ i=1,2,\ldots,n-m_u,\ \lambda > 0, \kappa = 0, \end{cases}$$

where m_u is the number of x_i values below the threshold value u.

10.7 Obtaining the Minimal from the Maximal GPD

The GPD_m can be obtained from the GPD_M using the following theorem.

Theorem 10.3 (Obtaining minima from maxima). *If the random variable $X \sim GPD_M(\lambda,\kappa)$, then $Y = -X \sim GPD_m(\lambda,\kappa)$. Similarly, if the random variable $X \sim GPD_m(\lambda,\kappa)$, then $Y = -X \sim GPD_M(\lambda,\kappa)$.*

Proof. Let $Y = -X$, then we have

$$\begin{aligned} F_Y(y) &= \Pr[Y \le y] = \Pr[-X \le y] = \Pr[X \ge -y] \\ &= 1 - \Pr[X < -y] = 1 - F_{GPD_M(\lambda,\kappa)}(-y) \\ &= (1 + \kappa y/\lambda)^{1/\kappa} = F_{GPD_m(\lambda,\kappa)}(y), \end{aligned}$$

which shows that $Y = -X \sim GPD_m(\lambda, \kappa)$. The converse result can be derived in a similar way. ■

Theorem 10.3 shows that, by using the change of variable $Y = -X$, a problem involving minimum can be reduced to a problem involving maxima. Hence, in the next section, we discuss estimation of the parameters and quantiles only for the maximal GPD.

10.8 Estimation for the GPD Families

We have seen in the previous sections that the GPD_M is used to model exceedances over a threshold u, while the GPD_m is used to model shortfalls under a threshold u. In this section, we use several methods to estimate the parameters and quantiles of the maximal GPD_M. The same methods can be used to obtain the corresponding estimates for the minimal GPD_m by changing the sign of the data. Theorem 10.3 suggests the following algorithm for estimating the parameters of the GPD_m:

1. Change sign of the data: $x_i \to -x_i, \quad i = 1, 2, \ldots, n$.
2. Estimate the parameters of the GPD_M (as explained in the following sections), say $\hat{\lambda}$ and $\hat{\kappa}$.
3. Propose the $\text{GPD}_m(\hat{\lambda}, \hat{\kappa})$ for given data on shortfalls under a threshold u.

10.8.1 The Maximum Likelihood Method

Point Estimation

In this section, we use the maximum likelihood method (see Section 5.1) to estimate the parameters and quantiles of the GPD_M, which is used to model exceedances over a threshold u. Taking the derivative of the cdf in (10.7), we obtain the pdf of the GPD_M,

$$f(x; \lambda, \kappa) = \begin{cases} \dfrac{1}{\lambda}\left(1 - \dfrac{\kappa x}{\lambda}\right)^{1/\kappa - 1}, & \left(1 - \dfrac{\kappa x}{\lambda}\right) \ge 0, \\ & \kappa \ne 0, \lambda > 0, \\ \dfrac{1}{\lambda}\, e^{-x/\lambda}, & x \ge 0, \kappa = 0, \lambda > 0. \end{cases} \tag{10.15}$$

For the case of $\kappa = 0$, the GPD_M is the exponential distribution and the MLE of λ can be easily obtained as $\hat{\lambda} = \bar{x}$, where $\bar{x}$ is the sample mean.

From (10.15), the log-likelihood function is

$$\ell(\lambda, \kappa) = -n \log \lambda + \left(\frac{1}{\kappa} - 1\right) \sum_{i=1}^{n} \log(1 - \kappa x_i/\lambda), \quad \kappa \neq 0, \tag{10.16}$$

and

$$\ell(\lambda) = -n \log \lambda - \lambda^{-1} \sum_{i=1}^{n} x_i, \quad \kappa = 0.$$

Thus, for $\kappa \neq 0$, the log-likelihood function can be made arbitrarily large by taking $\kappa > 1$ and λ/κ close to the maximum order statistic $x_{n:n}$. Consequently, the maximum likelihood estimators are taken to be the values $\hat{\lambda}$ and $\hat{\kappa}$, which yield a local maximum of $\ell(\lambda, \kappa)$. To find the local maximum, numerical methods are needed. For details, see Davison (1984), Davison and Smith (1990), DuMouchel (1983), Grimshaw (1993), Hosking and Wallis (1987), Hosking, Wallis, and Wood (1985), Smith (1985), and Smith and Weissman (1985).

The inverse of the information matrix is the asymptotic covariance matrix of the MLE of λ and κ. In the regular case, where $\kappa < 1/2$, the asymptotic covariance matrix is given by

$$\Sigma(\hat{\lambda}, \hat{\kappa}) = \frac{1}{n} \begin{bmatrix} 2\lambda^2(1-\kappa) & \lambda(1-\kappa) \\ \lambda(1-\kappa) & (1-\kappa)^2 \end{bmatrix}. \tag{10.17}$$

An estimate of this asymptotic covariance matrix is obtained by replacing the parameters by their MLEs,

$$\hat{\Sigma}(\hat{\lambda}, \hat{\kappa}) = \frac{1}{n} \begin{bmatrix} 2\hat{\lambda}^2(1-\hat{\kappa}) & \hat{\lambda}(1-\hat{\kappa}) \\ \hat{\lambda}(1-\hat{\kappa}) & (1-\hat{\kappa})^2 \end{bmatrix}. \tag{10.18}$$

Estimate of the pth quantile of the GPD is then given by

$$\hat{x}_p = \begin{cases} \hat{\lambda}\left[1 - (1-p)^{\hat{\kappa}}\right]/\hat{\kappa}, & \text{if } \hat{\kappa} \neq 0, \\ -\hat{\lambda} \log(1-p), & \text{if } \kappa = 0. \end{cases} \tag{10.19}$$

Remark 10.1 *The above numerical solutions used to obtain the MLE require an initial estimate $\theta_0 = \{\lambda_0, \kappa_0\}$. The exponential estimate of $\lambda = \bar{x}$ and $\kappa = 0$ can be used as initial estimates. A simple version of the EPM (see Section 10.8.4) can also be used as an initial starting point.*

Confidence Intervals for the Parameters

For $\kappa \neq 0$, confidence intervals for the parameters $\theta = \{\lambda, \kappa\}$ can be obtained using (5.43). We first obtain the MLE of θ by maximizing the log-likelihood function in (10.16). The inverse of the Fisher information matrix, evaluated at

$(\hat{\lambda}, \hat{\kappa})$, is the estimated covariance matrix of $(\hat{\lambda}, \hat{\kappa})$. The square root of the diagonal elements of this matrix are the standard errors, $(\hat{\sigma}_{\hat{\lambda}}, \hat{\sigma}_{\hat{\kappa}})$, of the estimates $(\hat{\lambda}, \hat{\kappa})$, respectively. Accordingly, the $(1-\alpha)100\%$ confidence intervals for the parameters are given by

$$\lambda \in \left(\hat{\lambda} \pm z_{\alpha/2}\, \hat{\sigma}_{\hat{\lambda}}\right) \quad \text{and} \quad \kappa \in \left(\hat{\kappa} \pm z_{\alpha/2}\, \hat{\sigma}_{\hat{\kappa}}\right). \tag{10.20}$$

Confidence Intervals for the Quantiles

We use the delta method to construct confidence intervals for the quantiles of the GPD_M in the same way as we did for the GEVD_M in Section 9.2.1. For $\kappa \neq 0$, $\theta = (\lambda, \kappa)$ and the gradient vector is

$$\nabla_\theta x_p = \begin{bmatrix} \dfrac{\partial x_p}{\partial \lambda} \\ \dfrac{\partial x_p}{\partial \kappa} \end{bmatrix} = \begin{bmatrix} \dfrac{1}{\kappa}\left[1-(1-p)^\kappa\right] \\ -\dfrac{\lambda}{\kappa^2}\left[1-(1-p)^\kappa\right] - \dfrac{\lambda}{\kappa}(1-p)^\kappa \log(1-p) \end{bmatrix},$$

where $\theta = \{\lambda, \kappa\}$. For $\kappa = 0$, $\theta = \lambda$ and the gradient becomes

$$\nabla_\theta x_p = \frac{\partial x_p}{\partial \lambda} = -\log(1-p), \tag{10.21}$$

which is a scalar. The estimated variance of $\hat{x}_p$ is

$$\hat{\sigma}^2_{\hat{x}_p} \approx \nabla^T_{\hat{\theta}} x_p \hat{\Sigma} \nabla_{\hat{\theta}} x_p, \tag{10.22}$$

where $\hat{\Sigma}$ is an estimate of the asymptotic variance-covariance matrix of $\hat{\theta}$, given by the inverse of the observed Fisher information matrix, and $\nabla_{\hat{\theta}} x_p$ is $\nabla_\theta x_p$ evaluated at $\hat{\theta}$.

Using (5.45), a $(1-\alpha)100\%$ confidence interval for x_p is then given by

$$x_p \in \left(\hat{x}_p \pm z_{\alpha/2} \hat{\sigma}_{\hat{x}_p}\right). \tag{10.23}$$

Example 10.1 (Maxima data examples). Table 10.1 shows the maximum likelihood estimates of the parameters of the maximal GPD and the associated average scaled absolute error (ASAE, defined in (9.64)) statistic values for some of the data sets in Chapter 1, for which the right tail is of interest.

It is interesting to compare the $\hat{\kappa}$ values of this table with those in Table 9.6. We obtain similar values for the wind data, moderate differences for the women, flood, and wave data, and large differences for the Bilbao and men data sets. Note that the larger the value of $\hat{\kappa}$ the larger the differences. This proves on one hand that the tail information (GPD) is not the same as the whole sample information for small sample sizes, and on the other hand that the ML method has some problems.

Table 10.2 gives the confidence intervals for some data sets. Note that this method is not able to give confidence intervals for all cases. ∎

Table 10.1: Threshold Values u, Maximum Likelihood Parameter Estimates and ASAE Values of the Maximal GPD for the Data Sets in Chapter 1 for Which the Right Tail Is of Interest.

ML Method				
Data Set	u	$\hat{\lambda}$	$\hat{\kappa}$	ASAE
Wind	36.82	6.52	–0.48	0.027
Bilbao	8.74	1.18	1.02	0.028
Men	104.0	3.44	1.37	0.080
Women	105.2	1.62	0.49	0.087
Flood	45.04	10.87	0.20	0.035
Wave	17.36	7.99	0.27	0.041

Table 10.2: Confidence Intervals for Some Quantiles.

Data Set	$x_{0.95}$	CI($x_{0.95}$)	$x_{0.99}$	CI($x_{0.99}$)
Wind	80.69	(28.57,132.8)	148.16	(–83.44,379.8)
Bilbao	9.84	–	9.89	–
Men	106.5	–	106.5	–
Women	107.72	(107.1,108.3)	108.1	(107.0,109.2)
Flood	69.52	(60.4,78.64)	77.72	(60.81,94.63)
Wave	33.81	(27.69,39.92)	38.49	(28.15,48.83)

10.8.2 The Method of Moments

The Method of Moments (MOM) estimators are (see Section 5.2)

$$\hat{\kappa}_{\text{MOM}} = \frac{\bar{x}^2 - \hat{S}^2}{2\hat{S}^2} \quad \text{and} \quad \hat{\lambda}_{\text{MOM}} = \frac{\bar{x}(\bar{x}^2/\hat{S}^2 + 1)}{2}, \tag{10.24}$$

where $\bar{x}$ and $\hat{S}^2$ are the sample mean and variance, respectively. Note that, for some cases, the moments may not exist.

10.8.3 The Probability Weighted Moments Method

To find the Probability-Weighted Moments (PWM) estimates for the parameters of the GPD_M, we use the α_s moments (see Section 5.3) and obtain

$$\alpha_s = M(1,0,s) \quad = \quad E[X\{1-F(X)\}^s] = \frac{\lambda}{(s+1)(s+1+\kappa)}, \tag{10.25}$$

which exist provided that $\kappa > -1$. Using $\alpha_0 = E(X)$ and α_1, we obtain

$$\begin{aligned} \lambda &= \frac{2\alpha_0\alpha_1}{\alpha_0 - 2\alpha_1}, \\ \kappa &= \frac{4\alpha_1 - \alpha_0}{\alpha_0 - 2\alpha_1}. \end{aligned} \tag{10.26}$$

The PWM are obtained by replacing α_0 and α_1 by estimators based on the observed sample moments. Accordingly, the PWM estimates are given by

$$\begin{aligned} \hat{\kappa}_{PWM} &= \frac{4t - \bar{x}}{\bar{x} - 2t}, \\ \hat{\lambda}_{PWM} &= \frac{2\bar{x}t}{\bar{x} - 2t}, \end{aligned} \tag{10.27}$$

where

$$t = \frac{1}{n}\sum_{i=1}^{n}(1 - p_{i:n})x_{i:n},$$

and $p_{i:n} = (i - 0.35)/n$.

The asymptotic covariance matrix is given by

$$\hat{\Sigma}(\hat{\lambda}, \hat{\kappa}) = \frac{v(\kappa)}{n}\begin{bmatrix} m_{11} & m_{12} \\ m_{21} & m_{22} \end{bmatrix}, \tag{10.28}$$

where

$$\begin{aligned} m_{11} &= \kappa^2(7 + 18\kappa + 11\kappa^2 + 2\kappa^3), \\ m_{12} &= m_{21} = \kappa(2 + \kappa)(2 + 6\kappa + 7\kappa^2 + 2\kappa^3), \\ m_{22} &= (1 + \kappa)(2 + \kappa)^2(1 + \kappa + 2\kappa^2), \end{aligned}$$

and

$$v(\kappa) = \frac{1}{(1 + 2\kappa)(3 + 2\kappa)}. \tag{10.29}$$

Example 10.2 (Maxima data examples). Table 10.3 shows the threshold values u, the PWM estimates of the parameters and the ASAE of the maximal GPD and the associated ASAE statistic values for the data sets in Chapter 1, for which the right tail is of interest. The results are comparable with those in Table 10.1. ■

10.8.4 The Elemental Percentile Method

Traditional methods of estimation (MLE and the moments-based methods) have problems because

1. The range of the distribution depends on the parameters: $x < \lambda/\kappa$, for $\kappa > 0$ and $x > 0$, for $\kappa < 0$. So, the MLE do not have the usual asymptotic properties.

Table 10.3: Threshold Values u, PWM Parameter Estimates and ASAE Values of the Maximal GPD for Some of the Data Sets in Chapter 1 for Which the Right Tail Is of Interest.

	PWM Method			
Data Set	u	$\hat{\lambda}$	$\hat{\kappa}$	ASAE
Wind	36.82	6.50	-0.44	0.035
Bilbao	8.74	1.20	1.05	0.029
Men	104.0	2.82	1.15	0.056
Women	105.2	1.55	0.49	0.087
Flood	45.04	11.02	0.22	0.036
Wave	17.36	8.30	0.32	0.043

2. The MLE requires numerical solutions.
3. For some samples, the likelihood may not have a local maximum. For $\kappa > 1$, the MLE do not exist (the likelihood can be made infinite).
4. When $\kappa < -1$, the mean and higher order moments do not exist. So, MOM and PWM do not exist when $\kappa < -1$.
5. The PWM estimators are good for cases where $-0.5 < \kappa < 0.5$. Outside this range of κ, the PWM estimates may not exist, and if they do exist their performance worsens as κ increases.

This leaves us with the elemental percentile method (EPM) for estimating the parameters and quantiles of the extreme models. The EPM is discussed in general in Section 5.4. Here, we use it to estimate the parameters of the GPD_M. Since the GPD_M has two parameters, we need two distinct order statistics. Let $x_{i:n}$ and $x_{j:n}$ be two distinct order statistics in a random sample of size n from $F_{GPD_M}(x; \lambda, \kappa)$. Then, equating the cdf evaluated at the observed order statistics to their corresponding quantile values, we obtain (see (3.86))

$$\begin{aligned} x_{i:n} &= \lambda \left[1 - (1 - p_{i:n})^{\kappa}\right] / \kappa, \\ x_{j:n} &= \lambda \left[1 - (1 - p_{j:n})^{\kappa}\right] / \kappa, \end{aligned} \tag{10.30}$$

where $p_{i:n} = i/(n+1)$. From the system (10.30), we obtain

$$x_{i:n}[1 - (1 - p_{j:n})^{\kappa}] = x_{j:n}[1 - (1 - p_{i:n})^{\kappa}]. \tag{10.31}$$

Equation (10.31) is a function of only one unknown κ, hence it can be easily solved for κ, using the bisection method (see Algorithm 10.1), obtaining an estimator of κ, $\hat{\kappa}(i, j)$. This estimator is then substituted in (10.30) to obtain a corresponding estimator of λ, $\hat{\lambda}(i, j)$, which is given by

$$\hat{\lambda}(i, j) = \frac{\hat{\kappa}(i, j)\, x_{j:n}}{1 - (1 - p_{j:n})^{\hat{\kappa}(i,j)}}. \tag{10.32}$$

Algorithm 10.1 (Solving Equation (10.31)).

- **Input:** The maximum order statistic $x_{n:n}$ and an elemental subset of two distinct order statistics, $x_{i:n}$ and $x_{j:n}$.

- **Output:** Initial estimate of κ.

For $i = 1, 2, \ldots, n-1$, compute

$$\delta_0 = \frac{x_{i:n}}{x_{j:n}} - \frac{\log(1-p_{i:n})}{\log(1-p_{j:n})}.$$

Then,

1. If $\delta_0 > 0$, use the bisection method on the interval

$$\left(0, \frac{\log(1 - x_{i:n}/x_{n:n})}{\log(1-p_{i:n})}\right)$$

to obtain a solution $\hat{\kappa}(i,n)$ of (10.31).

2. Otherwise, use the bisection method on the interval

$$\left(\frac{\log(x_{i:n}/x_{n:n})}{\log[(1-p_{i:n})/(1-p_{n:n})]}, 0\right)$$

to solve (10.31) and obtain $\hat{\kappa}(i,n)$.

The initial estimates are based on only two order statistics. To obtain the final estimates, we select a prespecified number, N, of elemental subsets each of size 2, either at random or using all possible subsets. For each of these elemental subsets, an elemental estimate of the parameters $\theta = \{\kappa, \lambda\}$ is computed. Let us denote these elemental estimates by $\hat{\theta}_1, \hat{\theta}_2, \ldots, \hat{\theta}_N$. The elemental estimates that are inconsistent with the data are discarded. These elemental estimates can then be combined, using some suitable (preferably robust) functions, to obtain an overall final estimate of θ. Examples of robust functions include the *median* (MED) and the *α-trimmed mean* (TM_α), where α indicates the percentage of trimming. Thus, a final estimate of $\theta = \{\kappa, \lambda\}$, can be defined as

$$\begin{aligned} \hat{\kappa}_{\text{MED}} &= \text{Median}(\hat{\kappa}_1, \hat{\kappa}_2, \ldots, \hat{\kappa}_N), \\ \hat{\lambda}_{\text{MED}} &= \text{Median}(\hat{\lambda}_1, \hat{\lambda}_2, \ldots, \hat{\lambda}_N), \end{aligned} \tag{10.33}$$

or

$$\begin{aligned} \hat{\kappa}_{\text{TM}_\alpha} &= \text{TM}_\alpha(\hat{\kappa}_1, \hat{\kappa}_2, \ldots, \hat{\kappa}_N), \\ \hat{\lambda}_{\text{TM}_\alpha} &= \text{TM}_\alpha(\hat{\lambda}_1, \hat{\lambda}_2, \ldots, \hat{\lambda}_N), \end{aligned} \tag{10.34}$$

where $\text{Median}(y_1, \ldots, y_N)$ is the median of the set of numbers $\{y_1, \ldots, y_N\}$, and $\text{TM}_\alpha(y_1, \ldots, y_N)$ is the mean obtained after trimming the $(\alpha/2)100\%$ largest and the $(\alpha/2)100\%$ smallest order statistics of $y_1, \ldots, y_N$.

To avoid estimates that are inconsistent with the data, it is better to estimate the upper end λ/κ instead of λ and, based on it, recover λ, that is, replace the last estimates in (10.33) by

$$\left(\hat{\lambda}/\hat{\kappa}\right)_{\text{MED}} = \text{Median}\left((\hat{\lambda}/\hat{\kappa})_{1,2,n}, (\hat{\lambda}/\hat{\kappa})_{1,3,n}, \ldots, (\hat{\lambda}/\hat{\kappa})_{1,n-1,n}\right)$$

and recover $\hat{\lambda}$ by means of

$$\hat{\lambda}_{\text{MED}} = \hat{\kappa}\ \ (\hat{\lambda}/\hat{\kappa})_{\text{MED}}.$$

Similarly, we replace the last estimates in (10.34) by

$$\left(\hat{\lambda}/\hat{\kappa}\right)_{\text{TM}_\alpha} = \text{TM}_\alpha\left((\hat{\lambda}/\hat{\kappa})_{1,2,n}, (\hat{\lambda}/\hat{\kappa})_{1,3,n}, \ldots, (\hat{\lambda}/\hat{\kappa})_{1,n-1,n}\right)$$

and recover $\hat{\lambda}$ by means of

$$\hat{\lambda}_{\text{TM}_\alpha} = \hat{\kappa}\ \ (\hat{\lambda}/\hat{\kappa})_{\text{TM}_\alpha}.$$

The quantile estimates are then obtained by replacing the parameter estimates in the quantile function in (10.19).

A Computationally Efficient Version of EPM

As per Remark 10.1, a computationally simple version of the EPM or the QLS can be used as an initial starting point for the numerical solution of the MLE. Here, the final estimates in (10.33) or (10.34) can be computed using only a small number of elemental subsets. Let $i = 1$ and compute $\hat{\lambda}_{1j}$ and $\hat{\kappa}_{1j}$, $j = 2, 3, \ldots, n$. These are only $n - 1$ estimates, which is much smaller than N. They can be combined to produce final estimates using (10.33) or (10.34). These estimates can then be used as starting values in the MLE algorithm.

Confidence Intervals

Since the estimates exist for any combination of parameter values, the use of sampling based methods such as the bootstrap methods (Efron (1979) and Diaconis and Efron (1974)) to obtain variances and confidence intervals is justified. The bootstrap sampling can be performed by drawing the data from the parametric cdf, $F_{GPD_M}(x; \hat{\lambda}, \hat{\kappa})$.

Example 10.3 (Maximum data sets). Table 10.4 shows the threshold values u, the EPM parameter estimates and the ASAE values of the maximal GPD for the data sets in Chapter 1 for which the right tail is of interest for both, the trimmed mean and the median versions. As you can see, they are very similar, though present some differences with those in Table 10.1, especially for the women and wave data sets.

■

Table 10.4: Threshold Values u, EPM Parameter Estimates and ASAE Values of the Maximal GPD for the Data Sets in Chapter 1 for Which the Right Tail Is of Interest.

EPM-TM Method				
Data Set	u	$\hat{\lambda}$	$\hat{\kappa}$	ASAE
Wind	36.82	6.16	–0.69	0.037
Bilbao	8.74	1.28	1.25	0.044
Men	104.0	2.76	1.11	0.055
Women	105.2	2.15	1.39	0.120
Flood	45.04	12.02	0.46	0.054
Wave	17.36	9.66	0.75	0.065
EPM-MED Method				
Data Set	u	$\hat{\lambda}$	$\hat{\kappa}$	ASAE
Wind	36.82	5.84	–0.75	0.048
Bilbao	8.74	1.28	1.16	0.035
Men	104.0	2.53	0.99	0.051
Women	105.2	2.22	1.25	0.122
Flood	45.04	11.31	0.30	0.041
Wave	17.36	9.86	0.53	0.056

10.8.5 The Quantile Least Squares Method

The quantile least squares method for the GPD estimates the parameters by minimizing the squares of the differences between the theoretical and the observed quantiles. Accordingly, the estimates of the parameters, for $\kappa \neq 0$, are obtained by solving the minimization problem

$$\underset{\lambda,\,\kappa}{\text{Minimize}} \quad \sum_{i=1}^{n} \left[x_{i:n} - \lambda \left[1 - (1 - p_{i:n})^{\kappa}\right] / \kappa\right]^2 , \tag{10.35}$$

and, for $\kappa = 0$, the function to be minimized is

$$\underset{\lambda}{\text{Minimize}} \quad \sum_{i=1}^{n} \left[x_{i:n} - \lambda \log(1 - p_{i:n})\right]^2 . \tag{10.36}$$

Example 10.4 (Maximal data sets). Table 10.5 shows the threshold values u, the QLS parameter estimates and the ASAE values of the maximal GPD for the data sets in Chapter 1 for which the right tail is of interest. The parameter estimates are closer to the ML estimates in Table 10.1 than those for the other methods.

■

Table 10.5: Threshold Values u, Quantile Least Squares Parameter Estimates and ASAE Values of the Maximal GPD for the Data Sets in Chapter 1 for Which the Right Tail Is of Interest.

LS Method				
Data Set	u	$\hat{\lambda}$	$\hat{\kappa}$	ASAE
Wind	36.82	6.54	−0.55	0.018
Bilbao	8.74	1.13	0.96	0.028
Men	104.0	2.62	1.04	0.053
Women	105.2	1.28	0.26	0.081
Flood	45.04	9.29	0.05	0.026
Wave	17.36	6.60	0.08	0.032

Example 10.5 (Minimal data sets). Table 10.6 shows the threshold values u, the parameter estimates and the ASAE values for all methods of the minimal GPD for the data sets in Chapter 1 for which the left tail is of interest. As expected, all the $\hat{\kappa}$ estimates are positive, suggesting a minimal Weibull domain of attraction. ■

10.9 Model Validation

In Section 6.3, we discussed the P-P and Q-Q plots as tools for model validation. The P-P plot is a scatter plot of the estimated versus the actual percentiles, that is, the scatter plot of

$$\hat{F}(x_{i:n}) \quad \text{versus} \quad p_{i:n}, \quad i = 1, 2, \ldots, n.$$

The Q-Q plot is a scatter plot of the estimated versus the observed quantiles:

$$\hat{F}^{-1}(p_{i:n}) \quad \text{versus} \quad x_{i:n}, \quad i = 1, 2, \ldots, n.$$

If the model fits the data well, the pattern of points in both plots would exhibit a 45-degree linear trend. Let us now examine the P-P and Q-Q plots for some of the data in Chapter 1.

Example 10.6 (Plots for the GPD$_M$ distribution). The P-P and Q-Q plots for some of the data sets in Chapter 1 for which the right tail is of interest are given in Figures 10.1 and 10.2 for the MLE and the QLS. The plots for the EPM-MED, EPM-TR, and PWM are similar. Some of the Q-Q plots show some lack of fit for large data values.

Note that the Q-Q plots reveal the lack of fit better than the P-P plots. This is because the natural scale of data is used instead of the probability scale $[0, 1]$. ■

Table 10.6: Threshold Values u, Parameter Estimates, and ASAE Values Obtained from Fitting the Minimal GPD to Some of the Data Sets in Chapter 1 for Which the Left Tail Is of Interest.

Data Set	u	$\hat{\lambda}$	$\hat{\kappa}$	ASAE
ML Method				
Epicenter	179.5	121.6	1.00	0.060
Chain	96.60	49.92	1.10	0.083
Insulation	1170.	285.5	0.64	0.064
Precipitation	45.40	13.31	0.82	0.030
EPM-TM Method				
Epicenter	179.5	90.66	0.15	0.168
Chain	96.60	30.83	0.45	0.046
Insulation	1170.	452.1	1.74	0.082
Precipitation	45.40	12.57	0.79	0.024
EPM-MED Method				
Epicenter	179.5	83.30	0.34	0.099
Chain	96.60	31.22	0.51	0.036
Insulation	1170.	462.2	1.62	0.082
Precipitation	45.40	12.73	0.76	0.027
PWM Method				
Epicenter	179.5	129.8	1.00	0.041
Chain	96.60	33.63	0.66	0.023
Insulation	1170.	269.3	0.61	0.058
Precipitation	45.40	11.87	0.70	0.021
LS Method				
Epicenter	179.5	106.0	0.68	0.061
Chain	96.60	32.82	0.61	0.024
Insulation	1170.	368.9	1.09	0.072
Precipitation	45.40	12.18	0.73	0.023

Example 10.7 (Plots for the GPD$_m$ distribution). The P-P and Q-Q plots for some of the data in Chapter 1 for which the left tail is of interest are given in Figures 10.3 and 10.4. For space-saving purposes, we give the plots for the MLE and the QLS. As can be seen from the graphs, the trend is reasonably

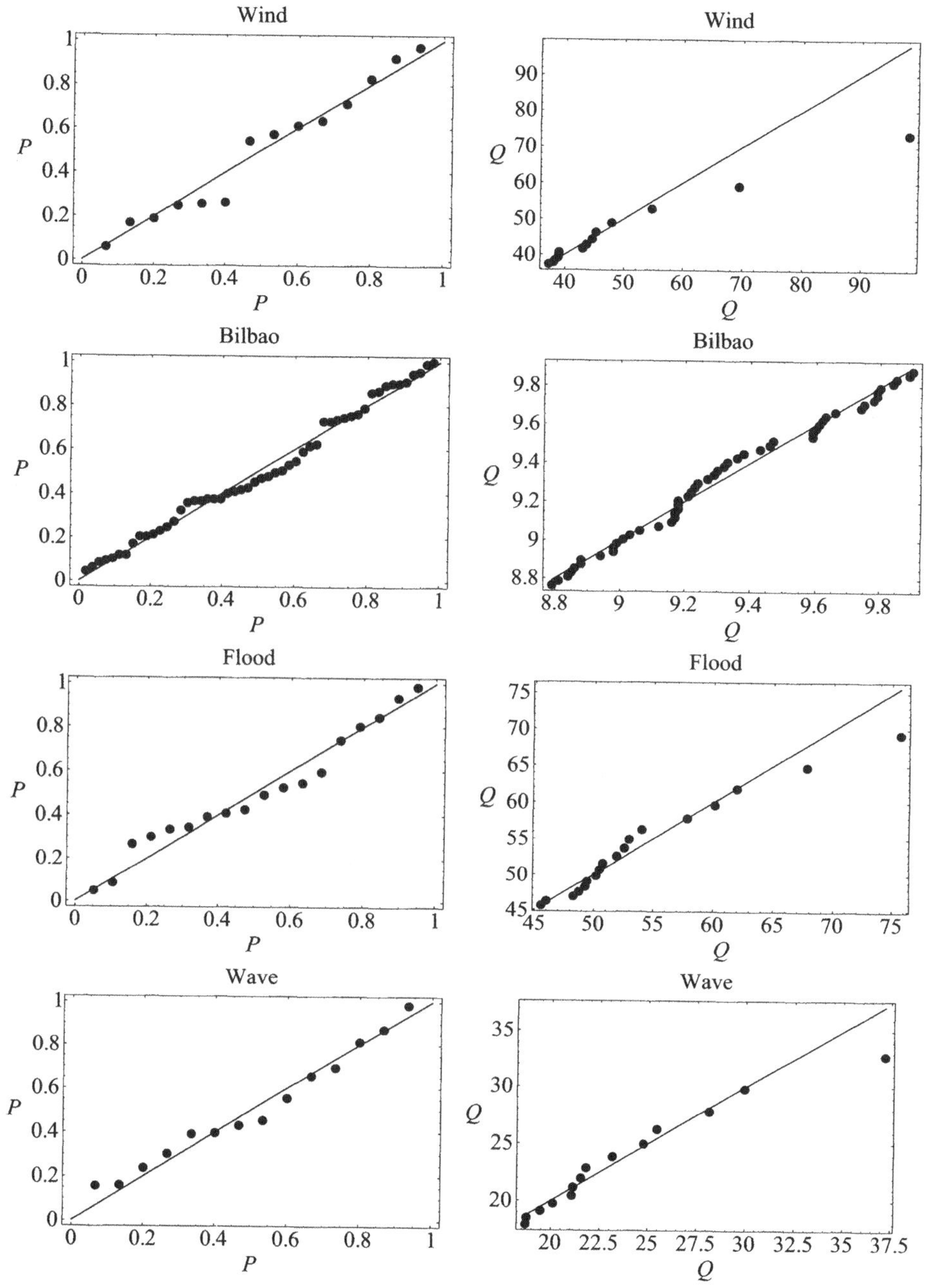

Figure 10.1: P-P and Q-Q plots obtained from fitting the maximal GPD to four data sets using the maximum likelihood method.

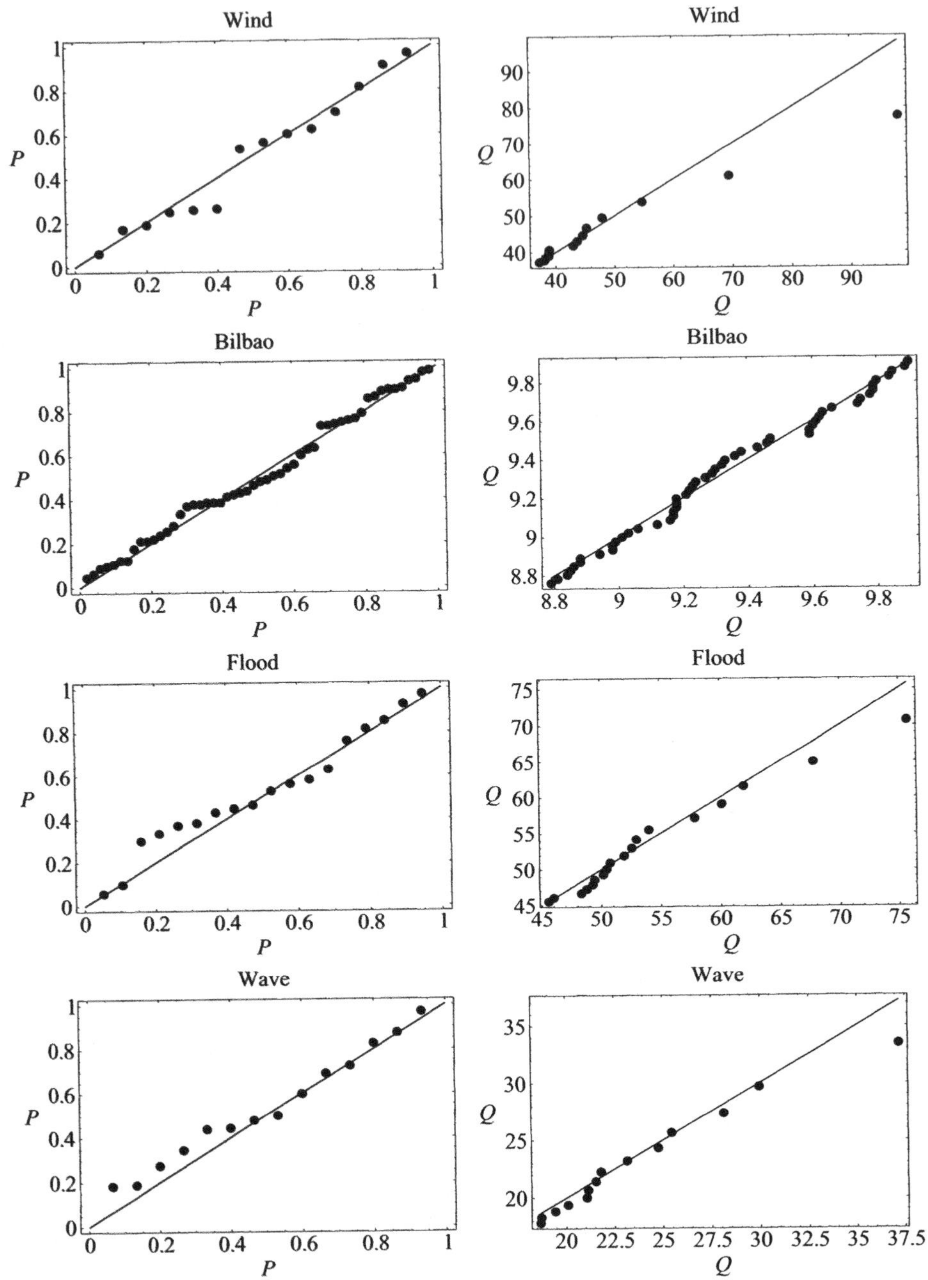

Figure 10.2: P-P and Q-Q plots obtained from fitting the maximal GPD to four data sets using the QLS method.

linear for all data sets. ∎

To further assess the quality of the GPD model, we can proceed as follows:

1. The estimates of λ are plotted versus the threshold values u. Note that the theoretical value of λ as a function of u is (see (10.8))

$$\lambda(u) = \lambda_0 - ku, \tag{10.37}$$

   where λ_0 is the value of λ associated with $u = 0$. Thus, we expect a linear trend in the above plot.

2. Provided that $\kappa > -1$, $u > 0$, and $\lambda - \kappa u > 0$, we have (see Yang (1978), Hall and Wellner (1981), and Davison and Smith (1990))

$$E(X - u|X > u) = \frac{\lambda - \kappa u}{1 + \kappa}. \tag{10.38}$$

   Accordingly, if the GPD is appropriate, the scatter plot of the mean observed excess over u versus u should resemble a straight line with a slope of $-\kappa/(1 + \kappa)$ and an intercept of $\lambda/(1 + \kappa)$. If the points in this scatter should show a strong linear relationship, then the GPD assumption should seem reasonable.

Example 10.8 (Bilbao data). Figure 10.5 shows the $\hat{\lambda}$, $\hat{\kappa}$, and $E[X - u|X \geq u]$ versus the threshold value u for the Bilbao data. It shows a clear linear trend for $\hat{\lambda}$ and $E[X - u|X \geq u]$ versus the threshold value u, and a relatively constant trend for $\hat{\kappa}$, indicating that the assumption of a GPD parent for these data is reasonable for $u > 8.8$ m.

Note that the slope of the $\hat{\lambda}$ versus the threshold value u is approximately $m = -1.1$, leading to an estimate $\hat{\kappa} = 1.1$ that is consistent with the ML estimates in Tables 10.1, 10.3, 10.4, and 10.5.

Finally, the slope of the $E[X - u|X \geq u]$ versus the threshold value u line is approximately $m = -0.45$, from which we get an estimate $\hat{\kappa} = m/(1-m) = 0.81$. ∎

10.10 Hypothesis Tests for the Domain of Attraction

Testing $H_0 : \kappa = 0$ versus $H_1 : \kappa > 0$ is equivalent to testing a Gumbel versus a Weibull domain of attraction. Similarly, testing $H_0 : \kappa = 0$ versus $H_1 : \kappa < 0$ is equivalent to testing a Gumbel versus a Fréchet domain of attraction. We can also test $H_0 : \kappa = 0$ versus $H_1 : \kappa \neq 0$.

To this end, we can:

1. Estimate $\hat{\kappa}$ for the exponential and the GPD models using the maximum likelihood method and then utilize the χ^2_1 statistic as described in Section 6.2.

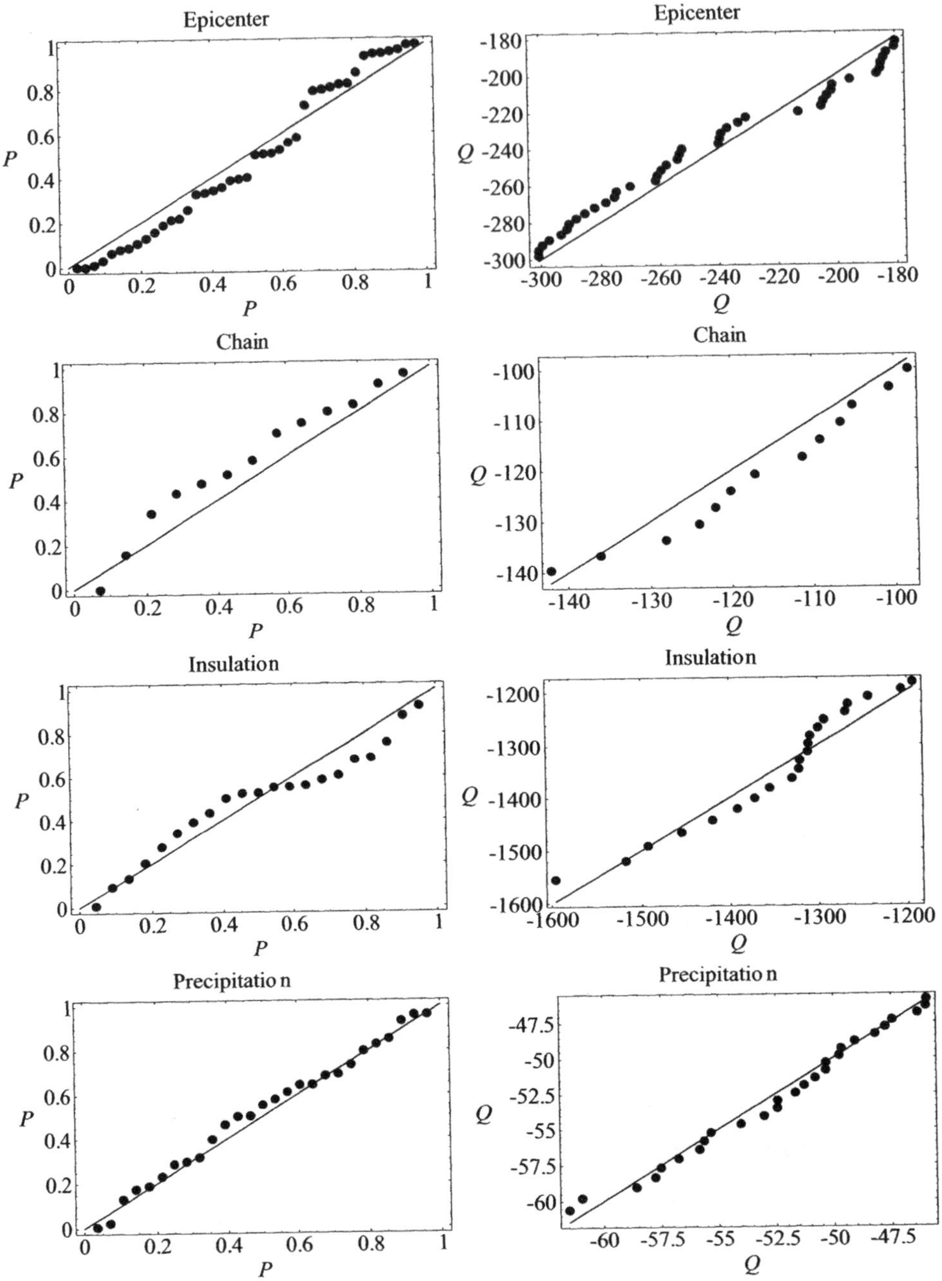

Figure 10.3: P-P and Q-Q plots obtained from fitting the minimal GPD to four data sets using the maximum likelihood method.

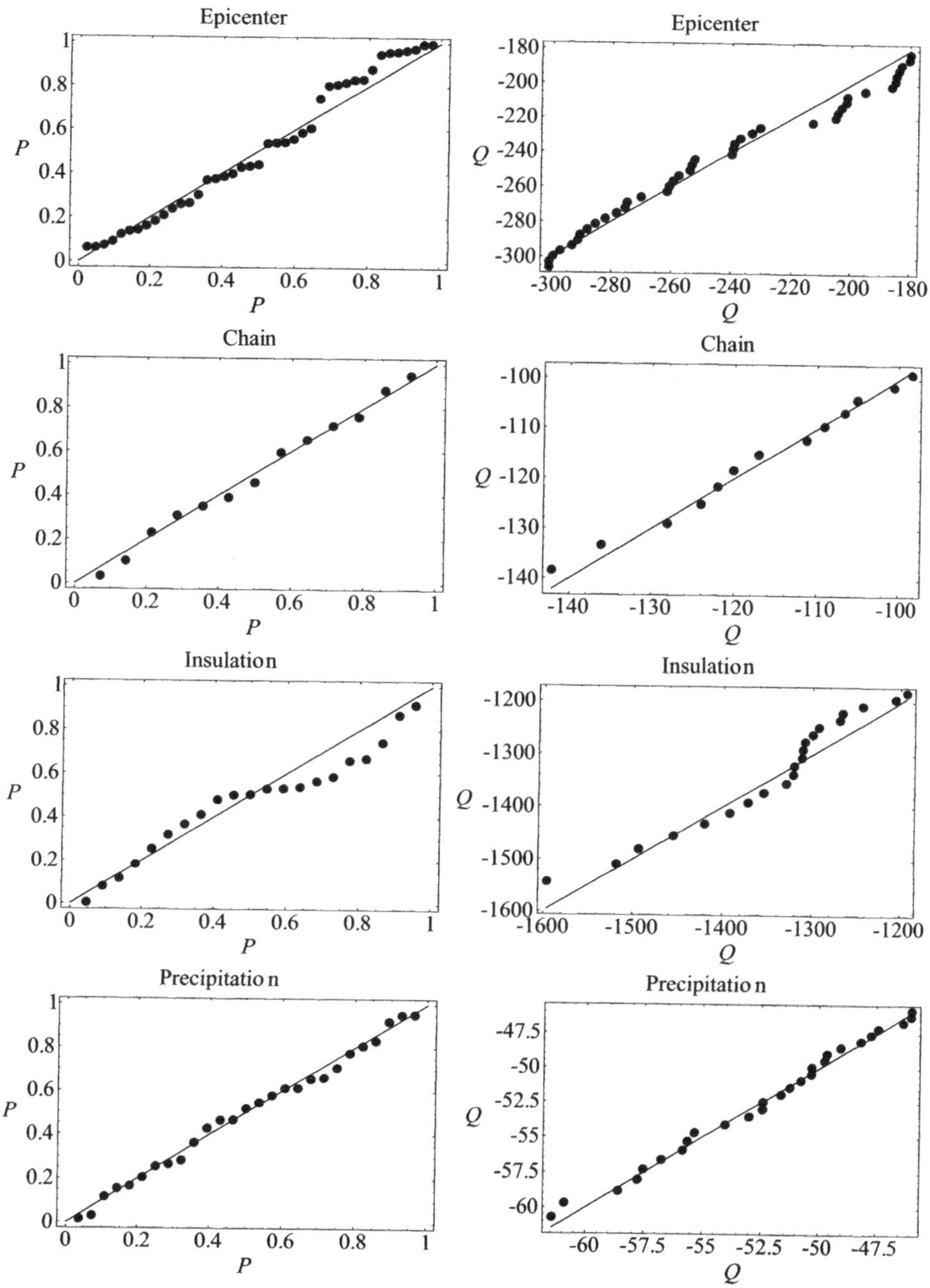

Figure 10.4: P-P and Q-Q plots obtained from fitting the minimal GPD to four data sets using the QLS method.

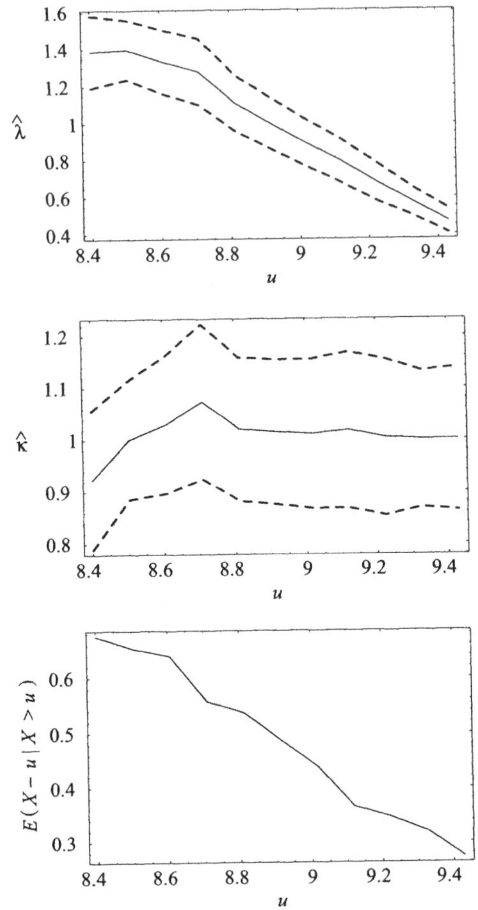

Figure 10.5: Plot of $\hat{\lambda}$, $\hat{\kappa}$ and $E(X - u|X \geq u)$ versus the threshold value u for the Bilbao data.

2. Use a confidence interval for κ and see if it contains $\kappa = 0$, and decide accordingly.

3. Fit a straight line to the λ versus u plot, described in Section 10.9, and test for null slope.

4. Fit a straight line to the $E[X - u|X \geq u]$ versus u plot, described in Section 10.9, and test for null slope.

Some recent interesting references related to hypothesis testing for extremes are Marohn (1998, 2000).

Example 10.9 (Testing the GPD model for the wave data). In this example, we apply the above four testing methods to the wave data:

1. For $u = 10$ m, the MLE of λ for the exponential model is $\hat{\lambda}_1 = 7.37$ with a log-likelihood $\ell_1 = -107.917$, and the MLE for the GPD model are $\hat{\lambda}_2 = 9.01$ and $\hat{\kappa}_2 = 0.222$ with a log-likelihood $\ell_2 = -107.151$. Since the difference $\ell_2 - \ell_1 = 0.766$ is smaller than the critical value $\chi_1(0.95) = 3.8415$, and has an associated p-value of 0.38, we conclude that a Gumbel maximal domain of attraction cannot be rejected.

2. For $u = 10$ m, the MLE of κ is $\hat{\kappa} = 0.222$ and a 95% confidence interval for κ is $(-0.065, 0.509)$. Since the value $\kappa = 0$ belongs to this interval, we conclude that a Gumbel maximal domain of attraction cannot be rejected.

3. Figure 10.6 shows the plots of $\hat{\lambda}$, $\hat{\kappa}$ and $E[X - u|X \geq u]$ versus the threshold value u for the wave data. It shows a linear trend for $\hat{\lambda}$ and $E[X - u|X \geq u]$ versus the threshold value u, and a relatively constant trend for $\hat{\kappa}$, indicating that the assumption of a GPD parent for these data is reasonable for $u > 10$ m.

   If we fit a regression straight line for $\hat{\lambda}$ versus the threshold value u using ten points ($u = 10, 11, \ldots, 19$), we get a regression line $\hat{\kappa} = 11.4 - 0.249u$, and a p value for the slope $p = 0.002$ that leads to the conclusion that a Gumbel maximal domain of attraction must be rejected. Note that the conclusion is now in contradiction with the previous ones. This is due to the reduced number of data points.

4. Finally, the slope of the $E[X - u|X \geq u]$ versus the threshold value u regression line is $m = -0.213$, from which we get $\hat{\kappa} = -m/(1+m) = -0.176$.

■

Exercises

10.1 Show that if $X \sim GPD_M(\lambda, \delta, \kappa)$, then $Y = -X \sim GPD_m(-\lambda, \delta, \kappa)$, that is, prove the second part of Theorem 10.3.

10.2 Show that if $X \sim GPD_m(-\lambda, \kappa)$, then $Y = -X \sim GPD_M(\lambda, \kappa)$, that is, prove the second part of Theorem 10.3.

10.3 Check if the $GPD_m(\lambda, \kappa)$ is also stable with respect to truncations from the right.

10.4 Discuss the stability of the maximal GPD with respect to simultaneous truncation from the right and the left.

10.5 Derive the formulas for confidence intervals of the quantiles of the GPD.

10.6 Fit a GPD to two data sets in Chapter 1 for different threshold values u and test the goodness of fit of the GPD by plotting λ and $E(X - u|X > u)$ versus u.

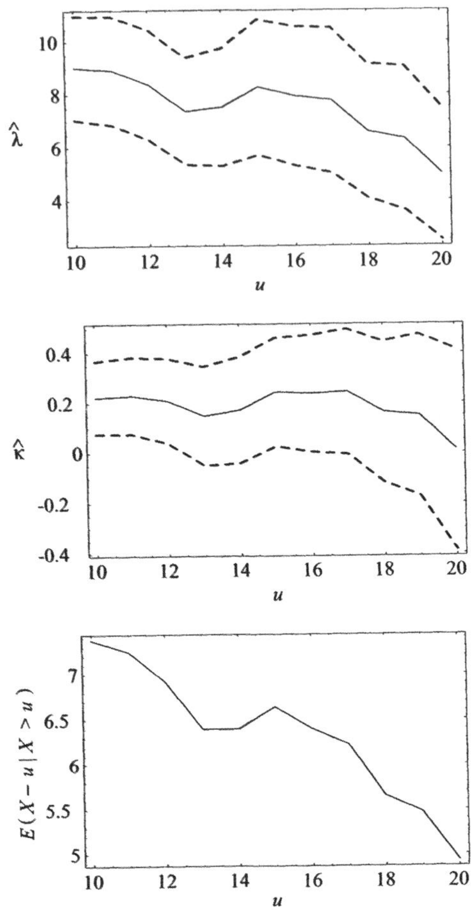

Figure 10.6: Plots of $\hat{\lambda}$, $\hat{\kappa}$, and $E(X - u|X \geq u)$ versus the threshold value u for the wave data.

10.7 Discuss how to make an adequate selection of the threshold value u in the generalized Pareto model in real cases.

10.8 Derive the formulas for the QLS estimates for the GEVD from (10.35) and (10.36) by differentiating these expressions with respect to the parameters. Are they closed or implicit formulas?

10.9 Discuss the existence of a probability paper plot for the GPD model.

10.10 Suggest a test of hypothesis for testing a Gumbel versus a GEVD domain of attraction using the GPD model.

Chapter 11

Multivariate Extremes

In Chapters 9 and 10, we dealt with the case of extremes of sequences of a single random variable. In this chapter, we study the case of extremes of sequences of multivariate data. Fortunately, extreme value theory is also well developed for the multivariate case and we can use this knowledge to solve many practical problems.

In this chapter, we answer the following questions:

1. Given a cdf, how can we know if it belongs to a domain of attraction and what the corresponding limit distribution is?

2. What are the sequences of normalizing constants that must be used to get such a limit?

3. Which multidimensional cdf functions $F(\mathbf{x})$ can appear as maximal limit distributions?

4. Given a set of data coming from a population with unknown cdf, how can we estimate an approximation to its maximal or minimal limit distribution?

5. Given a cdf $F(\mathbf{x})$, how can we know if it is a maximal limit distribution?

Unfortunately, the unique results for the univariate case of having a single family as the only possible limit break down. In the multivariate case, we have a much larger set of possibilities.

In Section 11.1, the extreme value m-dimensional problem is stated. The important tool of dependence function is introduced in Section 11.2. In Section 11.3, we give two alternative methods to find the limit distribution of a given cdf, and show how one can determine the sequences of normalizing constants that must be used to get such a limit. In Section 11.4, we give methods for determining which distributions are possible for a limit, and how one can determine whether a given cdf is a maximal limit distribution. In Section 11.5, we introduce some practically useful parametric bivariate models. In Section 11.6

we show how a set of data can be transformed to a set of data with Fréchet marginals. In Section 11.7, we give a multivariate version of the peaks over thresholds method. In Section 11.8, we discuss some methods for inference. Finally, some illustrative examples of applications are presented in Section 11.9.

Some other interesting discussions on multivariate extremes can be found in Abdous et al. (1999), Capéraà and Fougères (2000), Hall and Tajvidi (2000), Ker (2001), Nadarajah (2000), Peng (1999), Schlather (2001), Smith (1994), and Smith, Tawn, and Yuen (1990).

11.1 Statement of the Problem

Consider a sample of m-dimensional vectors $\mathbf{X}_1, \mathbf{X}_2, \ldots, \mathbf{X}_n$ of size n, coming from a population with cdf $F(\mathbf{x})$ and survival function $S(\mathbf{x})$, where the jth-component of $\mathbf{X}_i, i = 1, 2, \ldots, n$, is denoted by $X_{ij}, j = 1, 2, \ldots, m$. We use bold face letters to denote vectors of multivariate data.[1]

Let $\mathbf{Z}_n$ and $\mathbf{W}_n$ be the vectors of maxima and minima, respectively, that is, the vectors whose components are the respective maxima and minima of the components. The cdfs of $\mathbf{Z}_n$ and $\mathbf{W}_n$ are denoted by $H_n(\mathbf{x})$ and $L_n(\mathbf{x})$. Then, we have

$$\begin{aligned} H_n(\mathbf{x}) &= \Pr(\mathbf{Z}_n \le \mathbf{x}) \\ &= \Pr(\max(X_{11}, \ldots, X_{n1}) \le x_1, \ldots, \max(X_{1m}, \ldots, X_{nm}) \le x_m) \\ &= \Pr(X_{11} \le x_1, \ldots, X_{n1} \le x_1, \ldots, X_{1m} \le x_m, \ldots, X_{nm} \le x_m) \\ &= \Pr(X_{11} \le x_1, \ldots, X_{1m} \le x_m, \ldots, X_{n1} \le x_1, \ldots, X_{nm} \le x_m) \\ &= F(x_1, \ldots, x_m) \cdots F(x_1, \ldots, x_m) \\ &= F^n(x_1, \ldots, x_m) \end{aligned} \tag{11.1}$$

showing that $H_n(\mathbf{x}) = F^n(x_1, \ldots, x_m)$.

Similarly, it can be shown that the survival function is

$$S_n(\mathbf{x}) = \Pr(\mathbf{W}_n > \mathbf{x}) = S^n(x_1, \ldots, x_m), \tag{11.2}$$

where S is the corresponding survival function.

As we have done for the methods developed in Chapter 9, we discuss here the existence of vector sequences $\{\mathbf{a}_n\}, \{\mathbf{b}_n\}, \{\mathbf{c}_n\}$, and $\{\mathbf{d}_n\}$, such that $H_n(\mathbf{x})$ and $L_n(\mathbf{x})$ satisfy

$$\begin{aligned} \lim_{n\to\infty} H_n(\mathbf{a}_n + \mathbf{b}_n\mathbf{x}) &= \lim_{n\to\infty} \Pr(\mathbf{Z}_n \le \mathbf{a}_n + \mathbf{b}_n\mathbf{x}) \\ &= \lim_{n\to\infty} F^n(\mathbf{a}_n + \mathbf{b}_n\mathbf{x}) = H(\mathbf{x}), \end{aligned} \tag{11.3}$$

$$\lim_{n\to\infty} L_n(\mathbf{c}_n + \mathbf{d}_n\mathbf{x}) = \lim_{n\to\infty} \Pr(\mathbf{W}_n \le \mathbf{c}_n + \mathbf{d}_n\mathbf{x}) = L(\mathbf{x}), \tag{11.4}$$

[1]Some of the material in Sections 11.1–11.4 is reprinted from the book *Extreme Value Theory in Engineering*, by E. Castillo, Copyright © Academic Press (1988), with permission from Elsevier.

where $H(\mathbf{x})$ and $L(\mathbf{x})$ are nondegenerate distributions, and the product of vectors $\mathbf{xy}$ here is a component-wise product, that is, $\mathbf{xy} = (x_1y_1, \ldots, x_ny_n)$.

If (11.3) is satisfied, we say that $F(\mathbf{x})$ belongs to the maximal domain of attraction of $H(\mathbf{x})$. Similarly, if (11.4) is satisfied, we say that $F(\mathbf{x})$ belongs to the minimal domain of attraction of $L(\mathbf{x})$. In addition, if $H(\mathbf{x})$ satisfies (11.3), it is called a maximal limit distribution, and if $L(\mathbf{x})$ satisfies (11.4), it is called a minimal limit distribution.

11.2 Dependence Functions

In this section, the important concept of dependence function is introduced. Its importance in multivariate extreme value theory is due to the fact that it can be used to derive the limit distribution of a given multivariate cdf.

Definition 11.1 (Dependence function). *Let $F(\mathbf{x})$ be the cdf of an m-dimensional random variable with univariate marginals $F_i(x_i), i = 1, 2, \ldots, n$. We define the dependence function of $F(\mathbf{x})$, denoted $D_F(y_1, y_2, \ldots, y_m)$ as*

$$D_F(F_1(x_1), F_2(x_2), \ldots, F_m(x_m)) = F(x_1, x_2, \ldots, x_m), \tag{11.5}$$

which for increasing $F_i(x), i = 1, 2, \ldots, n$, becomes

$$D_F(y_1, y_2, \ldots, y_m) = F(F_1^{-1}(y_1), F_2^{-1}(y_2), \ldots, F_m^{-1}(y_m)). \tag{11.6}$$

Note that this function is defined on the unit hypercube $0 \le y_i \le 1, i = 1, 2, \ldots, m$.

We derive below the dependence functions for some well-known bivariate distributions.

Example 11.1 (Mardia's distribution). The cdf of Mardia's distribution is

$$F(x, y) = [\exp(x) + \exp(y) - 1]^{-1} + 1 - \exp(-x) - \exp(-y), \quad -1 \le \alpha \le 1,$$

from which we can derive the dependence function as follows:

$$\begin{aligned} D_F(u, v) &= F(F_1^{-1}(u), F_2^{-1}v)) = F(-\log(1-u), -\log(1-v)) \\ &= \left(\frac{1}{1-u} + \frac{1}{1-v} - 1\right)^{-1} + 1 - (1-u) - (1-v) \\ &= \left(\frac{1}{1-u} + \frac{1}{1-v} - 1\right)^{-1} - 1 + u + v. \end{aligned}$$

■

Example 11.2 (Morgenstern's distribution). The cdf of Morgenstern's distribution is

$$F(x, y) = e^{-x-y}\left\{1 + \alpha\left[1 - e^{-x}\right]\left[1 - e^{-y}\right]\right\} + 1 - e^{-x} - e^{-y},$$

and its dependence function is

$$\begin{aligned} D_F(u,v) &= F(-\log(1-u), -\log(1-v)) \\ &= (1-u)(1-v)(1+\alpha uv) - 1 + u + v \\ &= uv[1+\alpha(1-u)(1-v)]. \end{aligned}$$

■

Example 11.3 (Gumbel's type I distribution). The cdf of the Gumbel's type I distribution is

$$F(x,y) = \exp(-x-y+\theta xy) + 1 - \exp(-x) - \exp(-y),$$

and hence its dependence function is

$$\begin{aligned} D_F(u,v) &= F(-\log(1-u), -\log(1-v)) \\ &= \exp[\log(1-u) + \log(1-v) + \theta\log(1-u)\log(1-v)] \\ &\quad +1 - (1-u) - (1-v) \\ &= (1-u)(1-v)\exp[\theta\log(1-u)\log(1-v)] - 1 + u + v. \end{aligned}$$

■

Example 11.4 (Gumbel's type II distribution). The cdf of the Gumbel's type II distribution is

$$F(x,y) = \exp\left[-\left(x^m + y^m\right)^{1/m}\right] + 1 - \exp(-x) - \exp(-y)$$

and the corresponding dependence function becomes

$$\begin{aligned} D_F(u,v) &= F(-\log(1-u), -\log(1-v)) \\ &= \exp\{-[[-\log(1-u)]^m + [-\log(1-v)]^m]^{1/m}\} - 1 + u + v. \end{aligned}$$

■

Example 11.5 (Marshall-Olkin's distribution). The cdf of the Marshall-Olkin's distribution is

$$F(x,y) = \exp\left[-x-y-\lambda\max(x,y)\right] + 1 - \exp\left[-(1+\lambda)x\right] - \exp\left[-(1+\lambda)y\right],$$

from which the dependence function is derived as

$$\begin{aligned} D_F(u,v) &= F\left(\frac{-\log(1-u)}{1+\lambda}, \frac{-\log(1-v)}{1+\lambda}\right) \\ &= \left\{(1-u)(1-v)[1-\max(u,v)]^{\lambda}\right\}^{1/(1+\lambda)} - 1 + u + v. \end{aligned}$$

■

Example 11.6 (Independent bivariate exponential distribution). The cdf of the independent bivariate exponential distribution is

$$F(x,y) = (1-\exp(-x))(1-\exp(-y)).$$

Then, its dependence function can be written as

$$\begin{aligned} D_F(u,v) &= F(-\log(1-u), -\log(1-v)) \\ &= \{1-\exp[\log(1-u)]\}\{1-\exp[\log(1-v)]\} = uv. \end{aligned}$$

■

Tables 11.1 and 11.2 give a summary of some well-known bivariate distributions together with their cdfs, survival functions, limit distributions, dependence functions, and marginals.

The dependence function is one of the tools used to derive the limit distribution of a given multivariate cdf. This is discussed in the next section.

11.3 Limit Distribution of a Given CDF

In this section we give two alternative methods for obtaining the limit distributions of a given cdf. The first is based on the marginal limit distribution and also gives the normalizing sequences. The second uses the dependence functions.

11.3.1 Limit Distributions Based on Marginals

An easy way of obtaining the limit distribution of a multivariate parent and the associated normalizing sequences $\{\mathbf{a}_n\}$ and $\{\mathbf{b}_n\}$ is by means of the following theorem (see Galambos (1987), p. 290).

Theorem 11.1 (Convergence of marginal distributions). *Let $F_n(\mathbf{x})$ be a sequence of m-dimensional cdfs with univariate marginals $F_{in}(x_i)$. If $F_n(\mathbf{x})$ converges in distribution (weakly) to a nondegenerate continuous cdf $F(\mathbf{x})$, then $F_{in}(x_i)$ converges in distribution to the ith marginal $F_i(x_i)$ of $F(\mathbf{x})$ for $1 \le i \le m$.*

This theorem states that if the limit distribution of the m-dimensional variable exists, then the limit distributions of the marginals also exist and are the marginals of the limit distribution. This suggests using the sequences required for marginal convergence, as explained in Chapter 9, and use them to calculate the limit as m-dimensional sequences. This technique is illustrated below.

Example 11.7 (Mardia's distribution). Because the marginals of Mardia's distribution, $F_X(x) = 1-\exp(-x)$ and $F_Y(y) = 1-\exp(-y)$ are unit exponentials (see Example 9.4), we can choose

$$\mathbf{a}_n = (\log n, \log n) \quad \text{and} \quad \mathbf{b}_n = (1,1), \tag{11.7}$$

Table 11.1: Some Families of Bivariate Distributions Together with Their Marginals (F_X, F_Y), Survival Functions (S), Limit Distributions (H) and Dependence Functions (D_F).

Functions	Mardia's Distribution
$F(x,y)$	$[\exp(x)+\exp(y)-1]^{-1}+1-\exp(-x)-\exp(-y)$
$S(x,y)$	$[\exp(x)+\exp(y)-1]^{-1}$
$H(x,y)$	$\exp\left\{-\exp(-x)-\exp(-y)+[(\exp(x)+\exp(y)]^{-1}\right\}$
$D_F(u,v)$	$\dfrac{uv(u+v-2)}{uv-1}$
Marginals	$F_X(x)=1-\exp(-x)$ and $F_Y(y)=1-\exp(-y)$
Functions	**Morgenstern's Distribution**
$F(x,y)$	$\exp(-x-y)\{1+\alpha[1-\exp(-x)][1-\exp(-y)]\}$ $+1-\exp(-x)-\exp(-y)$
$S(x,y)$	$\exp(-x-y)\{1+\alpha[1-\exp(-x)][1-\exp(-y)]\}$
$H(x,y)$	$\exp\{-\exp(-x)-\exp(-y)\}$
$D_F(u,v)$	$uv[1+\alpha(u-1)(v-1)]$
Marginals	$F_X(x)=1-\exp(-x)$ and $F_Y(y)=1-\exp(-y)$
Functions	**Gumbel's Type I Distribution**
$F(x,y)$	$\exp(-x-y+\theta xy)+1-\exp(-x)-\exp(-y)$
$S(x,y)$	$\exp(-x-y+\theta xy)$
$H(x,y)$	$\exp\{-\exp(-x)-\exp(-y)\}$
$D_F(u,v)$	$(1-u)(1-v)\exp[\theta\log(1-u)\log(1-v)]-1+u+v$
Marginals	$F_X(x)=1-\exp(-x)$ and $F_Y(y)=1-\exp(-y)$
Functions	**Gumbel's Type II Distribution**
$F(x,y)$	$\exp\left[-(x^m+y^m)^{1/m}\right]+1-\exp(-x)-\exp(-y)$
$S(x,y)$	$\exp\left[-(x^m+y^m)^{1/m}\right]$
$H(x,y)$	$\exp\{-\exp(-x)-\exp(-y)\}$
$D_F(u,v)$	$\exp\left\{-[[-\log(1-u)]^m+[-\log(1-v)]^m]^{1/m}\right\}-1+u+v$
Marginals	$F_X(x)=1-\exp(-x)$ and $F_Y(y)=1-\exp(-y)$

Table 11.2: Some Families of Bivariate Distributions Together with Their Marginals (F_X, F_Y), Survival Functions (S), Limit Distributions (H) and Dependence Functions (D_F).

Functions	Marshall-Olkin's Distribution
$F(x,y)$	$\exp[-x-y-\lambda\max(x,y)]+1-\exp[-x(1+\lambda)]$ $-\exp[-y(1+\lambda)]$
$S(x,y)$	$\exp[-x-y-\lambda\max(x,y)]$
$H(x,y)$	$\exp\{-\exp(-x)-\exp(-y)\}$
$D_F(u,v)$	$\left\{(1-u)(1-v)[1-\max(u,v)]^{\lambda}\right\}^{1/(1+\lambda)}-1+u+v$
Marginals	$F_X(x)=1-\exp(-x(1+\lambda))$ and $F_Y(y)=1-\exp(-y(1+\lambda))$

Functions	Oakes-Manatunga's Distribution
$F(x,y)$	$\exp\{-[(\lambda_1 x)^{\kappa_1}+(\lambda_2 y)^{\kappa_2}]^{\alpha}\}$ $+1-\exp[-(\lambda_1 x)^{\kappa_1}]-\exp[-(\lambda_2 y)^{\kappa_2}]$
$S(x,y)$	$\exp\{-[(\lambda_1 x)^{\kappa_1}+(\lambda_2 y)^{\kappa_2}]^{\alpha}\}$
$D_F(u,v)$	$\exp\{-[-\log(1-u)-\log(1-v)]^{\alpha}\}-1+u+v$
Marginals	$F_X(x)=1-\exp[-(\lambda_1 x)^{\kappa_1}]$ and $F_Y(y)=1-\exp[-(\lambda_2 y)^{\kappa_2}]$

Functions	Frank's Distribution
$F(x,y)$	$\log_{\alpha}\left[1+\frac{(\alpha^x-1)(\alpha^y-1)}{\alpha-1}\right],\quad 0\le x,y\le 1,\quad \alpha>0$
$S(x,y)$	$1+\log_{\alpha}\left[1+\frac{(\alpha^x-1)(\alpha^y-1)}{\alpha-1}\right]-x-y$
$D_F(u,v)$	$\log_{\alpha}\left[1+\frac{(\alpha^u-1)(\alpha^v-1)}{\alpha-1}\right]$
Marginals	$F_X(x)=x$ and $F_Y(y)=y$

Functions	Farlie-Gumbel-Morgenstern's Distribution
$F(x,y)$	$xy[1+\alpha(1-x)(1-y)]$
$S(x,y)$	$1+xy[1+\alpha(1-x)(1-y)]-x-y$
$D_F(u,v)$	$uv[1+\alpha(1-u)(1-v)]$
Marginals	$F_X(x)=x$ and $F_Y(y)=y$

and hence

$$\begin{aligned}\lim_{n\to\infty} F^n(\mathbf{a}_n+\mathbf{b}_n\mathbf{x}) &= \lim_{n\to\infty} F^n(\log n + x_1, \log n + x_2)\\ &= \lim_{n\to\infty} \left\{[ne^{x_1}+ne^{x_2}-1]^{-1}+1-n^{-1}\left[e^{-x_1}+e^{-x_2}\right]\right\}^n\\ &= \exp\left[-e^{-x_1}-e^{-x_2}+(e^{x_1}+e^{x_2})^{-1}\right]\\ &= H_0(x_1)H_0(x_2)\exp\left\{[e^{x_1}+e^{x_2}]^{-1}\right\},\end{aligned}$$

where $H_0(x)$ is the cdf of the standard maximal Gumbel distribution. ■

Example 11.8 (Morgenstern's distribution). For the Morgenstern's distribution, we have

$$\mathbf{a}_n = (\log n, \log n) \quad \text{and} \quad \mathbf{b}_n = (1,1). \tag{11.8}$$

Then,

$$\begin{aligned}\lim_{n\to\infty} F^n(\mathbf{a}_n+\mathbf{b}_n\mathbf{x}) &= \lim_{n\to\infty} F^n(\log n + x_1, \log n + x_2)\\ &= \lim_{n\to\infty} F^n(\log[n\exp(x_1)], \log[n\exp(x_2)])\\ &= \lim_{n\to\infty}\left[\exp\left(-\sum_{i=1}^{2} x_i\right)\frac{1+\alpha\prod_{i=1}^{2}\left[1-n^{-1}\exp(-x_i)\right]}{n}\right.\\ &\qquad \left. +1-\frac{\sum_{i=1}^{2}\exp(-x_i))}{n}\right]^n\\ &= \exp\{-\exp(-x_1)-\exp(-x_2)\} = H_0(x_1)H_0(x_2),\end{aligned}$$

where $H_0(x)$ is the cdf of the standard maximal Gumbel distribution. ■

Example 11.9 (Gumbel's type I distribution). For the Gumbel's type I distribution, we have

$$\mathbf{a}_n = (\log n, \log n) \quad \text{and} \quad \mathbf{b}_n = (1,1). \tag{11.9}$$

Thus,

$$\begin{aligned}\lim_{n\to\infty} F^n(\mathbf{a}_n+\mathbf{b}_n\mathbf{x}) &= \lim_{n\to\infty} F^n(\log n + x_1, \log n + x_2)\\ &= \lim_{n\to\infty}\left[\frac{\exp[-x_1-x_2+\theta(\log n+x_1)(\log n+x_2)]}{n^2}\right.\\ &\qquad \left. +1-\frac{\exp(-x_1)+\exp(-x_2)}{n}\right]^n\\ &= \exp\{-\exp(-x_1)-\exp(-x_2)\} = H_0(x_1)H_0(x_2),\end{aligned}$$

where it has been taken into account that $\theta < 0$. ■

Example 11.10 (Gumbel's type II distribution). For the Gumbel's type II distribution, we have

$$\mathbf{a}_n = (\log n, \log n) \quad \text{and} \quad \mathbf{b}_n = (1, 1). \tag{11.10}$$

Thus,

$$\begin{aligned}
\lim_{n\to\infty} F^n(\mathbf{a}_n + \mathbf{b}_n\mathbf{x}) &= \lim_{n\to\infty} F^n(\log n + x_1, \log n + x_2) \\
&= \lim_{n\to\infty} \left(\exp[-[(\log n + x_1)^m + (\log n + x_2)^m]^{1/m}]\right. \\
&\quad \left. +1 - \frac{\exp(-x_1) + \exp(-x_2)}{n}\right)^n \\
&= \lim_{n\to\infty} \left(\exp(-2^{1/m}\log n) + 1 - \frac{\sum_{i=1}^{2} \exp(-x_i)}{n}\right)^n \\
&= \exp\{-\exp(-x_1) - \exp(-x_2)\} = H_0(x_1)H_0(x_2),
\end{aligned}$$

where it has been taken into account that $\theta < 0$. ∎

Example 11.11 (Marshall-Olkin's distribution). For the Marshall-Olkin distribution, we have

$$\begin{aligned}
\lim_{n\to\infty} F^n(\mathbf{a}_n + \mathbf{b}_n\mathbf{x}) &= \lim_{n\to\infty} F^n\left(\frac{\log n + x_1}{1+\lambda}, \frac{\log n + x_2}{1+\lambda}\right) \\
&= \lim_{n\to\infty} \left\{\exp\left[-\left(2\log n + \sum_{i=1}^{2} x_i + \frac{\lambda[\log n + \max(x_1, x_2)]}{1+\lambda}\right)\right.\right. \\
&\quad \left.\left. +1 - \frac{e^{-x_1} + e^{-x_2}}{n}\right]\right\}^n \\
&= \lim_{n\to\infty} \left[\frac{\exp\left(-\dfrac{x_1 + x_2 + \lambda\max(x_1, x_2)}{1+\lambda}\right)}{n^{(2+\lambda)/(1+\lambda)}}\right. \\
&\quad \left. +1 - \frac{\sum_{i=1}^{2} e^{-x_i}}{n}\right]^n \\
&= \exp\{-e^{-x_1} - e^{-x_2}\} = H_0(x_1)H_0(x_2).
\end{aligned}$$

∎

11.3.2 Limit Distributions Based on Dependence Functions

The following theorem (see Galambos (1987)) gives necessary and sufficient conditions for the maximum in a sequence of iid variables to have a limit distribution, and how to obtain the dependence function of these limits.

Theorem 11.2 (Domain of attraction of a given cdf). *Let $\mathbf{X}_1, \ldots, \mathbf{X}_n$ be iid m-dimensional random vectors with common cdf $F(\mathbf{x})$. Then, there are vectors $\mathbf{a}_n$ and $\mathbf{b}_n > \mathbf{0}$ such that $(\mathbf{Z}_n - \mathbf{a}_n)/\mathbf{b}_n$ converges in distribution to a nondegenerate distribution $H(\mathbf{x})$ if, and only if, each of its marginals belong to the maximal domain of attraction of some $H_\kappa(x)$ and if*

$$\lim_{n\to\infty} D_{F^n}(y_1^{1/n}, y_2^{1/n}, \ldots, y_m^{1/n}) = D_H(y_1, y_2, \ldots, y_m). \tag{11.11}$$

Note that $(\mathbf{Z}_n - \mathbf{a}_n)/\mathbf{b}_n$ here means componentwise difference and division.

This theorem enables one to determine if $F(\mathbf{x})$ belongs to the domain of attraction of $H(\mathbf{x})$ and if the maximum of a sequence has a maximal limit distribution.

Some examples are given below.

Example 11.12 (Mardia's distribution). For the Mardia's distribution, we have

$$\begin{aligned}
D_H(u,v) &= \lim_{n\to\infty} D_F^n(u^{1/n}, v^{1/n}) \\
&= \lim_{n\to\infty} \left[u^{1/n} + v^{1/n} - 1 + \left(\frac{1}{1-u^{1/n}} + \frac{1}{1-v^{1/n}} - 1 \right)^{-1} \right]^n \\
&= \lim_{n\to\infty} \left[u^{1/n} + v^{1/n} - 1 + \frac{(1-u^{1/n})(1-v^{1/n})}{1-(uv)^{1/n}} \right]^n \\
&= \lim_{n\to\infty} \left[u^{1/n} v^{1/n} \left(1 + \frac{(1-u^{1/n})(1-v^{1/n})}{1-(uv)^{1/n}} \right) \right]^n \\
&= uv \lim_{n\to\infty} \left[1 - \frac{\log u \log v}{n \log(uv)} \right] \\
&= uv \exp\left[\frac{-\log u \log v}{\log(uv)} \right] \\
&= uv \exp\left[-\left(\frac{1}{\log u} + \frac{1}{\log v} \right)^{-1} \right].
\end{aligned}$$

Consequently, the cdf of the limit distribution is

$$\begin{aligned}
H(x_1, x_2) &= D_H(H_0(x_1), H_0(x_2)) \\
&= H_0(x_1) H_0(x_2) \exp\left\{ [\exp(x_1) + \exp(x_2)]^{-1} \right\}.
\end{aligned} \tag{11.12}$$

Note that the joint distribution is not the product of its marginals. ■

Example 11.13 (Morgenstern's distribution). For the Morgenstern's distribution, we have

$$D_H(u,v) = \lim_{n\to\infty} D_F^n(u^{1/n}, v^{1/n})$$

$$\begin{aligned}
&= uv \lim_{n\to\infty} \left[1+\alpha(1-u^{1/n})(1-v^{1/n})\right]^n \\
&= uv \lim_{n\to\infty} \left[1+n\alpha(1-u^{1/n})(1-v^{1/n})\right] \\
&= uv \lim_{n\to\infty} \left(1+\frac{n\alpha[\log u \log v]}{n^2}\right) = uv,
\end{aligned}$$

where we have taken into account that

$$\text{if } u_n \to 0 \Rightarrow (1+u_n)^n \sim nu_n$$

and

$$\text{if } u_n \to 1 \Rightarrow 1-u_n \sim -\log u_n.$$

Consequently, the cdf of the limit distribution is

$$H(x_1,x_2) = D_H(H_0(x_1),H_0(x_2)) = H_0(x_1)H_0(x_2).$$

■

Example 11.14 (Gumbel's type I distribution). For the Gumbel's type I distribution, we have

$$\begin{aligned}
D_H(u,v) &= \lim_{n\to\infty} D_F^n(u^{1/n},v^{1/n}) \\
&= \lim_{n\to\infty} \Big\{(1-u^{1/n})(1-v^{1/n})\exp[\theta\log(1-u^{1/n})\log(1-v^{1/n})] \\
&\quad -1+u^{1/n}+v^{1/n}\Big\}^n \\
&= \lim_{n\to\infty} \left(\frac{\prod_{i=1}^{2}\log y_i \exp\left[\theta\prod_{i=1}^{2}\log(1-y_i^{1/n})\right]}{n^2} - 1 + \sum_{i=1}^{2} y_i^{1/n}\right)^n \\
&= \lim_{n\to\infty} \exp\left[n(u^{1/n}+v^{1/n}-2)\right] = \lim_{n\to\infty} \exp\left[\frac{u^{1/n}+v^{1/n}-2}{1/n}\right] \\
&= \lim_{n\to\infty} \exp\left[\sum_{i=1}^{2}(y_i^{1/n}\log y_i)\right] = \exp[\log(uv)] = uv.
\end{aligned}$$

Then, we have

$$H(x_1,x_2) = D_H(H_0(x_1),H_0(x_2)) = H_0(x_1)H_0(x_2).$$

■

Example 11.15 (Gumbel's type II distribution). For the Gumbel's type II distribution, we have

$$D_H(u,v) = \lim_{n\to\infty} D_F^n(u^{1/n},v^{1/n})$$

$$
\begin{aligned}
&= \lim_{n\to\infty}\left\{\exp\left\{-\left[\sum_{i=1}^{2}\left[-\log(1-y_i^{1/n})\right]^m\right]^{1/m}\right\}-1+\sum_{i=1}^{2}y_i^{1/n}\right\}^n \\
&= \lim_{n\to\infty}\left\{\exp\left\{-\left[-\log(1-v^{1/n})\right]\left[1+\left(\frac{\log(1-u^{1/n})}{\log(1-v^{1/n})}\right)^m\right]^{1/m}\right\}\right. \\
&\quad \left. -1+u^{1/n}+v^{1/n}\right\}^n \\
&= \lim_{n\to\infty}\exp[n(u^{1/n}+v^{1/n}-2)] = uv,
\end{aligned}
$$

and then, we get

$$H(x_1,x_2) = D_H(H_0(x_1),H_0(x_2)) = H_0(x_1)H_0(x_2).$$

■

Example 11.16 (Marshall-Olkin's distribution). For the Marshall-Olkin distribution, let $A(n) = -1+u^{1/n}+v^{1/n}$; we then have

$$
\begin{aligned}
D_H(u,v) &= \lim_{n\to\infty}\left[D_F(u^{1/n},v^{1/n})\right]^n \\
&= \lim_{n\to\infty}\left[\left\{\prod_{i=1}^{2}(1-y_i^{1/n})\left[1-\max(u^{1/n},v^{1/n})\right]^\lambda\right\}^{1/(1+\lambda)}+A(n)\right]^n \\
&= \lim_{n\to\infty}\left[\prod_{i=1}^{2}\log y_i\left(\frac{\left[1-\max(u^{1/n},v^{1/n})\right]^\lambda}{n^2}\right)^{1/(1+\lambda)}+A(n)\right]^n \\
&= \lim_{n\to\infty}\left[\left(\frac{\prod_{i=1}^{2}\log y_i\left\{\log\left[\max(u,v)\right]\right\}^\lambda}{n^3}\right)^{1/(1+\lambda)}+A(n)\right]^n \\
&= \lim_{n\to\infty}\exp\left[n\left(u^{1/n}+v^{1/n}-2\right)\right] = uv,
\end{aligned}
$$

which leads to

$$H(x_1,x_2) = D_H(H_0(x_1),H_0(x_2)) = H_0(x_1)H_0(x_2).$$

■

11.4 Characterization of Extreme Distributions

In this section, we first present a theorem, which will enable us to decide whether a given m-dimensional cdf can be a limit distribution of a maximal sequence of random variables (see Galambos (1987)). Then, we present two different approaches for the characterization of the family of all limit distributions arising from a maximal sequence: (a) Functional Equations Approach and (b) Point Process Approach.

11.4.1 Identifying Extreme Value Distributions

Theorem 11.3 (Maximal limit distributions). *An m-dimensional cdf $H(\mathbf{x})$ is a maximal limit distribution if, and only if, (a) its univariate marginals belong to the maximal domain of attraction of $H_\kappa(x)$ and (b) its dependence function $D_H(y_1, y_2, \ldots, y_m)$ satisfies the functional equation*

$$D_H^k(y_1^{1/k}, y_2^{1/k}, \ldots, y_m^{1/k}) = D_H(y_1, y_2, \ldots, y_m), \tag{11.13}$$

for any $k \geq 1$.

We illustrate this theorem by an example.

Example 11.17 (Mardia's distribution). The cdf of the Mardia's distribution can be a limit distribution because:

1. Its marginals belong to the maximal domain of attraction of $H_0(x)$; see (11.7).

2. Its dependence function satisfies (11.13):

$$\begin{aligned} D_H^k(y_1^{1/k}, y_2^{1/k}, \ldots, y_m^{1/k}) &= \left[y_1^{1/k} y_2^{1/k} \exp\left(-\frac{\log y_1^{1/k} \log y_2^{1/k}}{\log(y_1^{1/k} y_2^{1/k})} \right) \right]^k \\ &= y_1 y_2 \exp\left(-\frac{\log y_1 \log y_2}{\log(y_1 y_2)} \right) \\ &= D_H(y_1, y_2, \ldots, y_m). \end{aligned}$$

Table 11.3 gives some examples of two types of families of extreme value distributions, that is, they can be limit distributions of extremes of some cdfs.

A particular case of a Type A cdf is

$$F(x, y) = \exp\left[-\exp(-x) - \exp(-y) + \theta(\exp(x) + \exp(y))^{-1} \right], \quad 0 \leq \theta \leq 1,$$

and a particular case of a Type B cdf is

$$F(x, y) = \exp\left[-\left(\exp(-mx) + \exp(-my)\right)^{1/m} \right], \quad m \geq 1.$$

■

11.4.2 Functional Equations Approach

The first approach, for the characterization of the family of all limit distributions arising from a maximal sequence, is based on Theorem 11.3 and can be given as the following corollary:

Corollary 11.1 *Equation (11.13) is equivalent to*

$$H(x_1, x_2, \ldots, x_m) = \left[\prod_{i=1}^{m} H_0(x_i)\right]^{v(x_2-x_1, x_3-x_1, \ldots, x_m-x_1)}, \tag{11.14}$$

where $H_0(x) = \exp\{-\exp(-x)\}$, *and the function*

$$v(x_2 - x_1, x_3 - x_1, \ldots, x_m - x_1)$$

must be such that $H(x_1, x_2, \ldots, x_m)$ *becomes a proper pdf, but otherwise arbitrary.*

The main disadvantage of this approach is that it is not easy to give conditions for the function $v(\cdot)$ to lead to a valid cdf.

Example 11.18 (Mardia's distribution). Consider the limit distribution of Mardia's distribution in (11.12):

$$H(x_1, x_2) = H_0(x_1)H_0(x_2)\exp\left\{[\exp(x_1) + \exp(x_2)]^{-1}\right\}. \tag{11.15}$$

From (11.14) and (11.15) we get

$$\begin{aligned} H(x_1, x_2) &= H_0(x_1)H_0(x_2)\exp\left\{[\exp(x_1) + \exp(x_2)]^{-1}\right\} \\ &= (H_0(x_1)H_0(x_2))^{v(x_2-x_1)} \end{aligned}$$

and then

$$\begin{aligned} v(x_2 - x_1) &= 1 + \frac{\log\left(\exp\left\{[\exp(x_1) + \exp(x_2)]^{-1}\right\}\right)}{\log(H_0(x_1)H_0(x_2))} \\ &= 1 - \frac{[\exp(x_1) + \exp(x_2)]^{-1}}{\exp(-x_1) + \exp(-x_2)} \\ &= 1 - [2 + \exp(x_2 - x_1) + \exp(-(x_2 - x_1))]^{-1}. \end{aligned}$$

■

11.4.3 A Point Process Approach

The second approach, for the characterization of the family of all limit distributions arising from a maximal sequence, is based on a point process representation of multivariate extremes; following Coles and Tawn (1991, 1994), de Haan and Resnick (1977), de Haan (1985), Joe, Smith, and Weissman (1992), Resnick (1987), and Tawn (1988). This representation is useful for:

1. Deriving the nonhomogeneous Poisson process associated with extreme multivariate points.
2. Obtaining the general structure of multivariate extreme limit distributions.

3. Generalizing the exceedances over threshold model to n-dimensions.

We start with the nonhomogeneous Poisson process associated with extreme multivariate points, which is defined by the following theorem:

Theorem 11.4 (Point process representation). *Let $\mathbf{Z}_1$, $\mathbf{Z}_2$, ... be a sequence of iid random vectors on $\mathbb{R}_+^m$, whose cdf F is in the maximal domain of attraction of a multivariate extreme value distribution G and its marginal co!maximalmponents are identically distributed with a unit Fréchet distribution. Consider the point process $P_n = \{\mathbf{Z}_i/n | i = 1, 2, \ldots, n\}$. Then, P_n converges in distribution to a nonhomogeneous Poisson process P on $\mathbb{R}_+^m - \{\mathbf{0}\}$ when $n \to \infty$, with intensity measure*

$$\lambda(dr \times d\mathbf{w}) = m\frac{dr}{r^2}dS(\mathbf{w}), \tag{11.16}$$

where r_i and $\mathbf{w}_{ij}$ are the pseudo-polar and angular coordinates, that is,

$$r_i = \frac{1}{n}\sum_{j=1}^{m} Z_{ij}, \quad w_{ij} = \frac{Z_{ij}}{\sum\limits_{j=1}^{m} Z_{ij}}, \quad i = 1, 2, \ldots, n, \quad j = 1, 2, \ldots, m, \tag{11.17}$$

where Z_{ij} is the jth component of Z_i and S is a probability measure on the unit simplex

$$S_m = \left\{ (w_1, w_2, \ldots, w_m) \left| \sum_{j=1}^{m} w_j = 1, \quad w_j \geq 0, \quad j = 1, 2, \ldots, m \right. \right\} \tag{11.18}$$

that satisfies

$$\int_{S_m} w_j dS(\mathbf{w}) = 1/m, \quad j = 1, 2, \ldots, m. \tag{11.19}$$

Alternatively, one can use

$$\lambda^*(dr \times d\mathbf{w}) = m\frac{dr}{r^2}dS^*(\mathbf{w}), \tag{11.20}$$

where now $\mathbf{w} = (w_1, w_2, \ldots, w_{m-1})$ and

$$S_m^* = \left\{ (w_1, w_2, \ldots, w_{m-1}) \left| \sum_{j=1}^{m-1} w_j \leq 1, \quad w_j \geq 0, \quad j = 1, 2, \ldots, m-1 \right. \right\}$$

that satisfies

$$\int_{S_m^*} w_j dS^*(\mathbf{w}) = 1/m, \quad j = 1, 2, \ldots, m-1. \tag{11.21}$$

This theorem has a great practical importance because it says that for any sequence $\mathbf{Z}_1, \mathbf{Z}_2, \ldots$ there is a positive finite measure $S(\mathbf{w})$ satisfying condition (11.19) such that the associated Poisson nonhomogeneous process has intensity (11.16). Thus, given $S(\mathbf{w})$ we can calculate the probabilities of occurrence of Poisson processes in any given region A. The following corollary, which characterizes all possible multivariate limit distributions, is one of these applications.

Corollary 11.2 (Limit distributions for multivariate extremes.) *Any limit distribution of normalized component-wise maxima, with unit Fréchet marginals, has the form*

$$H(\mathbf{z}) = \exp\left[-v(\mathbf{z})\right], \tag{11.22}$$

where

$$v(\mathbf{z}) = m \int_{S_m} \max_{1 \le j \le m} (w_j / z_j) dS(\mathbf{w}) \tag{11.23}$$

for some $S(\mathbf{w})$ defined on (11.18) and satisfying (11.19).

This is a consequence of the limiting Poisson process P with intensity λ in (11.16). In fact, defining

$$A = \mathbb{R}_+^m - \left\{ \prod_{j=1}^{m} [0, z_j) \right\}, \tag{11.24}$$

as $n \to \infty$, we have that

$$\Pr\left[\frac{1}{n}\mathbf{Z}_i \notin A, i = 1, 2, \ldots, n\right] \to \exp\{-\Lambda(A)\}, \tag{11.25}$$

where

$$\begin{aligned} \Lambda(A) &= m \int_A \frac{dr}{r^2} dS(\mathbf{w}) = m \int_{S^m} \int_{r=\min_{1 \le j \le m}(z_j/w_j)}^{\infty} \frac{dr}{r^2} dS(\mathbf{w}) \\ &= m \int_{S^m} \max_{1 \le j \le m} (w_j / z_j) dS(\mathbf{w}). \end{aligned} \tag{11.26}$$

Then, we obtain the likelihood associated with a region A bounded away from the origin

$$\begin{aligned} L_A(\boldsymbol{\theta}; n^{-1}\mathbf{Z}_i) &= \exp\left[-\Lambda(A)\right] \prod_{i=1}^{n_A} \lambda(dr_i \times d\mathbf{w}_i) \\ &= \exp\left[-\Lambda(A)\right] \prod_{i=1}^{n_A} m \frac{dr_i}{r_i^2} dS(\mathbf{w}_i; \boldsymbol{\theta}), \end{aligned} \tag{11.27}$$

where $\boldsymbol{\theta}$ is the vector of parameters, and n_A is the number of data points in the set A.

Example 11.19 (Bivariate limits). For the bivariate limits, (11.23) becomes

$$\begin{aligned} v(z_1, z_2) &= 2\int_0^1 \max\left(\frac{w}{z_1}, \frac{1-w}{z_2}\right) dS^*(w) \\ &= 2\int_0^{\frac{z_1}{z_1+z_2}} \frac{1-w}{z_2} dS^*(w) + 2\int_{\frac{z_1}{z_1+z_2}}^1 \frac{w}{z_1} dS^*(w). \quad (11.28) \end{aligned}$$

Next, we apply the expression in (11.28) to several cases:

1. A positive finite measure that places a mass 1/2 on $w_1 = 0, w_1 = 1$ and $w_1 = 1, w_2 = 0$:

$$\begin{aligned} v(z_1, z_2) &= 1/z_1 + 1/z_2, \\ H(z_1, z_2) &= \exp\left[-\left(1/z_1 + 1/z_2\right)\right], \quad z_1, z_2 > 0. \end{aligned}$$

2. A positive finite measure that places mass 1 on $w_1 = w_2 = 0.5$:

$$\begin{aligned} v(z_1, z_2) &= \max(1/z_1, 1/z_2), \\ H(z_1, z_2) &= \exp\left[-\max(1/z_1, 1/z_2))\right], \quad z_1, z_2 > 0. \end{aligned}$$

3. A uniform positive mass of intensity 1:

$$\begin{aligned} v(z_1, z_2) &= \frac{\left(z_1^2 + z_1 z_2 + z_2^2\right)}{z_1 z_2 \left(z_1 + z_2\right)} \\ H(z_1, z_2) &= \exp\left[-\frac{\left(z_1^2 + z_1 z_2 + z_2^2\right)}{z_1 z_2 \left(z_1 + z_2\right)}\right], \quad z_1, z_2 > 0. \end{aligned}$$

■

Example 11.20 (Trivariate limits). For the trivariate limits, (11.23) becomes

$$\begin{aligned} v(z_1, z_2, z_3) &= 3\int_0^1\int_0^{1-w_1} \max\left(\frac{w_1}{z_1}, \frac{w_2}{z_2}, \frac{1-w_1-w_2}{z_3}\right) dS^*(w_1, w_2) \\ &= 3\int_0^1\int_{\max(0,1-w_1-w_1\frac{z_3}{z_1})}^{\min(1-w_1,\frac{z_2 w_1}{z_1})} \frac{w_1}{z_1}\, dS^*(w_1, w_2) \\ &\quad +3\int_0^1\int_0^{\min(\frac{z_2(1-w_1)}{z_2+z_3},1-w_1-w_1\frac{z_3}{z_1})} \frac{1-w_1-w_2}{z_3}\, dS^*(w_1, w_2). \end{aligned}$$

■

Example 11.21 (Multivariate limits). Consider a positive finite measure $S(\mathbf{w})$ that places mass 1 on the $(w_1, w_2, \ldots, w_m)$ points $(1, 0, \ldots, 0)$, $(0, 1, \ldots, 0)$, ..., $(0, 0, \ldots, 1)$. Then, from (11.23) we get

$$v(z_1, z_2, \ldots, z_m) = \left(1/z_1 + 1/z_2 + \ldots + 1/z_m\right),$$

that leads to

$$H(z_1, z_2, \ldots, z_m) = \exp[-(1/z_1 + 1/z_2 + \ldots + 1/z_m)], \quad z_1, \ldots, z_m > 0.$$

■

Corollaries 11.1 and 11.2 show that the set of possible limit distributions for two or more dimensions ($m > 1$) is much wider than the set of possible limit distributions for the one dimensional case, where all possible limit distributions could be included in a single parametric family (the von-Mises family). Thus, nonparametric methods are useful for $m \geq 2$. However, in practice, parametric families of limit distributions are used. Another possibility consists of using $F^m(\mathbf{x};\theta)$ for any family of cdfs $F(\mathbf{x};\theta)$, since for large m a limit distribution can be approximated.

11.5 Some Parametric Bivariate Models

In this section, we include some of the most important bivariate parametric models.

1. **Logistic model**: It corresponds to a density function in $0 \leq w \leq 1$:

$$S^*(w) = \frac{1}{2}(\alpha^{-1} - 1)\{w(1-w)\}^{-1-1/\alpha}\left\{w^{-1/\alpha} + (1-w)^{-1/\alpha}\right\}^{\alpha-2}$$

that leads to

$$\begin{aligned} v(z_1, z_2) &= \left(z_1^{-1/\alpha} + z_2^{-1/\alpha}\right)^{\alpha}, \quad z_1, z_2 \geq 0 \\ H(z_1, z_2) &= \exp\left\{-\left(z_1^{-1/\alpha} + z_2^{-1/\alpha}\right)^{\alpha}\right\}, \quad z_1, z_2 \geq 0. \end{aligned}$$

From this model with unit Fréchet marginals, we can obtain the more general model with cdf

$$F(x,y) = \exp\left\{-\left[\left(1-\kappa_1\frac{x-\lambda_1}{\delta_1}\right)^{1/\alpha\kappa_1} + \left(1-\kappa_2\frac{y-\lambda_2}{\delta_2}\right)^{1/\alpha\kappa_2}\right]^{\alpha}\right\},$$

which obviously has GEVD marginals.

An interesting particular case is the one with Gumbel marginals, that is,

$$F(x,y) = \exp\left\{-\left[\exp\left(\frac{\lambda_1 - x}{\delta_1\alpha}\right) + \exp\left(\frac{\lambda_2 - y}{\delta_2\alpha}\right)\right]^{\alpha}\right\}, \quad -\infty < x, y < \infty. \tag{11.29}$$

2. **Dirichlet model**: It corresponds to a density function in $0 \leq w \leq 1$:

$$S^*(w) = \frac{\alpha\beta\Gamma(\alpha+\beta+1)(\alpha w)^{\alpha-1}(\beta(1-w))^{\beta-1}}{2\Gamma(\alpha)\Gamma(\beta)(\alpha w + \beta(1-w))^{\alpha+\beta+1}} \tag{11.30}$$

that leads to

$$\begin{aligned} v(z_1,z_2) &= (z_1)^{-1}\left[1 - I_\beta\left(\frac{\alpha z_1}{\alpha z_1+\beta z_2};\alpha+1,\beta\right)\right] \\ &\quad +(z_2)^{-1}\left[I_\beta\left(\frac{\alpha z_1}{\alpha z_1+\beta z_2};\alpha,\beta+1\right)\right], \quad z_1,z_2\geq 0, \\ H(z_1,z_2) &= \exp\left\{-v(z_1,z_2)\right\}, \quad z_1,z_2\geq 0, \end{aligned}$$

where

$$I_\beta\left(x;\alpha,\beta\right) = \frac{\Gamma(\alpha+\beta)}{\Gamma(\alpha)\Gamma(\beta)}\int_0^x w^{\alpha-1}(1-w)^{\beta-1}dw$$

is the incomplete Beta function defined in (3.36).

3. **Bilogistic model**: It corresponds to a density function in $0 \leq w \leq 1$, that is,

$$S^*(w) = \frac{(1-\alpha)(1-u)u^{1-\alpha}}{2(1-w)w^2(\alpha(1-u)+\beta u)}, \quad 0 < w < 1, \tag{11.31}$$

and $0 < \alpha, \beta < 1$ which leads to

$$\begin{aligned} v(z_1,z_2) &= \int_0^1 \max\left(\frac{1-\alpha}{z_1 s^\alpha}, \frac{1-\beta}{z_2(1-s)^\beta}\right) ds, \quad z_1,z_2\geq 0, \\ H(z_1,z_2) &= \exp\left\{-\int_0^1 \max\left(\frac{1-\alpha}{z_1 s^\alpha}, \frac{1-\beta}{z_2(1-s)^\beta}\right) ds\right\}, \quad z_1,z_2\geq 0, \end{aligned}$$

where u is the root of the equation

$$(1-\alpha)(1-w)(1-u)^\beta - (1-\beta)wu^\alpha = 0.$$

11.6 Transformation to Fréchet Marginals

Theorem 11.4 and Corollary 11.2 are valid only for data with Fréchet marginals. Thus, to obtain the general class of bivariate limit distributions, we need to transform the data $\mathbf{x}$ to another variable $\mathbf{z}$ with Fréchet marginals. Several options are possible, depending on the type of data:

1. If the data $\mathbf{x}$ are extreme data (maxima or minima) and they approximately follow a GEVD, we can use the transformation

$$z_j(x_j) = \left[1 - \kappa_j\left(\frac{x_j-\lambda_j}{\delta_j}\right)\right]^{-1/\kappa_j}, \quad j = 1,2,\ldots,m, \tag{11.32}$$

which, for $\kappa_j = 0$, becomes

$$z_j(x_j) = \exp\left(\frac{x_j-\lambda_j}{\delta_j}\right).$$

Thus, for the general class of bivariate limit distributions, from (11.22) $H(\mathbf{x})$ becomes

$$\exp\left[-v\left(\left[1-\kappa_1\left(\frac{x_1-\lambda_1}{\delta_1}\right)\right]^{-1/\kappa_1},\ldots,\left[1-\kappa_m\left(\frac{x_m-\lambda_m}{\delta_m}\right)\right]^{-1/\kappa_m}\right)\right].$$

2. Another option is to use the empirical cdfs $F_j(x_j),\ j=1,2,\ldots,m$, using the equation

$$F_j(x_j)=F_{Z_j}(z_j)=\exp(-1/z_j)\Leftrightarrow z_j=z_j(x_j)=-\left[\log(F_j(x_j))\right]^{-1}.$$

This option is valid only for interpolating between sample values.

3. Finally, according to Pickands (1975), the exceedances over a high threshold follow a generalized Pareto distribution; Coles and Tawn (1991) suggest the marginal transformation $z_j = z_j(x_j)$, where

$$z_j=\begin{cases}-\left\{\log\left[1-p_j(1-\kappa_j(x_j-u_j)/\alpha_j)^{1/\kappa_j}\right]\right\}^{-1}, & \text{if} \quad x_j>u_j,\\ -\left[\log(F_j(x_j))\right]^{-1}, & \text{if} \quad x_j\le u_j,\end{cases} \tag{11.33}$$

where u_j are high threshold values, $p_j = 1 - F_j(u_j)$, and κ_j and α_j can be estimated from the marginal data. Note that this transformation is based on the empirical distribution for $x_j < u_j$ and on the generalized Pareto approximation for $x_j > u_j$. Thus, the two expressions in (11.33) come from the two equations

$$F_j(x_j)=F_{GPD(\alpha_j,\kappa_j)}(z_j) \quad \text{and} \quad F_j(x_j)=\exp(-1/z_j),$$

respectively.

11.7 Peaks Over Threshold Multivariate Model

In this section, we generalize the exceedances over a threshold model. Since we are in a multidimensional context, we need to generalize the concept of exceedances. For the univariate case, an exceedance of u is a value $z > u$. For the n-dimensional case, we need to define a family of functions $h(\mathbf{z};u)$ such that $\mathbf{z}$ is an n-dimensional exceedance if $h(\mathbf{z};u)\ge 0$. However, this family is not arbitrary, but must satisfy some conditions as:

$$A(u_1)=\{\mathbf{z}|h(\mathbf{z};u_1)\ge 0\}\subset A(u_2)=\{\mathbf{z}|h(\mathbf{z};u_2)\ge 0\}, \quad \forall\ u_1>u_2,$$

and

$$(\infty,\infty,\ldots,\infty)\in A(u)=\{\mathbf{z}|h(\mathbf{z};u)\ge 0\}, \quad \forall\ u.$$

If $h(\mathbf{z};u)=g(\mathbf{z})-u$ for some function $g(\cdot)$, the problem can be transformed into a unidimensional problem, since we can transform the initial data $\{\mathbf{z}_1,\mathbf{z}_2,\ldots,\mathbf{z}_n\}$ into $\{g(\mathbf{z}_1),g(\mathbf{z}_2),\ldots,g(\mathbf{z}_n)\}$ and use the univariate threshold model.

According to Theorem 11.4, in the multivariate case we have

$$\Lambda(A(u)) = m \int_{A(u)} \frac{dr}{r^2} dS(\mathbf{w}) \tag{11.34}$$

and then, the likelihood function becomes (see (8.2))

$$L(\boldsymbol{\theta}; \mathbf{z}) \propto \exp\{-\Lambda[A(u); \boldsymbol{\theta}]\} \prod_{i=1}^{N_{A(u)}} \lambda(\mathbf{z}_i/n; \boldsymbol{\theta}), \tag{11.35}$$

where $N_{A(u)}$ is the number of points falling in $A(u)$.

An interesting particular case arises when $h(\mathbf{z}; u) = (z_1 + z_2 + \ldots + z_n)/n - u$, that leads to $A(u) = \{\mathbf{z} | (z_1 + z_2 + \ldots, z_m)/n \geq u\}$ and to

$$\Lambda(A(u); \boldsymbol{\theta}) = m \int_{A(u)} \frac{dr}{r^2} dS(\mathbf{w}; \boldsymbol{\theta}) = m \int_{r=u}^{\infty} \frac{dr}{r^2} \int_{S_m} dS(\mathbf{w}; \boldsymbol{\theta}) = \frac{m}{u}, \tag{11.36}$$

where S_m is as defined in (11.18), which is constant with respect to $\boldsymbol{\theta}$. Then, the likelihood function becomes (see (8.2))

$$L(\boldsymbol{\theta}; \mathbf{z}) \propto \exp\{-\Lambda[A(u); \boldsymbol{\theta}]\} \prod_{i=1}^{N_{A(u)}} \lambda(\mathbf{z}_i/n; \boldsymbol{\theta}) \propto \prod_{i=1}^{N_{A(u)}} s(\mathbf{w}_i; \boldsymbol{\theta}), \tag{11.37}$$

where $\mathbf{w}_i$ was defined in (11.17), and $s(\mathbf{w}_i; \boldsymbol{\theta})$ is the parametric family being considered.

11.8 Inference

Three methods are available to estimate the parameters of multivariate models: the sequential method, the single step method, and the generalized method. These are explained below.

11.8.1 The Sequential Method

We fit the marginals in a first step, and we estimate the remaining parameters in a second step.

1. We plot the marginals on a maximal Gumbel probability paper and fit the adequate GEVD to each marginal.
2. We transform the data to unit Fréchet marginals using the transformation

$$\exp\left\{-\left[1 - \kappa\left(\frac{V_i - \lambda}{\delta}\right)\right]^{1/\kappa}\right\} = \exp\left(-\frac{1}{U_i}\right), \quad i = 1, 2, \ldots, n,$$

where $V_1, V_2, \ldots, V_n$ is the initial marginal sample and $U_1, U_2, \ldots, U_n$ is the transformed marginal sample. This transformation consists of equaling the maximal GEVD and the unit maximal Fréchet cdfs, to get

$$U_i = \left[1 - \kappa\left(\frac{V_i - \lambda}{\delta}\right)\right]^{-1/\kappa}, \tag{11.38}$$

that is (11.32).

3. We fit an extreme model, for example, the logistic distribution:

$$H(u_1, u_2) = \exp\left\{-\left(u_1^{-1/\alpha} + u_2^{-1/\alpha}\right)^{\alpha}\right\}, \tag{11.39}$$

using one or several of the following methods:

(a) **The weighted least squares method.** In this method, we

$$\underset{\theta}{\text{Minimize}} \quad \sum_{i=1}^{n} \frac{n}{p^{x_i y_i}(1 - p^{x_i y_i})} \left(F_{X,Y}(x_i, y_i; \theta) - p^{x_i y_i}\right)^2, \tag{11.40}$$

where

p^{xy} = proportion of points in the sample where $X \leq x$ and $Y \leq y$.

(b) **The direct maximum likelihood method.** It maximizes the log-likelihood associated with the density $f(x, y; \theta)$.

(c) **The point processes method.** There are two versions:

i. The point processes approach that maximizes the likelihood

$$L(\boldsymbol{\theta}; \mathbf{z}) \propto \prod_{i=1}^{n} s(\mathbf{w}_i; \boldsymbol{\theta}), \tag{11.41}$$

where $s(w; \boldsymbol{\theta})$ is the density of a given parametric family.

ii. A second one that maximizes

$$L(\boldsymbol{\theta}; \mathbf{z}) \propto \prod_{i=1}^{N_{A(u)}} s(\mathbf{w}_i; \boldsymbol{\theta}) = \prod_{i=1}^{u_{1i}+u_{2i}>k} s\left(\frac{u_{1i}}{u_{1i}+u_{2i}}; \boldsymbol{\theta}\right), \tag{11.42}$$

subject to $U_1 + U_2 > k$, where $s(w; \boldsymbol{\theta})$ is a given family of densities, and k is a threshold value.

11.8.2 The Single Step Method

We use the maximum likelihood method to fit all the parameters simultaneously. This is a better approach, but the statistical structure of the data can be hidden. Thus, even when using this method, we recommend to plot the marginal data on probability paper plots.

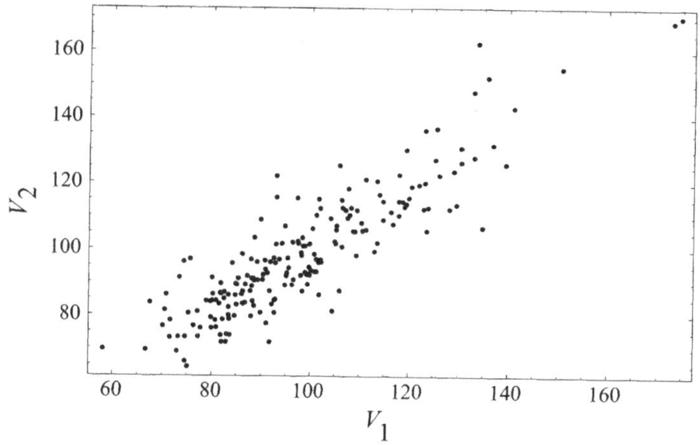

Figure 11.1: Maximum yearly wind speeds data V_1 and V_2 at two different locations during the last 200 years.

11.8.3 The Generalized Method

If we wish to improve our estimate, we can use the maximum likelihood method to fit a more general model, as for example, the logistic model that includes the κ_1 and κ_2 parameters:

$$F(x,y) = \exp\left\{-\left[\left(1-\kappa_1\frac{x-\lambda_1}{\delta_1}\right)^{1/\alpha\kappa_1} + \left(1-\kappa_2\frac{y-\lambda_2}{\delta_2}\right)^{1/\alpha\kappa_2}\right]^{\alpha}\right\}$$

and test using the χ^2 test if it is really an improvement of the previous method.

11.9 Some Multivariate Examples

In this section, we present three different examples to illustrate the methods described in Section 11.8.

11.9.1 The Yearly Maximum Wind Data

Consider the bivariate data (V_1, V_2) in Table 1.16 and Figure 11.1, that correspond to the yearly maximum wind speeds (in km/h) at two close locations. Since the locations are close, they have common weather conditions that result in a positive correlation, that becomes apparent from Figure 11.1.

If the marginal data is represented on a maximal Gumbel probability paper, the plots in Figure 11.2 are obtained, which reveal a very good fit and show that both data components V_1 and V_2 can be considered as maximal Gumbel not only on the right tail but on the whole range. This justifies fitting GEVD

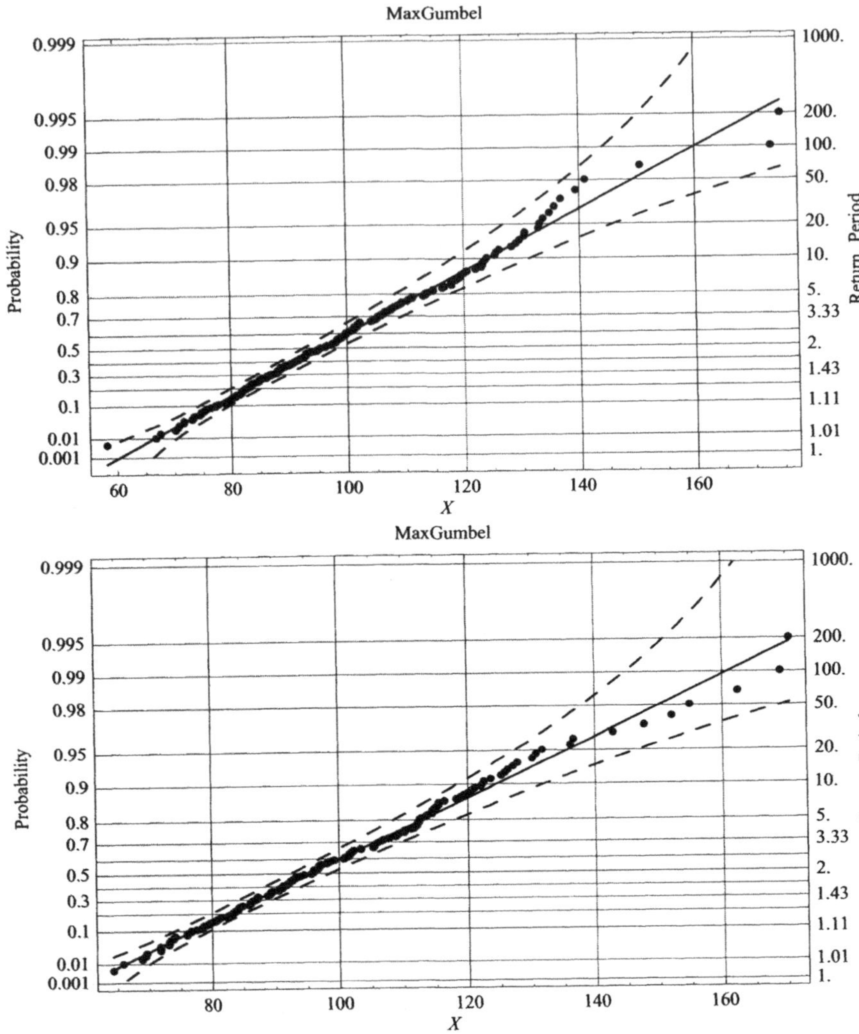

Figure 11.2: Yearly maximum wind data V_1 (upper plot) and V_2 (lower plot) at two different locations on a maximal Gumbel probability paper.

marginals and more precisely maximal Gumbel distributions. The maximum likelihood estimates are

$$\hat{\lambda}_1 = 89.89, \quad \hat{\delta}_1 = 15.08,$$
$$\hat{\lambda}_2 = 89.94, \quad \hat{\delta}_2 = 14.87.$$

To fit one of the extreme value models in Section 11.4.1 or use one of the point processes methods in Section 11.6, we need to transform the data V_1 and V_2 to max unit Fréchet marginals variables. To this end, we use the transformation

$$\exp\left[-\exp\left(\frac{\lambda - V_i}{\delta_i}\right)\right] = \exp\left(-\frac{1}{U_i}\right), \quad i = 1, 2,$$

that consists of equaling the maximal Gumbel and the maximal Fréchet cdfs to get

$$U_i = \exp\left(\frac{V_i - \lambda_i}{\delta_i}\right) \quad \Leftrightarrow V_i = \lambda_i + \delta_i \log U_i, \quad i = 1, 2.$$

Note that this is a special case of (11.38).

The transformed data once rounded is shown in Table 11.4, which can be used with the methods in Sections 11.4.1 and 11.6 to estimate the logistic distribution

$$H(u_1, u_2) = \exp\left\{-\left(u_1^{-1/\alpha} + u_2^{-1/\alpha}\right)^{\alpha}\right\}.$$

We have used the four different methods described in Section 11.8:

1. The weighted least squares method in (11.40) leads to an estimate $\hat{\alpha} = 0.286$.

2. The maximum likelihood method leads to $\hat{\alpha} = 0.329$.

3. The point processes approach method in (11.41) leads to the estimate $\hat{\alpha} = 0.279$.

4. The point processes approach method in (11.42) for $U_1 + U_2 > 10$ leads to the estimate $\hat{\alpha} = 0.329$.

Then, the cdf of the initial data (V_1, V_2) becomes

$$\begin{aligned} F(v_1, v_2) &= \Pr[V_1 \le v_1, V_2 \le v_2) \\ &= \Pr[\lambda_1 + \delta_1 \log U_1 \le v_1, \lambda_2 + \delta_2 \log U_2 \le v_2] \\ &= \Pr\left[U_1 \le \exp\left(\frac{v_1 - \lambda_1}{\delta_1}\right), U_2 \le \exp\left(\frac{v_2 - \lambda_2}{\delta_2}\right)\right] \\ &= \exp\left\{-\left[\exp\left(\frac{\lambda_1 - v_1}{\delta_1 \alpha}\right) + \exp\left(\frac{\lambda_2 - v_2}{\delta_2 \alpha}\right)\right]^{\alpha}\right\}, \quad (11.43) \end{aligned}$$

which obviously has maximal Gumbel marginals.

The previous estimation method is a sequential method because it first estimates the marginal parameters and then the dependence parameters.

Alternatively, we can estimate the five parameters $\lambda_1, \delta_1, \lambda_2, \delta_2, \alpha$ directly from the joint cdf (11.43), using the maximum likelihood method, we obtain the following estimates and log-likelihood:

$$\begin{aligned} \hat{\lambda}_1 &= 90.01, & \hat{\delta}_1 &= 15.13, \\ \hat{\lambda}_2 &= 89.88, & \hat{\delta}_2 &= 15.05, \\ \hat{\alpha} &= 0.326, & \ell &= -1566.2. \end{aligned}$$

Note that the estimates of both processes, the sequential and the joint methods, are very similar.

If we wish to improve our estimate, we can use the maximum likelihood method to fit the most general logistic model, that includes the κ parameters:

$$F(x,y) = \exp\left\{-\left[\left(1-\kappa_1\frac{x-\lambda_1}{\delta_1}\right)^{1/\alpha\kappa_1} + \left(1-\kappa_2\frac{y-\lambda_2}{\delta_2}\right)^{1/\alpha\kappa_2}\right]^{\alpha}\right\}, \tag{11.44}$$

which leads to

$$\begin{array}{lll} \hat{\lambda}_1 = 90.21, & \hat{\delta}_1 = 15.21, & \hat{\kappa}_1 = 0.019, \\ \hat{\lambda}_2 = 89.86, & \hat{\delta}_2 = 14.90, & \hat{\kappa}_2 = 0.001, \\ \hat{\alpha} = 0.329, & \ell = -1566.09. & \end{array}$$

Figure 11.3 shows the contours of the resulting pdf for both methods.

Using the likelihood ratio test, we can now test the null hypothesis H_0 versus the alternative hypothesis H_1, where

H_0 : The sample comes from the logistic model,
H_1 : The sample comes from the general logistic model.

Since the deviance function takes the value

$$D(\theta) = 2\left[\ell(\hat{\theta}|x) - \ell(\theta|x)\right] = 2(-1566.09 + 1566.20) = 0.22,$$

which leads to a p-value of 0.922, we conclude that the sample comes from the logistic model.

11.9.2 The Ocmulgee River Flood Data

Consider the bivariate data (Q_1, Q_2) in Table 1.15 and Figure 11.4 that corresponds to the yearly maximum floods of the Ocmulgee River data downstream at Macon and upstream at Hawkinsville from 1910 to 1949. Since the locations share the same river, it results in a positive correlation, that becomes apparent from the plot in Figure 11.4.

If the marginal data is plotted on a maximal Gumbel probability paper, we get the plots in Figure 11.5, which reveal a very good fit and show that the data can be considered as maximal Gumbel not only on the right tail but in the whole range. This justifies fitting a GEVD marginals and, more precisely, a maximal Gumbel distributions.

The parameter estimates for the maximal Gumbel marginal distributions corresponding to the maximum likelihood estimates:

$$\begin{array}{llll} \hat{\lambda}_1 & = & 26.378, & \hat{\delta}_1 & = & 17.042, \\ \hat{\lambda}_2 & = & 23.709, & \hat{\delta}_2 & = & 15.057. \end{array}$$

To fit one of the extreme value models in Section 11.4 or use one of the point processes methods described in Section 11.6, we need to transform the

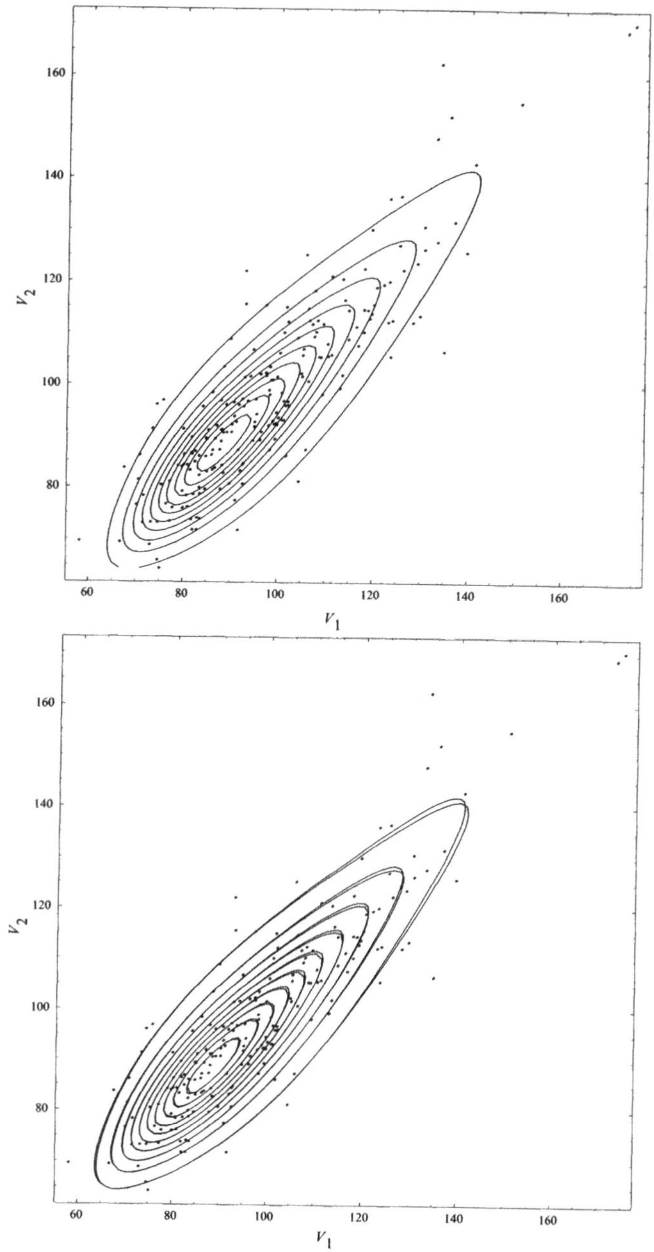

Figure 11.3: Contour plots of the joint density of (V_1, V_2) in (11.43) for the sequential (upper figure) and both the sequential and the joint estimation (11.44) of the parameters (they are almost indistinguishable).

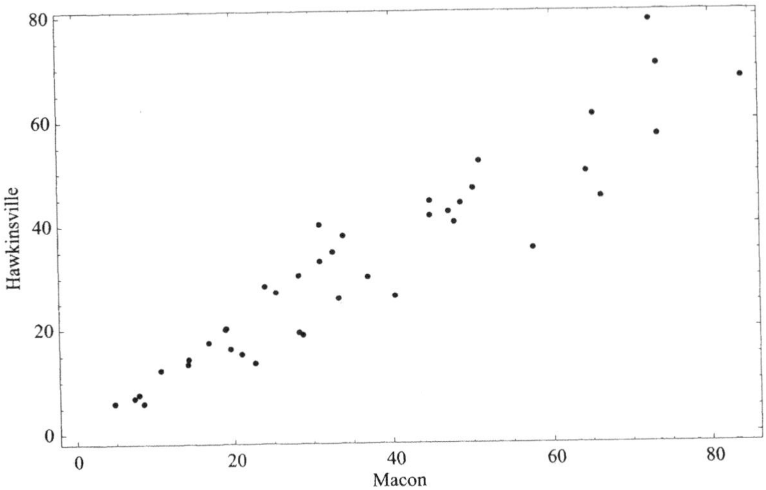

Figure 11.4: Yearly maximum floods of the Ocmulgee River data downstream at Macon and upstream at Hawkinsville from 1910 to 1949.

data Q_1 and Q_2 to max unit Fréchet marginals variables. To this end, we use the transformation

$$U_i = \exp\left(\frac{Q_i - \lambda_i}{\delta_i}\right) \quad \Leftrightarrow \quad Q_i = \lambda_i + \delta_i \log U_i, \quad i = 1, 2.$$

The marginal Gumbel estimates are

$$\hat{\lambda}_1 = 26.38, \quad \hat{\delta}_1 = 17.04, \quad \hat{\lambda}_2 = 23.71, \quad \hat{\delta}_2 = 15.06.$$

The transformed data is shown in Table 11.5, which can be used with the methods in Sections 11.4 and 11.6.

Finally, we estimate the logistic distribution

$$H(u_1, u_2) = \exp\left\{-\left(u_1^{-1/\alpha} + u_2^{-1/\alpha}\right)^{\alpha}\right\}$$

using the already described four different methods described in Section 11.8:

1. The weighted least squares method in (11.40) leads to the estimate $\hat{\alpha} = 0.00005$.

2. The maximum likelihood method leads to $\hat{\alpha} = 0.2314$.

3. The point processes approach method in (11.41) leads to the estimate $\hat{\alpha} = 0.210$.

4. The point processes approach method in (11.42) for $U_1 + U_2 > 5$ leads to the estimate $\hat{\alpha} = 0.257$.

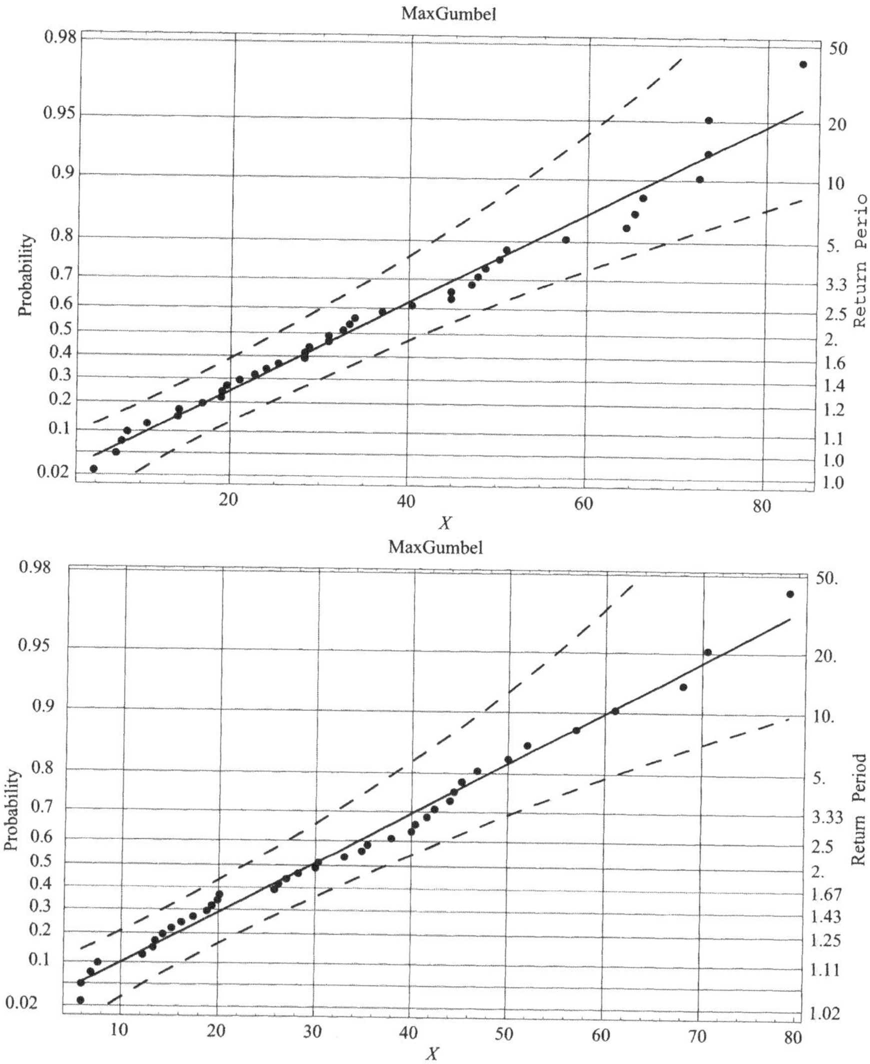

Figure 11.5: Yearly maximum floods of the Ocmulgee River data downstream at Macon and upstream at Hawkinsville from 1910 to 1949 on maximal Gumbel probability paper.

This ends the sequential estimation method with a first step in which we fit the marginals, and a second step in which we estimate the remaining parameters.

Then, the cdf of the initial data (Q_1, Q_2) becomes the model with maximal

Gumbel marginals (11.43):

$$F(q_1, q_2) = \exp\left\{-\left[\exp\left(\frac{\lambda_1 - q_1}{\delta_1 \alpha}\right) + \exp\left(\frac{\lambda_2 - q_2}{\delta_2 \alpha}\right)\right]^{\alpha}\right\}. \quad (11.45)$$

Alternatively, if we estimate the five parameters $\lambda_1, \delta_1, \lambda_2, \delta_2$, and α directly from the joint cdf (11.45), using the maximum likelihood method, we obtain the following estimates:

$$\hat{\lambda}_1 = 25.80, \quad \hat{\delta}_1 = 16.48, \quad \hat{\lambda}_2 = 23.46, \quad \hat{\delta}_2 = 14.63, \quad \hat{\alpha} = 0.237.$$

Note that the estimates of both processes are very similar. Figure 11.6 shows the contour lines for the pdf of the sequential and the joint estimation method.

Alternatively, we can use the maximum likelihood method to fit the most general logistic model (11.44) that includes the κ parameters. This leads to

$$\begin{aligned} \hat{\lambda}_1 &= 24.74, & \hat{\delta}_1 &= 15.89, & \hat{\kappa}_1 &= -0.274, \\ \hat{\lambda}_2 &= 22.66, & \hat{\delta}_2 &= 14.24, & \hat{\kappa}_2 &= -0.256, \\ \hat{\alpha} &= 0.195. \end{aligned}$$

and the pdf contours of the resulting model are shown in Figure 11.6.

11.9.3 The Maximum Car Speed Data

Consider the bivariate data (V_1, V_2) in Table 1.17 and Figure 11.7 that corresponds to the maximum weekend car speeds registered at two given locations 1 and 2, a highway and a mountain location, respectively, corresponding to 200 dry weeks and the first 1000 cars passing through the given locations.

If the marginal data is plotted on a maximal Gumbel probability paper, we get the plots in Figure 11.8, which reveal a very good fit for V_1 and show that these data can be considered as maximal Gumbel not only on the right tail but in the whole range. On the contrary, the plot associated with V_2 suggests a Weibull domain of attraction. Since the data reveal an asymptote approximately for $V_2 = 70$, we use this information to represent the data on a Weibull probability paper and obtain the plot in Figure 11.9, which reveal that these data can be considered as maximal Weibull not only on the right tail but in the whole range. This would justify fitting a maximal Gumbel distribution for V_1 and a maximal Weibull distribution for V_2. However, we shall fit a GEVD for both.

The maximum likelihood parameter estimates of the GEVD for V_1 are

$$\hat{\lambda}_1 = 90.509, \quad \hat{\delta}_1 = 15.07, \quad \text{and} \quad \hat{\kappa}_1 = 0.03.$$

Similarly, the maximum likelihood parameter estimates of the GEVD for V_2 are

$$\hat{\lambda}_2 = 58.24, \quad \hat{\delta}_2 = 5.84, \quad \text{and} \quad \hat{\kappa}_2 = 0.52.$$

If we estimate the five parameters $\lambda_1, \delta_1, \lambda_2, \delta_2$, and α directly from the joint cdf (11.45), using the maximum likelihood method for the GEVD-logistic model,

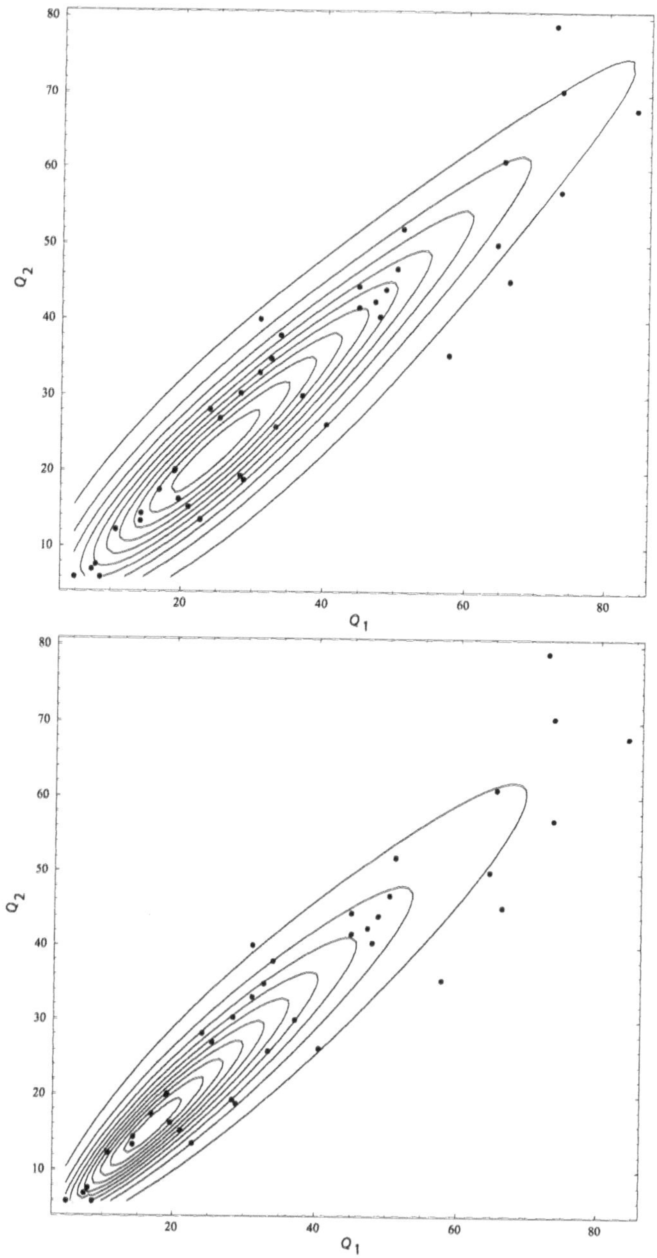

Figure 11.6: Contour plots of the joint density of (Q_1, Q_2) for the sequential (11.45) and the joint estimation of the parameters.

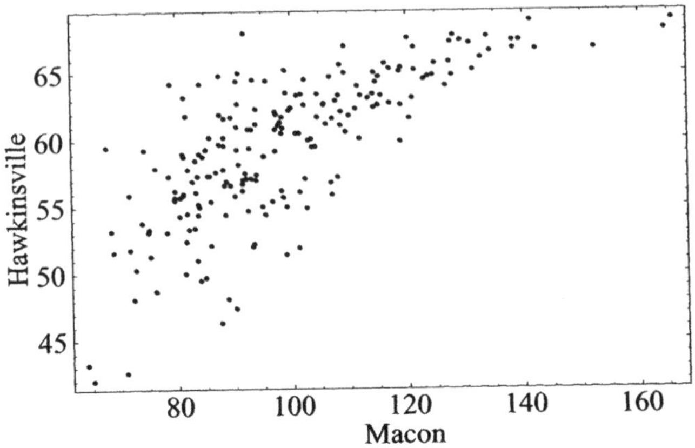

Figure 11.7: Data corresponding to the maximum weekend car speeds registered at two given locations 1 and 2 in 200 dry weeks.

we obtain the following estimates:

$$\begin{array}{llll}\hat{\lambda}_1 = 90.49, & \hat{\delta}_1 = 15.17, & \hat{\kappa}_1 = 0.022, & \\ \hat{\lambda}_2 = 58.07, & \hat{\delta}_2 = 5.73, & \hat{\kappa}_2 = 0.487, & \hat{\alpha}_2 = 0.428.\end{array}$$

Figure 11.10 shows the data and the contour lines of the pdf for this model, which shows a good fit.

Exercises

11.1 Show that for the case $m = 4$, the $v(z_1, z_2, z_3, z_4)/4$ function (the expression (11.23)) becomes

$$\begin{aligned}
&\int_0^1 \int_0^{1-w_1} \int_0^{1-w_1-w_2} \max\left(\frac{w_1}{z_1}, \frac{w_2}{z_2}, \frac{w_3}{z_3}, \frac{1-w_1-w_2-w_3}{z_4}\right) dS^*(\mathbf{w}) \\
&= \int_0^1 \int_0^{\min(1-w_1, \frac{z_2 w_1}{z_1})} \int_{\max(0, 1-w_1-w_2-\frac{z_4 w_1}{z_1})}^{\min(1-w_1-w_2, \frac{z_3 w_1}{z_1})} \frac{w_1}{z_1}\, dS^*(\mathbf{w}) \\
&+ \int_0^1 \int_{\frac{z_2 w_1}{z_1}}^{1-w_1} \int_{\max(0, 1-w_1-w_2-\frac{z_4 w_2}{z_2})}^{\min(1-w_1-w_2, \frac{z_3 w_2}{z_2})} \frac{w_2}{z_2}\, dS^*(\mathbf{w}) \\
&+ \int_0^1 \int_0^{1-w_1} \int_{\max(0, \frac{z_3 w_1}{z_1}, \frac{z_3 w_2}{z_2}, \frac{(1-w_1-w_2) z_3}{z_3+z_4})}^{1-w_1-w_2} \frac{w_3}{z_3}\, dS^*(\mathbf{w}) \\
&+ \int_0^1 \int_0^{1-w_1} \int_0^{a} \frac{1-w_1-w_2-w_3}{z_4}\, dS^*(\mathbf{w}),
\end{aligned}$$

where $dS^*(\mathbf{w})$ represents $dS^*(w_1, w_2, w_3)$ and

$$a = \min\left(1 - w_1 - w_2 - \frac{z_4 w_1}{z_1}, 1 - w_1 - w_2 - \frac{z_4 w_2}{z_2}, \frac{(1 - w_1 - w_2) z_3}{z_3 + z_4}\right).$$

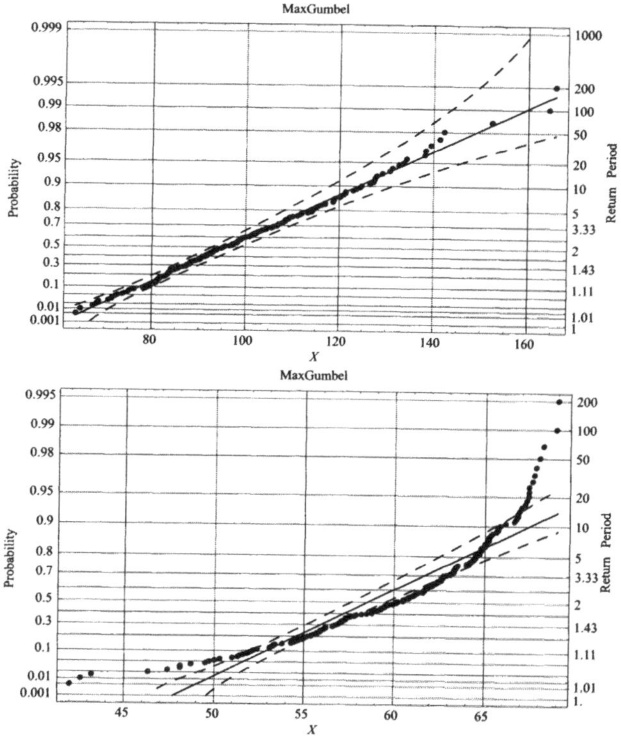

Figure 11.8: Data corresponding to the maximum weekend car speeds registered at two given locations 1 and 2 in 200 dry weeks on maximal Gumbel probability paper.

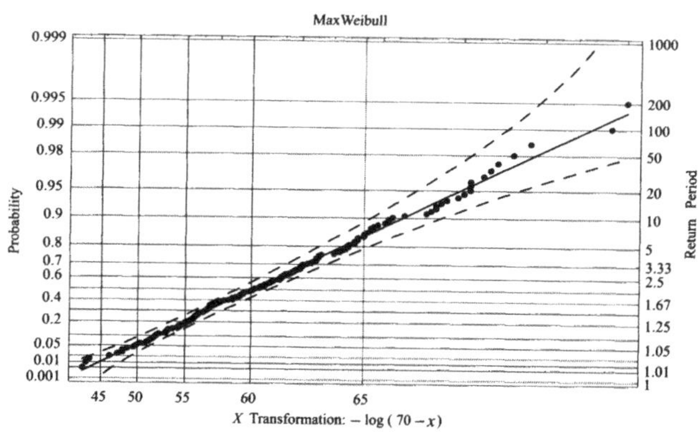

Figure 11.9: Data V_2 on maximal Weibull probability paper.

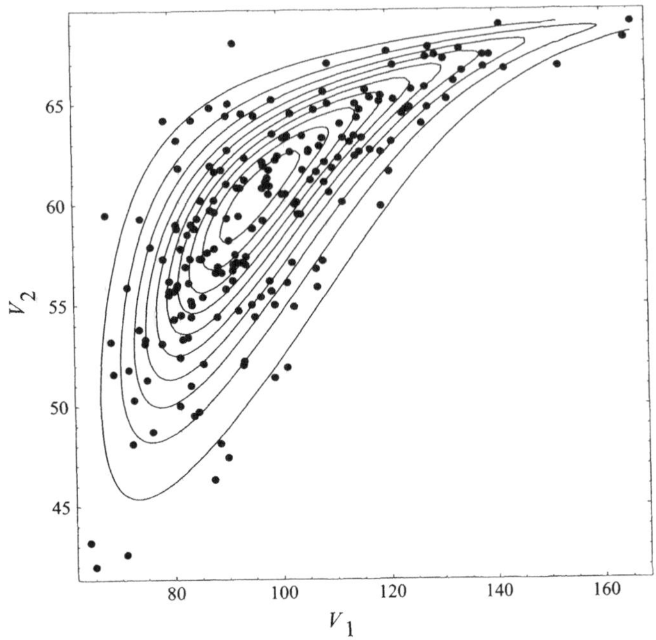

Figure 11.10: Contour plots of the joint density of the resulting model for (V_1, V_2).

11.2 Consider the following bivariate cdfs:

1. Ali-Mikhail-Haq Distribution:

$$F(x,y) = \frac{xy}{1-\alpha(1-x)(1-y)}, \quad 0 \le x, y \le 1,$$

where $-1 \le \alpha \le 1$;

2. Pareto-Copula:

$$F(x,y) = \left(x^{-1/c} + y^{-1/c} - 1\right)^{-c}, \quad 0 \le x, y \le 1,$$

where $c > 0$;

3. Gumbel-Hougaard-Copula:

$$F(x,y) = \exp\left\{-\left[(-\log x)^{\alpha} + (-\log y)^{\alpha}\right]^{1/\alpha}\right\}, \quad 0 \le x, y \le 1,$$

where $\alpha \ge 1$;

4. Bivariate Logistic Model:

$$F(x,y) = \frac{1}{1+e^{-x}+e^{-y}+\theta e^{-x-y}}, \quad -\infty < x, y < \infty,$$

where $0 \le \theta \le 2$.

For each of these cdfs, obtain:

(a) The marginals.

(b) The dependence function.

(c) The survival function.

(d) The limit distribution.

11.3 Use Theorem 11.1 to obtain the limit distribution of the:

(a) Frank distribution.

(b) Farlie-Gumbel-Morgenstern distribution.

11.4 Use Theorem 11.2 to solve the previous exercise.

11.5 Obtain the function $v(x_2 - x_1)$ in Corollary 11.1 corresponding to the Mardia distribution of Example 11.12.

11.6 Find the transformation required to obtain Fréchet marginals for Weibull and Gumbel models.

11.7 Given the Farlie-Gumbel-Morgenstern distribution

$$xy\left[1 + \alpha(1-x)(1-y)\right],$$

using the two methods described in this chapter:

(a) Obtain its dependence function.

(b) Obtain its maximal limit distribution based on marginals.

(c) Obtain its maximal limit distribution based on dependence functions.

11.8 Given the Frank distribution

$$\log_\alpha\left[1 + \frac{(\alpha^x - 1)(\alpha^y - 1)}{\alpha - 1}\right], \quad 0 \le x, y \le 1, \quad \alpha > 0,$$

using the two methods described in this chapter:

(a) Obtain its dependence function.

(b) Obtain its maximal limit distribution based on marginals.

(c) Obtain its maximal limit distribution based on dependence functions.

11.9 Can the Frank distribution be a limit distribution for maxima?

11.10 Discuss the convenience of the different alternatives given in Section 11.6 for transforming to unit Fréchet marginals in practical cases.

11.11 Choose a real engineering bivariate case involving extremes and suggest how to use the peak over threshold method.

Table 11.3: Some Families of Bivariate Extreme Value Distributions, where $F^e(x)$ Refers to Univariate Extreme Value Distributions.

Functions	Type A Distributions
$F(x,y)$	$F_X^e(x)F_Y^e(y)\exp\left\{-\theta\left[\frac{1}{\log F_X^e(x)}+\frac{1}{\log F_Y^e(y)}\right]^{-1}\right\}$
$S(x,y)$	$F_X^e(x)F_Y^e(y)\exp\left\{-\theta\left[\frac{1}{\log F_X^e(x)}+\frac{1}{\log F_Y^e(y)}\right]^{-1}\right\}$
	$+1-F_X^e(x)-F_Y^e(y)$
$H(x,y)$	$F_X^e(x)F_Y^e(y)\exp\left\{-\theta\left[\frac{1}{\log F_X^e(x)}+\frac{1}{\log F_Y^e(y)}\right]^{-1}\right\}$
$D_F(u,v)$	$uv\exp\left\{-\theta\left[\frac{1}{\log u}+\frac{1}{\log v}\right]^{-1}\right\}$
Marginals	$F_X^e(x)$; $F_Y^e(y))$
Functions	**Type B Distributions**
$F(x,y)$	$\exp\left\{-\left[(-\log F_X^e(x))^m+(-\log F_Y^e(y))^m\right]^{1/m}\right\}$
$S(x,y)$	$\exp\left\{-\left[(-\log F_X^e(x))^m+(-\log F_Y^e(y))^m\right]^{1/m}\right\}$
	$+1-F_X^e(x)-F_Y^e(y)$
$H(x,y)$	$\exp\left\{-\left[(-\log F_X^e(x))^m+(-\log F_Y^e(y))^m\right]^{1/m}\right\}$
$D_F(u,v)$	$\exp\left\{-\left[(-\log u)^m+(-\log v)^m\right]^{1/m}\right\}$
Marginals	$F_X^e(x)$; $F_Y^e(y))$
Functions	**Example of Type A Distributions**
$F(x,y)$	$\exp\{-\exp(-x)-\exp(-y)+\theta/(\exp(x)+\exp(y))\}$, $(0\le\theta\le 1)$
Functions	**Example of Type B Distributions**
$F(x,y)$	$\exp\left\{-\left[\exp(-mx)+\exp(-my)\right]^{1/m}\right\}$, $m\ge 1$
Functions	**Logistic Model**
$F(x,y)$	$\exp\left\{-\left(x^{1/\alpha}+y^{-1/\alpha}\right)^{\alpha}\right\}$, $x,y>0$, $\alpha\in(0,1]$
Functions	**Mixed Model**
$S(x,y)$	$\exp\left\{-(x+y)+\frac{\theta xy}{x+y}\right\}$
Functions	**Asymmetric Nonexchangeable Mixed Model**
$S(x,y)$	$\exp\left\{-(x+y)+xy\left[x(\theta+\phi)+y(2\phi+\theta)\right](x+y)^{-2}\right\}$
Functions	**Asymmetric Nonexchangeable Logistic Model**
$S(x,y)$	$\exp\left\{-(1-\theta)x-(1-\phi)y-\left[(\theta x)^r+(\phi y)^r\right]^{1/r}\right\}$

Table 11.4: Fréchet Marginals (U_1, U_2) for the Yearly Maximum Wind Speed Data.

v_1	u_2	u_1	u_2	u_1	u_2	u_1	u_2	u_1	u_2
2.12	1.59	55.98	78.90	2.19	5.39	9.53	4.56	0.78	0.83
1.38	0.94	5.18	5.18	0.60	0.47	1.55	1.05	1.71	1.15
0.27	0.41	2.08	1.24	3.37	4.53	2.12	1.24	20.85	65.36
4.75	7.81	0.53	0.78	1.88	0.96	0.90	1.08	10.60	23.05
0.36	1.50	0.55	0.68	5.92	3.28	0.83	1.14	1.21	1.47
0.59	0.98	0.65	0.51	1.09	1.20	0.55	0.40	2.76	2.89
1.06	1.30	0.65	0.60	2.23	4.47	0.51	0.49	0.99	3.55
1.93	1.16	8.98	4.44	1.52	0.94	1.13	0.30	4.11	2.90
3.46	2.83	13.91	4.84	0.41	0.41	7.36	5.58	2.71	2.28
3.87	3.35	0.70	0.50	1.82	2.09	0.74	1.08	1.93	2.18
0.28	0.77	0.59	0.34	0.43	0.55	9.35	2.85	2.25	1.58
29.62	35.20	1.28	1.59	3.65	4.32	0.78	0.64	1.02	1.05
1.74	1.72	0.71	0.76	6.94	4.65	1.90	1.18	4.78	2.23
0.87	0.83	4.08	8.07	3.01	4.53	0.48	0.68	1.63	5.50
1.66	2.20	0.52	0.39	7.08	14.89	17.55	12.75	1.53	2.25
26.64	11.15	3.92	2.85	0.33	0.32	7.61	7.01	274.8	223.8
0.88	1.08	0.81	0.96	1.97	1.27	2.12	3.88	3.62	1.73
0.33	0.24	3.24	6.69	0.30	0.46	18.26	130.7	3.21	3.68
19.64	3.04	0.88	1.59	5.18	3.55	0.61	0.81	1.12	0.83
0.58	0.80	22.42	16.58	0.72	0.63	8.41	7.25	0.65	0.77
0.59	0.30	0.63	0.30	0.23	0.66	0.65	0.47	6.75	5.18
11.03	8.81	1.14	1.50	1.92	1.36	0.45	0.39	7.12	4.94
2.57	3.63	1.72	1.83	0.54	0.46	0.51	0.67	3.13	4.32
0.71	0.96	0.91	2.47	17.32	48.95	0.88	0.93	0.38	0.18
1.03	1.16	2.78	3.13	1.20	0.54	6.41	5.21	0.61	0.71
1.43	1.62	1.16	0.64	2.26	1.50	0.91	0.61	1.85	1.17
0.22	0.25	0.34	1.10	0.68	1.46	2.97	5.28	0.91	1.04
1.41	1.20	0.38	0.53	1.82	1.22	2.01	2.96	0.72	0.95
3.30	3.90	2.19	1.49	0.81	0.66	0.59	0.79	0.99	0.54
14.67	15.5	6.41	3.90	0.89	0.68	0.57	0.64	1.22	8.63
1.21	0.70	2.63	0.56	5.76	4.18	6.41	8.93	1.19	1.63
2.20	0.77	2.06	1.77	0.52	1.10	0.43	0.33	0.52	0.68
1.65	2.30	1.77	2.09	1.40	1.13	9.10	22.28	4.63	1.88
14.77	11.53	0.63	0.34	2.99	2.06	1.23	5.58	0.12	0.25
0.95	1.50	1.08	0.43	0.77	0.78	1.75	2.47	10.46	12.17
2.89	0.84	1.20	0.69	2.87	10.71	1.38	3.11	0.39	1.59
4.91	5.92	1.75	0.83	1.06	1.55	13.37	9.61	0.78	1.77
247.2	203.6	9.10	7.55	0.30	0.32	1.32	2.19	0.36	0.20
2.74	2.19	0.88	0.50	0.28	0.56	0.36	0.32	0.96	1.05
12.68	4.44	3.55	2.81	1.45	1.33	1.22	2.16	0.66	0.34

Table 11.5: Fréchet Marginals (U_1, U_2) for the Flood at Macon and Hawkinsville.

v_1	u_2	u_1	u_2	u_1	u_2	u_1	u_2	u_1	u_2
1.15	0.72	0.35	0.31	2.95	3.95	4.24	6.55	0.28	0.31
0.65	0.79	3.51	3.03	0.94	1.24	0.49	0.54	1.31	2.95
10.35	4.17	1.86	1.52	3.68	3.85	1.12	1.55	0.73	0.57
14.97	39.33	1.12	0.75	0.34	0.34	3.37	3.46	15.79	22.37
9.31	5.73	0.40	0.47	0.67	0.61	0.65	0.78	0.57	0.66
0.81	0.51	9.81	11.9	1.50	1.15	1.31	1.85	1.55	2.57
0.49	0.50	0.33	0.33	15.79	9.12	2.95	3.28	4.05	4.63
2.28	1.18	6.25	2.17	1.44	2.09	0.87	1.35	29.4	18.94

Appendix A

Statistical Tables

1. Standard Normal Distribution
2. Student's t-Distribution
3. Chi-Square Distribution
4. Snedecor $F_{(m,n)}$ Distribution (the 0.75 Quantiles)
5. Snedecor $F_{(m,n)}$ Distribution (the 0.95 Quantiles)
6. Snedecor $F_{(m,n)}$ Distribution (the 0.99 Quantiles)

Table A.1: The cdf $\Phi(z)$ of the Standard Normal $N(0,1)$ Random Variable.

z	0.00	0.01	0.02	0.03	0.04	0.05	0.06	0.07	0.08	0.09
0.0	0.5000	0.5040	0.5080	0.5120	0.5160	0.5199	0.5239	0.5279	0.5319	0.5359
0.1	0.5398	0.5438	0.5478	0.5517	0.5557	0.5596	0.5636	0.5675	0.5714	0.5753
0.2	0.5793	0.5832	0.5871	0.5910	0.5948	0.5987	0.6026	0.6064	0.6103	0.6141
0.3	0.6179	0.6217	0.6255	0.6293	0.6331	0.6368	0.6406	0.6443	0.648	0.6517
0.4	0.6554	0.6591	0.6628	0.6664	0.6700	0.6736	0.6772	0.6808	0.6844	0.6879
0.5	0.6915	0.695	0.6985	0.7019	0.7054	0.7088	0.7123	0.7157	0.7190	0.7224
0.6	0.7257	0.7291	0.7324	0.7357	0.7389	0.7422	0.7454	0.7486	0.7517	0.7549
0.7	0.7580	0.7611	0.7642	0.7673	0.7704	0.7734	0.7764	0.7794	0.7823	0.7852
0.8	0.7881	0.7910	0.7939	0.7967	0.7995	0.8023	0.8051	0.8078	0.8106	0.8133
0.9	0.8159	0.8186	0.8212	0.8238	0.8264	0.8289	0.8315	0.8340	0.8365	0.8389
1.0	0.8413	0.8438	0.8461	0.8485	0.8508	0.8531	0.8554	0.8577	0.8599	0.8621
1.1	0.8643	0.8665	0.8686	0.8708	0.8729	0.8749	0.8770	0.8790	0.8810	0.8830
1.2	0.8849	0.8869	0.8888	0.8907	0.8925	0.8944	0.8962	0.8980	0.8997	0.9015
1.3	0.9032	0.9049	0.9066	0.9082	0.9099	0.9115	0.9131	0.9147	0.9162	0.9177
1.4	0.9192	0.9207	0.9222	0.9236	0.9251	0.9265	0.9279	0.9292	0.9306	0.9319
1.5	0.9332	0.9345	0.9357	0.9370	0.9382	0.9394	0.9406	0.9418	0.9429	0.9441
1.6	0.9452	0.9463	0.9474	0.9484	0.9495	0.9505	0.9515	0.9525	0.9535	0.9545
1.7	0.9554	0.9564	0.9573	0.9582	0.9591	0.9599	0.9608	0.9616	0.9625	0.9633
1.8	0.9641	0.9649	0.9656	0.9664	0.9671	0.9678	0.9686	0.9693	0.9699	0.9706
1.9	0.9713	0.9719	0.9726	0.9732	0.9738	0.9744	0.9750	0.9756	0.9761	0.9767
2.0	0.9772	0.9778	0.9783	0.9788	0.9793	0.9798	0.9803	0.9808	0.9812	0.9817
2.1	0.9821	0.9826	0.983	0.9834	0.9838	0.9842	0.9846	0.9850	0.9854	0.9857
2.2	0.9861	0.9864	0.9868	0.9871	0.9875	0.9878	0.9881	0.9884	0.9887	0.9890
2.3	0.9893	0.9896	0.9898	0.9901	0.9904	0.9906	0.9909	0.9911	0.9913	0.9916
2.4	0.9918	0.9920	0.9922	0.9925	0.9927	0.9929	0.9931	0.9932	0.9934	0.9936
2.5	0.9938	0.9940	0.9941	0.9943	0.9945	0.9946	0.9948	0.9949	0.9951	0.9952
2.6	0.9953	0.9955	0.9956	0.9957	0.9959	0.9960	0.9961	0.9962	0.9963	0.9964
2.7	0.9965	0.9966	0.9967	0.9968	0.9969	0.9970	0.9971	0.9972	0.9973	0.9974
2.8	0.9974	0.9975	0.9976	0.9977	0.9977	0.9978	0.9979	0.9979	0.9980	0.9981
2.9	0.9981	0.9982	0.9982	0.9983	0.9984	0.9984	0.9985	0.9985	0.9986	0.9986
3.0	0.9987	0.9987	0.9987	0.9988	0.9988	0.9989	0.9989	0.9989	0.9990	0.9990
3.1	0.999	0.9991	0.9991	0.9991	0.9992	0.9992	0.9992	0.9992	0.9993	0.9993
3.2	0.9993	0.9993	0.9994	0.9994	0.9994	0.9994	0.9994	0.9995	0.9995	0.9995
3.3	0.9995	0.9995	0.9995	0.9996	0.9996	0.9996	0.9996	0.9996	0.9996	0.9997
3.4	0.9997	0.9997	0.9997	0.9997	0.9997	0.9997	0.9997	0.9997	0.9997	0.9998
3.5	0.9998	0.9998	0.9998	0.9998	0.9998	0.9998	0.9998	0.9998	0.9998	0.9998
3.6	0.9998	0.9998	0.9999	0.9999	0.9999	0.9999	0.9999	0.9999	0.9999	0.9999

Table A.2: The $1-\alpha$ Quantiles of the Student t_n Distribution.

n	$\alpha = 0.1$	0.05	0.025	0.01	0.005
1	3.078	6.314	12.706	31.821	63.657
2	1.886	2.920	4.303	6.965	9.925
3	1.638	2.353	3.182	4.541	5.841
4	1.533	2.132	2.776	3.747	4.604
5	1.476	2.015	2.571	3.365	4.032
6	1.440	1.943	2.447	3.143	3.707
7	1.415	1.895	2.365	2.998	3.499
8	1.397	1.860	2.306	2.896	3.355
9	1.383	1.833	2.262	2.821	3.250
10	1.372	1.812	2.228	2.764	3.169
11	1.363	1.796	2.201	2.718	3.106
12	1.356	1.782	2.179	2.681	3.055
13	1.350	1.771	2.160	2.650	3.012
14	1.345	1.761	2.145	2.624	2.977
15	1.341	1.753	2.131	2.602	2.947
16	1.337	1.746	2.120	2.583	2.921
17	1.333	1.740	2.110	2.567	2.898
18	1.330	1.734	2.101	2.552	2.878
19	1.328	1.729	2.093	2.539	2.861
20	1.325	1.725	2.086	2.528	2.845
21	1.323	1.721	2.080	2.518	2.831
22	1.321	1.717	2.074	2.508	2.819
23	1.319	1.714	2.069	2.500	2.807
24	1.318	1.711	2.064	2.492	2.797
25	1.316	1.708	2.060	2.485	2.787
26	1.315	1.706	2.056	2.479	2.779
27	1.314	1.703	2.052	2.473	2.771
28	1.313	1.701	2.048	2.467	2.763
29	1.311	1.699	2.045	2.462	2.756
30	1.310	1.697	2.042	2.457	2.750
40	1.303	1.684	2.021	2.423	2.704
60	1.296	1.671	2.000	2.390	2.660
120	1.289	1.658	1.980	2.358	2.617
∞	1.282	1.645	1.960	2.327	2.576

Table A.3: The $1-\alpha$ Quantiles of the χ^2_n Distribution.

n	$\alpha = 0.99$	0.98	0.95	0.90	0.8	0.7	0.5
1	0.000157	0.000628	0.00393	0.015791	0.064185	0.1485	0.4550
2	0.0201	0.040405	0.10259	0.21072	0.44629	0.7133	1.3863
3	0.1148	0.1848	0.3518	0.5844	1.0052	1.4237	2.3660
4	0.2971	0.4294	0.7107	1.0636	1.6488	2.1947	3.3567
5	0.5543	0.7520	1.1455	1.6103	2.3425	2.9999	4.3515
6	0.8721	1.1344	1.6354	2.2041	3.0701	3.8276	5.3481
7	1.2390	1.5643	2.1673	2.8331	3.8223	4.6713	6.3458
8	1.6465	2.0325	2.7326	3.4895	4.5936	5.5274	7.3441
9	2.0879	2.5324	3.3251	4.1682	5.3801	6.3933	8.3428
10	2.5582	3.0591	3.9403	4.8652	6.1791	7.2672	9.3418
11	3.0535	3.6087	4.5748	5.5778	6.9887	8.1479	10.341
12	3.5706	4.1783	5.2260	6.3038	7.8073	9.0343	11.340
13	4.1069	4.7654	5.8919	7.0415	8.6339	9.9257	12.340
14	4.6604	5.3682	6.5706	7.7895	9.4673	10.821	13.339
15	5.2293	5.9849	7.2609	8.5468	10.307	11.721	14.339
16	5.8122	6.6142	7.9616	9.3122	11.152	12.624	15.338
17	6.4078	7.2550	8.6718	10.085	12.002	13.531	16.338
18	7.0149	7.9062	9.3905	10.865	12.857	14.440	17.338
19	7.6327	8.5670	10.117	11.651	13.716	15.352	18.338
20	8.2604	9.2367	10.851	12.443	14.578	16.266	19.337
21	8.8972	9.9146	11.591	13.240	15.445	17.182	20.337
22	9.5425	10.600	12.338	14.041	16.314	18.101	21.337
23	10.196	11.293	13.091	14.848	17.187	19.021	22.337
24	10.856	11.992	13.848	15.659	18.062	19.943	23.337
25	11.524	12.697	14.611	16.473	18.940	20.867	24.337
26	12.198	13.409	15.379	17.292	19.820	21.792	25.336
27	12.879	14.125	16.151	18.114	20.703	22.719	26.336
28	13.565	14.847	16.928	18.939	21.588	23.647	27.336
29	14.256	15.574	17.708	19.768	22.475	24.577	28.336
30	14.953	16.306	18.493	20.599	23.364	25.508	29.336

n	$\alpha = 0.3$	0.2	0.1	0.05	0.03	0.02	0.01
1	1.0742	1.6424	2.7055	3.8415	4.7093	5.4119	6.6349
2	2.4079	3.2189	4.6052	5.9915	7.0131	7.8240	9.2103
3	3.6649	4.6416	6.2514	7.8147	8.9473	9.8374	11.345
4	4.8784	5.9886	7.7794	9.4877	10.712	11.668	13.277
5	6.0644	7.2893	9.2364	11.070	12.375	13.388	15.086
6	7.2311	8.5581	10.645	12.592	13.968	15.033	16.812
7	8.3834	9.8032	12.017	14.067	15.509	16.622	18.475
8	9.5245	11.030	13.362	15.507	17.010	18.168	20.090
9	10.656	12.242	14.684	16.919	18.480	19.679	21.666
10	11.781	13.442	15.987	18.307	19.922	21.161	23.209
11	12.899	14.631	17.275	19.675	21.342	22.618	24.725
12	14.011	15.812	18.549	21.026	22.742	24.054	26.217
13	15.119	16.985	19.812	22.362	24.125	25.472	27.688
14	16.222	18.151	21.064	23.685	25.493	26.873	29.141
15	17.322	19.311	22.307	24.996	26.848	28.259	30.578
16	18.418	20.465	23.542	26.296	28.191	29.633	32.000
17	19.511	21.615	24.769	27.587	29.523	30.995	33.409
18	20.601	22.760	25.989	28.869	30.845	32.346	34.805
19	21.689	23.900	27.204	30.144	32.158	33.687	36.191
20	22.775	25.038	28.412	31.410	33.462	35.020	37.566
21	23.858	26.171	29.615	32.671	34.759	36.343	38.932
22	24.939	27.301	30.813	33.924	36.049	37.659	40.289
23	26.018	28.429	32.007	35.172	37.332	38.968	41.638
24	27.096	29.553	33.196	36.415	38.609	40.270	42.980
25	28.172	30.675	34.382	37.652	39.880	41.566	44.314
26	29.246	31.795	35.563	38.885	41.146	42.856	45.642
27	30.319	32.912	36.741	40.113	42.407	44.140	46.963
28	31.391	34.027	37.916	41.337	43.662	45.419	48.278
29	32.461	35.139	39.087	42.557	44.913	46.693	49.588
30	33.530	36.250	40.256	43.773	46.160	47.962	50.892

Table A.4: The 0.75 Quantiles of the Snedecor $F_{(m,n)}$ Distribution.

n	$m = 1$	2	3	4	5	6	7	8	9	10
1	5.828	7.500	8.200	8.581	8.820	8.983	9.102	9.192	9.263	9.320
2	2.571	3.000	3.153	3.232	3.280	3.312	3.335	3.353	3.366	3.377
3	2.024	2.280	2.356	2.390	2.409	2.422	2.430	2.436	2.441	2.445
4	1.807	2.000	2.047	2.064	2.072	2.077	2.079	2.08	2.081	2.082
5	1.692	1.853	1.884	1.893	1.895	1.894	1.894	1.892	1.891	1.890
6	1.621	1.762	1.784	1.787	1.785	1.782	1.779	1.776	1.773	1.771
7	1.573	1.701	1.717	1.716	1.711	1.706	1.701	1.697	1.693	1.690
8	1.538	1.657	1.668	1.664	1.658	1.651	1.645	1.640	1.635	1.631
9	1.512	1.624	1.632	1.625	1.617	1.609	1.602	1.596	1.591	1.586
10	1.491	1.598	1.603	1.595	1.585	1.576	1.569	1.562	1.556	1.551
11	1.475	1.577	1.580	1.570	1.560	1.550	1.542	1.535	1.528	1.523
12	1.461	1.560	1.561	1.550	1.539	1.529	1.520	1.512	1.505	1.500
13	1.450	1.545	1.545	1.534	1.521	1.511	1.501	1.493	1.486	1.480
14	1.440	1.533	1.532	1.519	1.507	1.495	1.485	1.477	1.470	1.463
15	1.432	1.523	1.520	1.507	1.494	1.482	1.472	1.463	1.456	1.449
16	1.425	1.514	1.510	1.497	1.483	1.471	1.460	1.451	1.443	1.437
17	1.419	1.506	1.502	1.487	1.473	1.460	1.450	1.441	1.433	1.426
18	1.413	1.499	1.494	1.479	1.464	1.452	1.441	1.431	1.423	1.416
19	1.408	1.493	1.487	1.472	1.457	1.444	1.432	1.423	1.414	1.407
20	1.404	1.487	1.481	1.465	1.450	1.437	1.425	1.415	1.407	1.399
21	1.400	1.482	1.475	1.459	1.444	1.430	1.419	1.409	1.400	1.392
22	1.396	1.477	1.470	1.454	1.438	1.424	1.413	1.402	1.394	1.386
23	1.393	1.473	1.466	1.449	1.433	1.419	1.407	1.397	1.388	1.380
24	1.390	1.470	1.462	1.445	1.428	1.414	1.402	1.392	1.383	1.375
25	1.387	1.466	1.458	1.441	1.424	1.410	1.398	1.387	1.378	1.370
26	1.384	1.463	1.454	1.437	1.420	1.406	1.393	1.383	1.374	1.366
27	1.382	1.460	1.451	1.433	1.417	1.402	1.390	1.379	1.370	1.361
28	1.380	1.457	1.448	1.430	1.413	1.399	1.386	1.375	1.366	1.358
29	1.378	1.455	1.445	1.427	1.410	1.395	1.383	1.372	1.362	1.354
30	1.376	1.452	1.443	1.424	1.407	1.392	1.380	1.369	1.359	1.351
40	1.363	1.435	1.424	1.404	1.386	1.371	1.357	1.345	1.335	1.327
60	1.349	1.419	1.405	1.385	1.366	1.349	1.335	1.323	1.312	1.303
120	1.336	1.402	1.387	1.365	1.345	1.328	1.313	1.300	1.289	1.279
∞	1.328	1.393	1.377	1.354	1.333	1.315	1.300	1.286	1.275	1.250

n	$m = 11$	12	15	20	24	30	40	60	120	∞
1	9.367	9.406	9.493	9.581	9.625	9.670	9.714	9.759	9.804	9.850
2	3.386	3.393	3.410	3.426	3.435	3.443	3.451	3.459	3.468	3.480
3	2.448	2.450	2.455	2.460	2.463	2.465	2.467	2.470	2.472	2.473
4	2.082	2.083	2.083	2.083	2.083	2.082	2.082	2.082	2.081	2.081
5	1.889	1.888	1.885	1.882	1.880	1.878	1.876	1.874	1.872	1.870
6	1.769	1.767	1.762	1.757	1.754	1.751	1.748	1.744	1.741	1.738
7	1.687	1.684	1.678	1.671	1.667	1.663	1.659	1.655	1.650	1.647
8	1.627	1.624	1.617	1.609	1.604	1.600	1.595	1.589	1.584	1.580
9	1.582	1.579	1.570	1.561	1.556	1.551	1.545	1.539	1.533	1.528
10	1.547	1.543	1.534	1.523	1.518	1.512	1.506	1.499	1.492	1.487
11	1.518	1.514	1.504	1.493	1.487	1.481	1.474	1.466	1.459	1.454
12	1.495	1.490	1.480	1.468	1.461	1.454	1.447	1.439	1.431	1.426
13	1.475	1.470	1.459	1.447	1.440	1.432	1.425	1.416	1.408	1.402
14	1.458	1.453	1.441	1.428	1.421	1.414	1.405	1.397	1.387	1.381
15	1.443	1.438	1.426	1.413	1.405	1.397	1.389	1.380	1.370	1.363
16	1.431	1.426	1.413	1.399	1.391	1.383	1.374	1.365	1.354	1.348
17	1.420	1.414	1.401	1.387	1.379	1.370	1.361	1.351	1.341	1.334
18	1.410	1.404	1.391	1.376	1.368	1.359	1.350	1.340	1.328	1.321
19	1.401	1.395	1.382	1.367	1.358	1.349	1.339	1.329	1.317	1.310
20	1.393	1.387	1.374	1.358	1.349	1.340	1.330	1.319	1.307	1.300
21	1.386	1.380	1.366	1.350	1.341	1.332	1.322	1.311	1.298	1.290
22	1.379	1.374	1.359	1.343	1.334	1.324	1.314	1.303	1.290	1.282
23	1.374	1.368	1.353	1.337	1.327	1.318	1.307	1.295	1.282	1.274
24	1.368	1.362	1.347	1.331	1.321	1.311	1.300	1.289	1.275	1.267
25	1.363	1.357	1.342	1.325	1.316	1.306	1.294	1.282	1.269	1.260
26	1.359	1.352	1.337	1.320	1.311	1.300	1.289	1.277	1.263	1.254
27	1.354	1.348	1.333	1.315	1.306	1.295	1.284	1.271	1.257	1.248
28	1.350	1.344	1.329	1.311	1.301	1.291	1.279	1.266	1.252	1.243
29	1.347	1.340	1.325	1.307	1.297	1.286	1.275	1.262	1.247	1.237
30	1.343	1.337	1.321	1.303	1.293	1.282	1.270	1.257	1.242	1.233
40	1.319	1.312	1.295	1.276	1.265	1.253	1.240	1.225	1.208	1.197
60	1.294	1.287	1.269	1.248	1.236	1.223	1.208	1.191	1.172	1.158
120	1.270	1.262	1.243	1.220	1.207	1.192	1.175	1.156	1.131	1.113
∞	1.255	1.247	1.227	1.190	1.180	1.160	1.140	1.120	1.080	1.000

Table A.5: The 0.95 Quantiles of the Snedecor $F_{(m,n)}$ Distribution.

n	$m=1$	2	3	4	5	6	7	8	9	10
1	161.4	199.5	215.7	224.6	230.2	234.0	236.8	238.9	240.5	241.9
2	18.51	19.00	19.16	19.25	19.30	19.33	19.35	19.37	19.38	19.40
3	10.13	9.552	9.277	9.117	9.013	8.941	8.887	8.845	8.812	8.786
4	7.709	6.944	6.591	6.388	6.256	6.163	6.094	6.041	5.999	5.964
5	6.608	5.786	5.409	5.192	5.050	4.950	4.876	4.818	4.772	4.735
6	5.987	5.143	4.757	4.534	4.387	4.284	4.207	4.147	4.099	4.060
7	5.591	4.737	4.347	4.120	3.972	3.866	3.787	3.726	3.677	3.637
8	5.318	4.459	4.066	3.838	3.687	3.581	3.500	3.438	3.388	3.347
9	5.117	4.256	3.863	3.633	3.482	3.374	3.293	3.230	3.179	3.137
10	4.965	4.103	3.708	3.478	3.326	3.217	3.135	3.072	3.020	2.978
11	4.844	3.982	3.587	3.357	3.204	3.095	3.012	2.948	2.896	2.854
12	4.747	3.885	3.490	3.259	3.106	2.996	2.913	2.849	2.796	2.753
13	4.667	3.806	3.411	3.179	3.025	2.915	2.832	2.767	2.714	2.671
14	4.600	3.739	3.344	3.112	2.958	2.848	2.764	2.699	2.646	2.602
15	4.543	3.682	3.287	3.056	2.901	2.790	2.707	2.641	2.588	2.544
16	4.494	3.634	3.239	3.007	2.852	2.741	2.657	2.591	2.538	2.494
17	4.451	3.592	3.197	2.965	2.810	2.699	2.614	2.548	2.494	2.450
18	4.414	3.555	3.160	2.928	2.773	2.661	2.577	2.510	2.456	2.412
19	4.381	3.522	3.127	2.895	2.740	2.628	2.544	2.477	2.423	2.378
20	4.351	3.493	3.098	2.866	2.711	2.599	2.514	2.447	2.393	2.348
21	4.325	3.467	3.072	2.840	2.685	2.573	2.488	2.420	2.366	2.321
22	4.301	3.443	3.049	2.817	2.661	2.549	2.464	2.397	2.342	2.297
23	4.279	3.422	3.028	2.796	2.640	2.528	2.442	2.375	2.320	2.275
24	4.260	3.403	3.009	2.776	2.621	2.508	2.423	2.355	2.300	2.255
25	4.242	3.385	2.991	2.759	2.603	2.490	2.405	2.337	2.282	2.236
26	4.225	3.369	2.975	2.743	2.587	2.474	2.388	2.321	2.265	2.220
27	4.210	3.354	2.960	2.728	2.572	2.459	2.373	2.305	2.250	2.204
28	4.196	3.340	2.947	2.714	2.558	2.445	2.359	2.291	2.236	2.190
29	4.183	3.328	2.934	2.701	2.545	2.432	2.346	2.278	2.223	2.177
30	4.171	3.316	2.922	2.690	2.534	2.421	2.334	2.266	2.211	2.165
40	4.085	3.232	2.839	2.606	2.449	2.336	2.249	2.180	2.124	2.077
60	4.001	3.150	2.758	2.525	2.368	2.254	2.167	2.097	2.040	1.993
120	3.920	3.072	2.680	2.447	2.290	2.175	2.087	2.016	1.959	1.910
∞	3.840	3.000	2.600	2.370	2.210	2.100	2.010	1.940	1.880	1.830

n	$m=11$	12	15	20	24	30	40	60	120	∞
1	243.0	243.9	245.9	248.0	249.1	250.1	251.1	252.2	253.3	254.3
2	19.40	19.41	19.43	19.45	19.45	19.46	19.47	19.48	19.49	19.50
3	8.763	8.745	8.703	8.660	8.639	8.617	8.594	8.572	8.549	8.536
4	5.936	5.912	5.858	5.803	5.774	5.746	5.717	5.688	5.658	5.630
5	4.704	4.678	4.619	4.558	4.527	4.496	4.464	4.431	4.398	4.360
6	4.027	4.000	3.938	3.874	3.841	3.808	3.774	3.740	3.705	3.670
7	3.603	3.575	3.511	3.445	3.410	3.376	3.340	3.304	3.267	3.230
8	3.313	3.284	3.218	3.150	3.115	3.079	3.043	3.005	2.967	2.930
9	3.102	3.073	3.006	2.936	2.900	2.864	2.826	2.787	2.748	2.710
10	2.943	2.913	2.845	2.774	2.737	2.700	2.661	2.621	2.580	2.540
11	2.818	2.788	2.719	2.646	2.609	2.570	2.531	2.490	2.448	2.400
12	2.717	2.687	2.617	2.544	2.505	2.466	2.426	2.384	2.341	2.300
13	2.635	2.604	2.533	2.459	2.420	2.380	2.339	2.297	2.252	2.210
14	2.565	2.534	2.463	2.388	2.349	2.308	2.266	2.223	2.178	2.130
15	2.507	2.475	2.403	2.328	2.288	2.247	2.204	2.160	2.114	2.070
16	2.456	2.425	2.352	2.276	2.235	2.194	2.151	2.106	2.059	2.010
17	2.413	2.381	2.308	2.230	2.190	2.148	2.104	2.058	2.011	1.960
18	2.374	2.342	2.269	2.191	2.150	2.107	2.063	2.017	1.968	1.920
19	2.340	2.308	2.234	2.155	2.114	2.071	2.026	1.980	1.930	1.880
20	2.310	2.278	2.203	2.124	2.082	2.039	1.994	1.946	1.896	1.840
21	2.283	2.25	2.176	2.096	2.054	2.010	1.965	1.916	1.866	1.810
22	2.259	2.226	2.151	2.071	2.028	1.984	1.938	1.889	1.838	1.780
23	2.236	2.204	2.128	2.048	2.005	1.961	1.914	1.865	1.813	1.760
24	2.216	2.183	2.108	2.027	1.984	1.939	1.892	1.842	1.790	1.730
25	2.198	2.165	2.089	2.007	1.964	1.919	1.872	1.822	1.768	1.710
26	2.181	2.148	2.072	1.990	1.946	1.901	1.853	1.803	1.749	1.690
27	2.166	2.132	2.056	1.974	1.930	1.884	1.836	1.785	1.731	1.670
28	2.151	2.118	2.041	1.959	1.915	1.869	1.820	1.769	1.714	1.650
29	2.138	2.104	2.027	1.945	1.901	1.854	1.806	1.754	1.698	1.640
30	2.126	2.092	2.015	1.932	1.887	1.841	1.792	1.740	1.683	1.620
40	2.038	2.003	1.924	1.839	1.793	1.744	1.693	1.637	1.577	1.510
60	1.952	1.917	1.836	1.748	1.700	1.649	1.594	1.534	1.467	1.390
120	1.869	1.834	1.750	1.659	1.608	1.554	1.495	1.429	1.352	1.250
∞	1.821	1.750	1.670	1.570	1.520	1.460	1.390	1.320	1.220	1.000

Table A.6: The 0.99 Quantiles of the Snedecor $F_{(m,n)}$ Distribution.

n	$m = 1$	2	3	4	5	6	7	8	9	10
1	4052.	4999.	5403.	5625.	5764.	5859.	5928.	5981.	6022.	6056.
2	98.50	99.00	99.17	99.25	99.30	99.33	99.36	99.37	99.39	99.40
3	34.12	30.82	29.46	28.71	28.24	27.91	27.67	27.49	27.35	27.23
4	21.20	18.00	16.69	15.98	15.52	15.21	14.98	14.80	14.66	14.55
5	16.26	13.27	12.06	11.39	10.97	10.67	10.46	10.29	10.16	10.05
6	13.75	10.92	9.780	9.148	8.746	8.466	8.260	8.102	7.976	7.874
7	12.25	9.547	8.451	7.847	7.460	7.191	6.993	6.840	6.719	6.620
8	11.26	8.649	7.591	7.006	6.632	6.371	6.178	6.029	5.911	5.814
9	10.56	8.022	6.992	6.422	6.057	5.802	5.613	5.467	5.351	5.257
10	10.04	7.559	6.552	5.994	5.636	5.386	5.200	5.057	4.942	4.849
11	9.646	7.206	6.217	5.668	5.316	5.069	4.886	4.744	4.632	4.539
12	9.330	6.927	5.953	5.412	5.064	4.821	4.640	4.499	4.388	4.296
13	9.074	6.701	5.739	5.205	4.862	4.620	4.441	4.302	4.191	4.100
14	8.862	6.515	5.564	5.035	4.695	4.456	4.278	4.140	4.030	3.939
15	8.683	6.359	5.417	4.893	4.556	4.318	4.142	4.004	3.895	3.805
16	8.531	6.226	5.292	4.773	4.437	4.202	4.026	3.890	3.780	3.691
17	8.400	6.112	5.185	4.669	4.336	4.102	3.927	3.791	3.682	3.593
18	8.285	6.013	5.092	4.579	4.248	4.015	3.841	3.705	3.597	3.508
19	8.185	5.926	5.010	4.500	4.171	3.939	3.765	3.631	3.523	3.434
20	8.096	5.849	4.938	4.431	4.103	3.871	3.699	3.564	3.457	3.368
21	8.017	5.780	4.874	4.369	4.042	3.812	3.640	3.506	3.398	3.310
22	7.945	5.719	4.817	4.313	3.988	3.758	3.587	3.453	3.346	3.258
23	7.881	5.664	4.765	4.264	3.939	3.710	3.539	3.406	3.299	3.211
24	7.823	5.614	4.718	4.218	3.895	3.667	3.496	3.363	3.256	3.168
25	7.770	5.568	4.675	4.177	3.855	3.627	3.457	3.324	3.217	3.129
26	7.721	5.526	4.637	4.140	3.818	3.591	3.421	3.288	3.182	3.094
27	7.677	5.488	4.601	4.106	3.785	3.558	3.388	3.256	3.149	3.062
28	7.636	5.453	4.568	4.074	3.754	3.528	3.358	3.226	3.12	3.032
29	7.598	5.420	4.538	4.045	3.725	3.499	3.330	3.198	3.092	3.005
30	7.562	5.390	4.51	4.018	3.699	3.473	3.304	3.173	3.067	2.979
40	7.314	5.179	4.313	3.828	3.514	3.291	3.124	2.993	2.888	2.801
60	7.077	4.977	4.126	3.649	3.339	3.119	2.953	2.823	2.718	2.632
120	6.851	4.787	3.949	3.480	3.174	2.956	2.792	2.663	2.559	2.472
∞	6.630	4.610	3.780	3.320	3.020	2.880	2.640	2.510	2.410	2.320

n	$m = 11$	12	15	20	24	30	40	60	120	∞
1	6083.	6106.	6157.	6209.	6235.	6261.	6287.	6313.	6339.	6366 .
2	99.41	99.42	99.43	99.45	99.46	99.47	99.47	99.48	99.49	99.50
3	27.13	27.05	26.87	26.69	26.60	26.50	26.41	26.32	26.22	26.13
4	14.45	14.37	14.20	14.02	13.93	13.84	13.75	13.65	13.56	13.46
5	9.963	9.888	9.722	9.553	9.466	9.379	9.291	9.202	9.112	9.020
6	7.790	7.718	7.559	7.396	7.313	7.229	7.143	7.057	6.969	6.880
7	6.538	6.469	6.314	6.155	6.074	5.992	5.908	5.824	5.737	5.650
8	5.734	5.667	5.515	5.359	5.279	5.198	5.116	5.032	4.946	4.460
9	5.178	5.111	4.962	4.808	4.729	4.649	4.567	4.483	4.398	4.310
10	4.772	4.706	4.558	4.405	4.327	4.247	4.165	4.082	3.996	3.910
11	4.462	4.397	4.251	4.099	4.021	3.941	3.860	3.776	3.690	3.600
12	4.220	4.155	4.010	3.858	3.780	3.701	3.619	3.535	3.449	3.360
13	4.025	3.960	3.815	3.665	3.587	3.507	3.425	3.341	3.255	3.170
14	3.864	3.800	3.656	3.505	3.427	3.348	3.266	3.181	3.094	3.000
15	3.730	3.666	3.522	3.372	3.294	3.214	3.132	3.047	2.959	2.870
16	3.616	3.553	3.409	3.259	3.181	3.101	3.018	2.933	2.845	2.750
17	3.519	3.455	3.312	3.162	3.084	3.003	2.92	2.835	2.746	2.650
18	3.434	3.371	3.227	3.077	2.999	2.919	2.835	2.749	2.660	2.570
19	3.360	3.297	3.153	3.003	2.925	2.844	2.761	2.674	2.584	2.590
20	3.294	3.231	3.088	2.938	2.859	2.778	2.695	2.608	2.517	2.420
21	3.236	3.173	3.030	2.880	2.801	2.720	2.636	2.548	2.457	2.360
22	3.184	3.121	2.978	2.827	2.749	2.667	2.583	2.495	2.403	2.310
23	3.137	3.074	2.931	2.781	2.702	2.620	2.535	2.447	2.354	2.260
24	3.094	3.032	2.889	2.738	2.659	2.577	2.492	2.403	2.310	2.210
25	3.056	2.993	2.850	2.699	2.620	2.538	2.453	2.364	2.270	2.170
26	3.021	2.958	2.815	2.664	2.585	2.503	2.417	2.327	2.233	2.130
27	2.988	2.926	2.783	2.632	2.552	2.470	2.384	2.294	2.198	2.100
28	2.959	2.896	2.753	2.602	2.522	2.440	2.354	2.263	2.167	2.006
29	2.931	2.868	2.726	2.574	2.495	2.412	2.325	2.234	2.138	2.030
30	2.906	2.843	2.700	2.549	2.469	2.386	2.299	2.208	2.111	2.010
40	2.727	2.665	2.522	2.369	2.288	2.203	2.114	2.019	1.917	1.800
60	2.559	2.496	2.352	2.198	2.115	2.028	1.936	1.836	1.726	1.600
120	2.399	2.336	2.192	2.035	1.950	1.860	1.763	1.656	1.533	1.380
∞	2.307	2.180	2.040	1.880	1.790	1.700	1.590	1.470	1.320	1.000

Bibliography

Abdous, B., Ghoudi, K., and Khoudraji, A. (1999). Non-parametric estimation of the limit dependence function of multivariate extremes. *Extremes*, 2(3):245–268.

Anderson, C. W. and Coles, S. G. (2002). The largest inclusions in a piece of steel. *Extremes*, 5(3):237–252 (2003).

Andrä, W. and Saul, R. (1974). Versuche mit bündeln aus parallelen drähten und litzen für die nordbrücke mannheim-ludwigshafen und das zeltdach in münchen. *Die Bautechnik*, 9, 10, 11:289–298, 332–340, 371–373.

Andrä, W. and Saul, R. (1979). Die festigkeit insbesondere dauerfertigkeit langer paralleldrahtbündel. *Die Bautechnik*, 4:128–130.

Ang, A. H. S. (1973). Structural risk analysis and reliability-based design. *Journal of the Structural Division, ASCE*, 99:1891–1910.

Arena, F. (2002). Ocean wave statistics for engineering applications. *Rendiconti del Circolo Matematico di Palermo. Serie II. Supplemento*, 70(I):21–52. The Fourth International Conference in "Stochastic Geometry, Convex Bodies, Empirical Measures and Applications to Engineering Science," Vol. I, Tropea, 2001.

Arnold, B. C., Balakrishnan, N., and Nagaraja, H. N. (1992). *A First Course in Order Statistics*. John Wiley & Sons, New York.

Arnold, B. C., Castillo, E., and Sarabia, J. M. (1996). Modelling the fatigue life of longitudinal elements. *Naval Research Logistics Quarterly*, 43:885–895.

Aziz, P. M. (1956). Application of the statistical theory of extremes values to the analysis of maximum pit depth for aluminium. *Corrosion*, 12:35–46.

Balakrishnan, N., ed. (1992). *Handbook of the Logistic Distribution*. Marcel Dekker, New York.

Balakrishnan, N. and Basu, A. P., eds. (1995). *The Exponential Distribution: Theory, Methods and Applications*. Gordon and Breach Science Publishers, Newark, New Jersey.

Balakrishnan, N. and Chan, P. S. (1994). Asymptotic best linear unbiased estimation for the log-gamma distribution. *Sankhyā, B*, 56:314–322.

Balakrishnan, N. and Chan, P. S. (1995a). Maximum likelihood estimation for the log-gamma distribution under type-II censored samples and associated inference. In N. Balakrishnan, ed., *Recent Advances in Life-Testing and Reliability*, pp. 409–437. CRC Press, Boca Raton, Florida.

Balakrishnan, N. and Chan, P. S. (1995b). Maximum likelihood estimation for the three-parameter log-gamma distribution under type-II censoring. In N. Balakrishnan, ed., *Recent Advances in Life-Testing and Reliability*, pp. 439–453. CRC Press, Boca Raton, Florida.

Balakrishnan, N. and Chan, P. S. (1998). Log-gamma order statistics and linear estimation of parameters. In N. Balakrishnan and C. R. Rao, eds., *Handbook of Statistics*, vol. 17, pp. 61–83. North-Holland, Amsterdam.

Balakrishnan, N. and Cohen, A. C. (1991). *Order Statistics and Inference: Estimation Methods*. Academic Press, San Diego, California.

Balakrishnan, N. and Nevzorov, V. B. (2003). *A Primer on Statistical Distributions*. John Wiley & Sons, New York.

Balakrishnan, N. and Rao, C. R. (1997). A note on the best linear unbiased estimation based on order statistics. *The American Statistician*, 51:181–185.

Balakrishnan, N. and Rao, C. R., eds. (1998a). *Order Statistics: Applications*, vol. 17. North-Holland, Amsterdam.

Balakrishnan, N. and Rao, C. R., eds. (1998b). *Order Statistics: Theory and Methods*, vol. 16. North-Holland, Amsterdam.

Barlow, R. E. (1972). Averaging time and maxima for air pollution concentrations. In *Proceedings of the Sixth Berkeley Symposium on Mathematical Statistics and Probability (University of California, Berkeley, California, 1970/1971), Volume VI: Effects of Pollution on Health*, pp. 433–442. University California Press, Berkeley, California.

Barlow, R. E. and Singpurwalla, N. D. (1974). Averaging time and maxima for dependent observations. In *Proceedings of the Symposium on Statistical Aspects of Air Quality Data*.

Barnett, V. (1975). Probability plotting methods and order statistics. *Applied Statistics*, 24:95–108.

Basu, A. P. (1971). Bivariate failure rate. *Journal of the American Statistical Association*, 66:103–104.

Batdorf, S. B. (1982). Tensile strength of unidirectionally reinforced composites-I. *Journal of Reinforced Plastics and Composites*, 1:153–164.

Batdorf, S. B. and Ghaffanian, R. (1982). Tensile strength of unidirectionally reinforced composites-II. *Journal of Reinforced Plastics and Composites*, 1:165–176.

Battjes, J. A. (1977). Probabilistic aspects of ocean waves. In *Proceedings of the Seminar on Safety of Structures Under Dynamic Loading*. University of Trondheim, Trondheim, Norway.

Beard, L. R. (1962). *Statistical Methods in Hidrology*. U.S. Army Corps of Engineers, Sacramento, California.

Benson, M. A. (1968). Uniform flood-frequency estimating methods for federal agencies. *Water Resources Research*, 4:891–908.

Birnbaum, Z. W. and Saunders, S. C. (1958). A statistical model for life-length of materials. *Journal of the American Statistical Association*, 53:151–159.

Block, H. W. and Basu, A. P. (1974). A continuous bivariate exponential extension. *Journal of the American Statistical Association*, 69:1031–1037.

Blom, G. (1958). *Statistical Estimates and Transformed Beta-Variables*. John Wiley & Sons, New York.

Bogdanoff, J. L. and Schiff, A. (1972). Earthquake effects in the safety and reliability analysis of engineering structures. In A. M. Freudenthal, ed., *International Conference on Structural Safety and Reliability*, pp. 147–148. Pergamon Press, Washington, D.C.

Borgman, L. E. (1963). Risk criteria. *Journal of Waterway, Port, Coastal, and Ocean Engineering, ASCE*, 89:1–35.

Borgman, L. E. (1970). Maximum wave height probabilities for a random number of random intensity storms. In *Proceedings of the 12th Conference on Coastal Engineering*. Washington, D.C.

Borgman, L. E. (1973). Probabilities of highest wave in hurricane. *Journal of Waterway, Port, Coastal, and Ocean Engineering, ASCE*, 99:185–207.

Bretchneider, C. C. (1959). Wave variability and wave spectra for wind generated gravity waves. Technical Memo. 118, U. S. Beach Erosion Board, Washington, D.C.

Bryant, P. J. (1983). Cyclic gravity waves in deep water. *Australian Mathematical Society Journal, B*, 25(1):2–15.

Bühler, H. and Schreiber, W. (1957). Lösung eineger aufgaben der dauerfestigkeit mit dem treppenstufen-verfahren. *Archiv für eisenhüttenwessen*, 28:153–156.

Capéraà, P. and Fougères, A. L. (2000). Estimation of a bivariate extreme value distribution. *Extremes*, 3(4):311–329.

Castillo, E. (1988). *Extreme Value Theory in Engineering*. Academic Press, San Diego, California.

Castillo, E. (1994). Extremes in engineering applications. In J. Galambos, J. Lechner, and E. Simiu, eds., *Extreme Value Theory and Applications*, pp. 15–42. Kluwer Academic Publishers, Dordrecht, The Netherlands.

Castillo, E., Ascorbe, A., and Fernández-Canteli, A. (1983a). Static progressive failure in multiple tendons: A statistical approach. In *The 44th Session of the International Statistical Institute*, vol. 45.1. Madrid, Spain.

Castillo, E., Fernández-Canteli, A., Ascorbe, A., and Mora, E. (1983b). The Box-Jenkins model and the progressive fatigue failure of large parallel elements stay-endons. In *Statistical Extremes and Applications*. ASI-NATO.

Castillo, E., Fernández-Canteli, A., Ascorbe, A., and Mora, E. (1984a). Aplicación de los modelos de series temporales al análisis estadístico de la resistencia de tendones de puentes atirantados. *Anales de Ingeniería Mecánica*, 2:379–382.

Castillo, E., Fernández-Canteli, A., Esslinger, V., and T̈hurlimann (1985). Statistical models for fatigue analysis of wires, strands and cables. In *IABSE Proceedings*, vol. 82, pp. 1–140.

Castillo, E., Fernández-Canteli, A., Ruiz-Tolosa, J. R., and Sarabia, J. M. (1990). Statistical models for analysis of fatigue life of long elements. *Journal of Engineering Mechanics*, 116:1036–1049.

Castillo, E., Galambos, J., and Sarabia, J. M. (1989). The selection of the domain of attraction of an extreme value distribution from a set of data. In *Extreme Value Theory*, vol. 51 of *Lecture Notes in Statistics*, pp. 181–190. Springer-Verlag, New York.

Castillo, E. and Hadi, A. S. (1995a). A method for estimating parameters and quantiles of distributions of continuous random variables. *Computational Statistics and Data Analysis*, 20:421–439.

Castillo, E. and Hadi, A. S. (1995b). Modeling lifetime data with application to fatigue models. *Journal of the American Statistical Association*, 90:1041–1054.

Castillo, E. and Hadi, A. S. (1995c). Parameter and quantile estimation for the generalized extreme-value distribution. *Environmetrics*, 5:417–432.

Castillo, E. and Hadi, A. S. (1997). Fitting the generalized Pareto distribution to data. *Journal of the American Statistical Association*, 92:1609–1620.

Castillo, E., Losada, M. A., Mínguez, R., Castillo, C., and Baquerizo, A. (2004). Optimal engineering design method that combines safety factors and failure probabilities: Application to rubble-mound breakwaters. *Journal of Waterways, Ports, Coastal and Ocean Engineering*, 130:77–88.

Castillo, E., Luceño, A., Montalbán, A., and Fernández-Canteli, A. (1987). A dependent fatigue lifetime model. *Communications in Statistics, Theory and Methods*, 16(4):1181–1194.

Castillo, E., Pérez, A. V. U., Gutiérrez-Solana, F., and Fernández-Canteli, A. (1984b). Estudio de la relación existente entre dos métodos clásicos del análisis de la resistencia a fatiga de elementos estructurales. *Anales de Ingenieria Mecánica*, 2(1):150–156.

Castillo, E. and Sarabia, J. M. (1992). Engineering analysis of extreme value data. *Journal of Waterway, Port, Coastal, and Ocean Engineering, ASCE*, 118:129–146.

Castillo, E. and Sarabia, J. M. (1994). Extreme value analysis of wave heights. *Journal of Research of the National Institute of Standards and Technology*, 99:445–454.

Castillo, E., Sarabia, J. M., and Hadi, A. S. (1997). Fitting continuous bivariate distributions to data. *Journal of the Royal Statistical Society, D*, 46:355–369.

Cavanie, A., Arhan, M., and Ezraty, R. (1976). A statistical relationship between individual heights and periods of storm waves. In *Proceedings BOSS'76*, pp. 354–360. Trondheim, Norway.

Chakrabarti, S. K. and Cooley, R. P. (1977). Statistical distributions of periods and heights of ocean waves. *Journal of Geophysical Research*, 82:1363–1368.

Chan, P. S. and Balakrishnan, N. (1995). The infeasibility of probability weighted moments estimation of some generalized distributions. In N. Balakrishnan, ed., *Recent Advances in Life-Testing and Reliability*, vol. 1, pp. 565–573. CRC Press, Boca Raton, Florida.

Chow, V. T. (1951). A general formula for hydrologic frequency analysis. *Eos Transactions AGU*, 32:231–237.

Chow, V. T. (1964). *Handbook of Applied Hydrology*. McGraw-Hill, New York.

Coleman, B. D. (1956). Time dependence of mechanical breakdown phenomena. *Journal of Applied Physics*, 27:862–866.

Coleman, B. D. (1957a). A stochastic process model for mechanical breakdown. *Transactions of the Society of Rheology*, 1:153–168.

Coleman, B. D. (1957b). Time dependence of mechanical breakdown in bundles of fibers I. *Journal of Applied Physics*, 28:1058–1064.

Coleman, B. D. (1958a). On the strength of classical fibres bundles. *International Journal of Mechanics of Solids*, 7:60–70.

Coleman, B. D. (1958b). Statistics and time dependence of mechanical breakdown in fibers. *Journal of Applied Physics*, 29:968–983.

Coleman, B. D. (1958c). Time dependence of mechanical breakdown in bundles of fibers III. *Transactions of the Society of Rheology*, 2:195–218.

Coles, S. (2001). *An Introduction to Statistical Modeling of Extreme Values*. Springer-Verlag, London, England.

Coles, S. G. and Tawn, J. A. (1991). Modelling extreme multivariate events. *Journal of the Royal Statistical Society, B*, 53:377–392.

Coles, S. G. and Tawn, J. A. (1994). Statistical methods for multivariate extremes: an application to structural design (with discussion). *Journal of the Royal Statistical Society, A*, 43:1–48.

Court, A. (1953). Wind extremes as design factors. *Journal of the Franklin Institute*, 256:39–55.

Cunnane, C. (1978). Unbiased plotting positions: A review. *Journal of Hydrology*, 37:205–222.

Davenport, A. G. (1968a). The dependence of wind loads upon meteorological parameters. In *Proceedings Wind Effects on Building and Structures*, vol. 1. University of Toronto Press, Toronto, Canada.

Davenport, A. G. (1968b). The relationship of wind structure to wind loading. In *Proceedings of the International Seminar on Wind Building and Structures*. University of Toronto Press, Toronto, Canada.

Davenport, A. G. (1972). Structural safety and reliability under wind action. In *International Conference on Structural Safety and Reliability*, pp. 131–145. Pergamon Press, Washington, D.C.

Davenport, A. G. (1978). Wind structure and wind climate. In *Int. Res. Seminar on Safety of Structures Under Dynamic Loading*, pp. 238–256. Trondheim, Norway.

David, H. A. and Nagaraja, H. N. (2003). *Order Statistics*. John Wiley & Sons, New York, 3rd ed.

Davison, A. C. (1984). Modelling excesses over high thresholds, with an application. In *Statistical Extremes and Applications NATO ASI Series*, pp. 461–482. D. Reidel, Dordrecht.

Davison, A. C. and Smith, R. L. (1990). Models for exceedances over high thresholds. *Journal of the Royal Statistical Society, B*, 52:393–442.

de Haan, L. (1985). Extremes in higher dimensions: the model and some statistics. In *Proceedings of the 45th Session of the International Statistical Institute, Vol. 4*, vol. 51, pp. 185–192. (with discussion).

de Haan, L. and de Ronde, J. (1998). Sea and wind: multivariate extremes at work. *Extremes*, 1(1):7–45.

de Haan, L. and Resnick, S. I. (1977). Limit theory for multivariate sample extremes. *Z. Wahrscheinlichkeitstheorie und Verw. Gebiete*, 40(4):317–337.

de Haan, L. and Sinha, A. K. (1999). Estimating the probability of a rare event. *The Annals of Statistics*, 27(2):732–759.

Dengel, D. (1971). Einige grundlegende gesichtspunkte für die olanung undauswertung von dauerschwingversuchen. *Materialprüfung*, 13:145–151.

Diaconis, P. and Efron, B. (1974). Computer intensive methods in statistics. *Scientific American*, 248:116–130.

DiCiccio, T. J. (1987). Approximate inference for the generalized gamma distribution. *Technometrics*, 29:32–39.

Draper, L. (1963). Derivation of a design wave from instrumental records of sea waves. In *Proceedings of the Institute of Civil Engineers*, vol. 26, pp. 291–303. London, England.

Drees, H., de Haan, L., and Li, D. (2003). On large deviation for extremes. *Statistics and Probability Letters*, 64(1):51–62.

Duebelbeiss, E. (1979). Dauerfertigkeitsversuche mit einem midifizierten treppenstufenverfahren. *Materialprüfung*, 16:240–244.

DuMouchel, W. (1983). Estimating the stable index σ in order to measure tail thickness. *The Annals of Statistics*, 11:1019–1036.

Dupuis, D. J. and Field, C. A. (1998). A comparison of confidence intervals for generalized extreme-value distributions. *Journal of Statistical Computation and Simulation*, 61(4):341–360.

Earle, M. D., Effermeyer, C. C., and Evans, D. J. (1974). Height-period joint probabilities in hurricane camille. *Journal of Waterway, Port, Coastal, and Ocean Engineering, ASCE*, 3:257–264.

Efron, B. (1979). Bootstrap methods: Another look at the jackknife. *The Annals of Statistics*, 7:1–26.

Eldredge, G. G. (1957). Analysis of corrosion pitting by extreme-value statistics and its application to oil well tubing caliper surveys. *Corrosion*, 13:67–76.

Embrechts, P., Klüppelberg, C., and Mikosch, T. (1997). *Modelling Extremal Events*. Springer-Verlag, Berlin.

Endicott, H. S. and Weber, H. K. (1956). Extremal nature of dielectric breakdown-effect of sample size. In *Symposium on Minimum Property Values of Electrical Insulating Materials. ASTM Special Technical Publication*, vol. 188, pp. 5–11. Philadelphia, PA.

Endicott, H. S. and Weber, H. K. (1957). Electrode area effect for the impulse breakdown of transformer oil. *AIEE Transactions. Part III: Power Apparatus and Systems*, 76:393–398.

Epstein, B. (1954). Truncated life tests in the exponential case. *Annals of Mathematical Statististics*, 25:555–564.

Epstein, B. and Sobel, M. (1954). Some theorems relevant to life testing from an exponential distribution. *Annals of Mathematical Statistics*, 25:373–381.

Evans, M., Hastings, N., and Peacock, B. (2000). *Statistical Estimates and Transformed Beta-Variables*. John Wiley & Sons, New York.

Fei, H. L., Lu, X. W., and Xu, X. L. (1998). Procedures for testing outlying observations with a Weibull or extreme-value distribution. *Acta Mathematicae Applicatae Sinica. Yingyong Shuxue Xuebao*, 21(4):549–561.

Fernández-Canteli, A. (1982). Statistical interpretation of the miner-number using an index of probability of total damage. In *IABSE Colloquium Fatigue of Steel and Concrete Structures*. Lausanne.

Fernández-Canteli, A., Esslinger, V., and Thurlimann, B. (1984). Ermüdungsfestigkeit von bewehrungs und spannstählen. Tech. rep., ETH Zürich.

Ferro, C. A. T. and Segers, J. (2003). Inference for clusters of extreme values. *Journal of the Royal Statistical Society, B*, 65(2):545–556.

Fisher, R. A. and Tippett, L. H. C. (1928). Limiting forms of the frequency distributions of the largest or smallest member of a sample. *Procedings of the Cambridge Philosophical Society*, 24:180–190.

Freudenthal, A. M. (1975). Reliability assessment of aircraft structures based on probabilistic interpretation of the sactter factor. In *AFML-TR-74-198, Air Force Materials Laboratory, Wright-Patterson AFB.*

Freund, J. (1961). A bivariate extension of the exponential distribution. *Journal of the American Statistical Association*, 56:971–977.

Gabriel, K. (1979). Anwendung von statistischen methoden und wahrscheinlichkeitsbetrachtungen auf das verhalten von bündeln und seilen als zugglieder aus vielenund langen drähten. In *Weitgespannte flačhentragwerke 2, International Symposium*. Stuttgart.

Galambos, J. (1978). *The Asymptotic Theory of Extreme Order Statistics*. John Wiley & Sons, New York.

Galambos, J. (1987). *The Asymptotic Theory of Extreme Order Statistics*. Robert E. Krieger, Malabar, Florida, 2nd ed.

Galambos, J. (1994). The development of the mathematical theory of extremes in the past half century. *Rossiĭskaya Akademiya Nauk. Teoriya Veroyatnosteĭ i ee Primeneniya*, 39(2):272–293.

Galambos, J. (1995). *Advanced Probability Theory*. Marcel Dekker, New York, 2nd ed.

Galambos, J. (1998). Univariate extreme value theory and applications. In N. Balakrishnan and C. R. Rao, eds., *Order Statistics: Theory and Methods*, vol. 16 of *Handbook of Statistics*, pp. 315–333. North-Holland, Amsterdam.

Galambos, J. (2000). The classical extreme value model: mathematical results versus statistical inference. In *Statistics for the 21st Century*, vol. 161 of *Statistical Textbooks Monograhs*, pp. 173–187. Marcel Dekker, New York.

Galambos, J. (2001). Point estimators and tests of hypothesis for the parameters of the classical extreme value distribution functions. *Journal of Applied Statistical Science*, 10(1):1–8.

Galambos, J. and Macri, N. (2000). The life length of humans does not have a limit. *Journal of Applied Statistical Science*, 9(2):253–263.

Galambos, J. and Macri, N. (2002). Classical extreme value model and prediction of extreme winds. *Journal of Structural Engineeing*, 125:792–794.

Galambos, J. and Simonelli, I. (1996). *Bonferroni-Type Inequalities with Applications*. Springer-Verlag, New York.

Glen, A. G. and Leemis, L. M. (1997). The arctangent survival distribution. *Journal of Quality Technology*, 29:205–210.

Glynn, P. W. and Whitt, W. (1995). Heavy-traffic extreme-value limits for queues. *Operations Research Letters*, 18(3):107–111.

Gomes, M. I. and de Haan, L. (1999). Approximation by penultimate extreme value distributions. *Extremes*, 2(1):71–85.

Gomes, M. I. and Oliveira, O. (2001). The bootstrap methodology in statistics of extremes-choice of the optimal sample fraction. *Extremes*, 4(4):331–358 (2002).

Gómez-Corral, A. (2001). On extreme values of orbit lengths in $M/G/1$ queues with constant retrial rate. *OR Spektrum. Quantitative Approaches in Management*, 23(3):395–409.

Goodknight, R. G. and Russel, T. L. (1963). Investigation of statistics of wave heights. *Journal of Waterway, Port, Coastal, and Ocean Engineering, ASCE*, 89:29–54.

Greenwood, J. A., Landwehr, J. M., Matalas, N. C., and Wallis, J. R. (1979). Probability weighted moments: Definition and relation to parameters of several distributions expressible in inverse form. *Water Resources Research*, 15:1049–1054.

Grigoriu, M. (1984). Estimate of extreme winds from short records. *Journal of the Structural Engineering Division, ASCE*, 110:1467–1484.

Grimshaw, S. D. (1993). Computing maximum likelihood estimates for the generalized Pareto distribution. *Technometrics*, 35:185–191.

Gringorten, I. I. (1963). A plotting rule for extreme probability paper. *Journal of Geophysical Research*, 68:813–814.

Grover, H. J. (1966). Fatigue of aircraft structures. Tech. Rep. 01-1A-13, NAVAIR United States Government Office, Washington, D.C.

Guillou, A. and Hall, P. (2001). A diagnostic for selecting the threshold in extreme value analysis. *Journal of the Royal Statistical Society, B*, 63(2):293–305.

Gumbel, E. J. (1961). Bivariate logistic distributions. *Journal of the American Statistical Association*, 56:335–349.

Gumbel, E. J. (1964). Technische anwerdung der statistischen theorie der extremwerte. *Schweizer Archiv*, 30:33–47.

Gumbel, E. J. and Goldstein, N. (1964). Analysis of empirical bivariate extremal distributions. *Journal of the American Statistical Association*, 59:794–816.

Günbak, A. R. (1978). Statistical analysis of 21 wave records off the danish white sand coast during one storm in january 1976. In *Division of Port and Ocean engineering*. The University of Trondheim, Trondheim, Norway.

Gupta, V. K., Duckstein, L., and Peebles, R. W. (1976). On the joint distribution of the largest flood and its time occurrence. *Water Resources Research*, 12:295–304.

Hajdin, N. (1976). Vergleich zwischen den paralleldrahtseilen und verschlossenen seilen am beispiel der eisenbahnscharägseilbrücke über die save in Belgrad. *IVBH Vorbericht zum*, 10:471–475.

Hall, P. and Tajvidi, N. (2000). Distribution and dependence-function estimation for bivariate extreme-value distributions. *Bernoulli*, 6(5):835–844.

Hall, P. and Wang, J. Z. (1999). Estimating the end-point of a probability distribution using minimum-distance methods. *Bernoulli*, 5(1):177–189.

Hall, W. J. and Wellner, J. (1981). Mean residual life. In M. Csörgo, J. N. K. R. D. A. Dawson, and A. K. M. E. Saleh, eds., *Statistics and Related Topics*, pp. 169–184. North-Holland, Amsterdam.

Harter, H. L. (1977). *A Survey of the Literature on the Size Effect on Material Strength.* Air Force Flight Dynamics Laboratory/FBRD, Air Force Systems Command, United States Air Force, Wright-Patterson Air Force Base, Ohio. AFFDL-TR-77-11.

Harter, H. L. (1978a). A bibliography of extreme value theory. *International Statistical Review*, 46:279–306.

Harter, H. L. (1978b). A chronological annotated bibliography on order statistics. In *U.S. Government Printing Office*, vol. Pre-1950. Washington, D.C.

Harter, H. L. (1984). Another look at plotting positions. *Communications in Statistics, Theory and Methods*, 13:1613–1633.

Hasofer, A. M. (1972). Wind-load design based on statistics. *Proceedings of the Instution of Civil Engineers*, 51:69–82.

Hasofer, A. M. (1979). Extreme value theory: a review of approaches. In *Applications of Statistics and Probability in Soil and Structural Engineering (Proceedings of the Third International Conference*, vol. III, pp. 40–53. Unisearch, Kensington.

Hasofer, A. M. and Sharpe, K. (1969). The analysis of wind gusts. *Australian Meteorological Magazine*, 17:198–214.

Helgason, T. and Hanson, J. M. (1976). Fatigue strenght of high-yield reinforcing bars. Tech. rep., American Association of State Highway and Transportation Officials.

Hershfield, D. M. (1962). Extreme rainfall relationship. *Journal of the Hydraulics Division, ASCE*, 6:73–92.

Hill, L. R. and Schmidt, P. L. (1948). Insulation breakdown as a function of area. *Electrical Engineering*, 67:76.

Hosking, J. R. M. (1984). Testing whether the shape parameter is zero in the generalized extreme-value distribution. *Biometrika*, 71:367–374.

Hosking, J. R. M. and Wallis, J. R. (1987). Parameter and quantile estimation for the generalized Pareto distribution. *Technometrics*, 29:339–349.

Hosking, J. R. M., Wallis, J. R., and Wood, E. F. (1985). Estimation of the generalized extreme-value distribution by the method of probability-weighted moments. *Technometrics*, 27:251–261.

Houmb, O. G. and Overvik, J. (1977). On the statistical properties of 115 wave records from the Norwegian continental shelf. Tech. rep., The University of Trondheim, Trondheim, Norway.

Joe, H., Smith, R. L., and Weissman, I. (1992). Bivariate threshold methods for extremes. *Journal of the Royal Statistical Society, B*, 54:171–183.

Johnson, N. L., Kotz, S., and Balakrishnan, N. (1994). *Continuous Univariate Distributions, Volume 1*. John Wiley & Sons, New York, 2nd ed.

Johnson, N. L., Kotz, S., and Balakrishnan, N. (1995). *Continuous Univariate Distributions, Volume 2*. John Wiley & Sons, New York, 2nd ed.

Johnson, N. L., Kotz, S., and Balakrishnan, N. (1997). *Discrete Multivariate Distributions, Volume 2*. John Wiley & Sons, New York.

Johnson, N. L., Kotz, S., and Kemp, A. W. (1992). *Univariate Discrete Distributions*. John Wiley & Sons, New York, 2nd ed.

Kang, S. and Serfozo, R. F. (1997). Maxima of sojourn times in acyclic Jackson queueing networks. *Computers and Operations Research and their Application to Problems of World Concern*, 24(11):1085–1095.

Karr, A. F. (1976). Two extreme value processes arising in hydrology. *Journal of Applied Probability*, 13(1):190–194.

Ker, A. P. (2001). On the maximum of bivariate normal random variables. *Extremes*, 4(2).

Kirby, W. (1969). On the random occurrences of major floods. *Water Resources Research*, 5:778–784.

Kotz, S., Balakrishnan, N., and Johnson, N. L. (2000). *Continuous Multivariate Distributions, Volume 1*. John Wiley & Sons, New York, 2nd ed.

Kotz, S. and Nadarajah, S. (2000). *Extreme Value Distributions*. Imperial College Press, London, England.

Larsen, R. I. (1969). A new mathematical model of air pollutant concentration averaging time and frequency. *Journal of the Air Pollution and Control Association*, 19:24–30.

Lawless, J. F. (1980). Inference in the generalized gamma and log-gamma distribution. *Technometrics*, 22:67–82.

Lawless, J. F. (2003). *Statistical Models and Methods for Lifetime Data*. John Wiley & Sons, New York, 2nd ed.

Leadbetter, M. R. (1995). On high level exceedance modeling and tail inference. *Journal of Statistical Planning and Inference*, 45(1-2):247–260.

Leadbetter, M. R., Lindgren, G., and Rootzén, H. (1983). *Extremes and Related Properties of Random Sequences and Processes*. Springer-Verlag, New York.

Lévi, R. (1949). Calculs probabilistes de la securité des constructions. *Annales des Ponts et Chaussées*, 119:493–539.

Liao, M. and Shimokawa, T. (1999). A new goodness-of-fit test for type-I extreme-value and 2-parameter Weibull distributions with estimated parameters. *Journal of Statistical Computation and Simulation*, 64(1):23–48.

Lighthill, J. (1999). Ocean spray and the thermodynamics of tropical cyclones. *Journal of Engineering Mathematics*, 35(1-2):11–42.

Lindgren, G. and Rootzén, H. (1987). Extreme values: theory and technical applications. *Scandinavian Journal of Statistics*, 14(4):241–279.

Logan, K. H. (1936). Soil corrosion studies: Rates of loss of weight and pitting of ferrous specimens. *Journal of Research of the National Bureau of Standards*, 16:431–466.

Logan, K. H. and Grodsky, V. A. (1931). Soil corrosion studies, 1930: Rates of corrosion and pitting of bare ferrous speciments. *Journal of Research of the National Bureau of Standards*, 7:1–35.

Longuet-Higgins, M. S. (1952). On the statistical distribution of the heights of sea waves. *Journal of Marine Research*, 9:245–266.

Longuet-Higgins, M. S. (1975). On the joint distribution of the periods and amplitudes of sea waves. *Journal of Geophysical Research*, 80:2688–2694.

Lu, J. C. and Peng, L. (2002). Likelihood based confidence intervals for the tail index. *Extremes*, 5(4):337–352 (2003).

Maennig, W. W. (1967). Untersuchungen zur planung und auswertung von dauerschwingversuchen an stahl in bereichen der zei und der dauerfestigkeit. Tech. rep., Düsseldorf, Germany.

Maennig, W. W. (1970). Bemerkungen zur beurteilung des dauerschwingverhaltens von stahl und einige untersuchungen zur bestimmung des dauerfestigkeitsbereichs. *Materialprüfung*, 12:124–131.

Mann, N. R., Schafer, R. E., and Singpurwalla, N. D. (1974). *Methods for Statistical Analysis of Reliability and Life Data*. John Wiley & Sons, New York.

Marohn, F. (1998). Testing the Gumbel hypothesis via the pot-method. *Extremes*, 1(2):191–213.

Marohn, F. (2000). Testing extreme value models. *Extremes*, 3(4):363–384.

Marshall, A. W. and Olkin, I. (1967). A multivariate exponential distribution. *Journal of the American Statistical Association*, 62:30–44.

Matalas, N. C. and Wallis, J. R. (1973). Eureka! It fits a pearson type III distribution. *Water Resources Research*, 9:282–291.

Matthys, G. and Beirlant, J. (2003). Estimating the extreme value index and high quantiles with exponential regression models. *Statistica Sinica*, 13(3):853–880.

McCormick, W. P. and Park, Y. S. (1992). Approximating the distribution of the maximum queue length for $M/M/s$ queues. In *Queueing and Related Models*, vol. 9, pp. 240–261. Oxford University Press, New York.

Mendenhall, W. (1958). A bibliography on life testing and related topics. *Biometrika*, 45:521–543.

Midlarsky, M. I. (1989). A distribution of extreme inequality with applications to conflict behavior: a geometric derivation of the Pareto distribution. *Mathematical and Computer Modelling*, 12(4-5):577–587.

Mistéth, E. (1973). Determination of the critical loads considering the anticipated durability of structures. *Acta Technica Academiae Scientiarum Hungaricae*, 74:21–38.

Mistéth, E. (1974). Dimensioning of structures for flood discharge according to the theory of probability. *Acta Technica Academiae Scientiarum Hungaricae*, 76:107–127.

Morrison, J. E. and Smith, J. A. (2001). Scaling properties of flood peaks. *Extremes*, 4(1):5–22.

Moses, F. (1974). Reliability of structural systems. *Journal of the Structural Division, ASCE*, 100:1813–1820.

Murzewski, J. (1972). Optimization of structural safety for extreme load and strenght distributions. *Archiwum Inzynierii Ladowej*, 18:573–583.

Mustafi, C. K. (1963). Estimation of parameters of the extreme value distribution with limited type of primary probability distribution. *Calcutta Statistical Association Bulletin*, 12:47–54.

Nadarajah, S. (2000). Approximations for bivariate extreme values. *Extremes*, 3(1):87–98.

Nadarajah, S. (2003). Reliability for extreme value distributions. *Mathematical and Computer Modelling*, 37(9-10):915–922.

Nelson, W. (2004). *Applied Life Data Analysis*. John Wiley & Sons, New York.

North, M. (1980). Time-dependent stochastic model of floods. *Journal of the Hydrological Division, ASCE*, HY5:649–665.

Onorato, M., Osborne, A. R., and Serio, M. (2002). Extreme wave events in directional, random oceanic sea states. *Physics of Fluids*, 14(4):L25–L28.

Peng, L. (1999). Estimation of the coefficient of tail dependence in bivariate extremes. *Statistics and Probability Letters*, 43(4):399–409.

Phoenix, S. L. (1978). The asymptotic time to failure of a mechanical system of parallel members. *SIAM Journal of Applied Mathematics*, 34:227–246.

Phoenix, S. L. and Smith, R. L. (1983). A comparison of probabilistic techniques for the strength of fibrous materials under local load-sharing among fibers. *International Journal of Solids Structures*, 19:479–496.

Phoenix, S. L. and Tierney, L. J. (1983). A statistical model for the time dependent failure of unidirectional composite materials under local elastic load-sharing among fibers. *Engineering Fracture Mechanics*, 18:193–215.

Phoenix, S. L. and Wu, E. M. (1983). Statistics for the time dependent failure of kevlar-49/epoxy composites: micromechanical modeling and data interpretation. *Mechanics of Composite Materials*, pp. 135–215.

Pickands, J., III (1975). Statistical inference using extreme order statistics. *Annals of Statistics*, 3:119–131.

Prentice, R. L. (1974). A log gamma model and its maximum likelihood estimation. *Biometrika*, 61:539–544.

Prescott, P. and Walden, A. T. (1980). Maximum likelihood estimation of the parameters of the generalized extreme-value distribution. *Biometrika*, 67:723–724.

Prot, M. (1949a). La securité. *Annales of Ponts and Chaussées*, 119:19–49.

Prot, M. (1949b). Statistique et sécurité. *Revue de Métallurgie*, 46:716–718.

Prot, M. (1950). Vues nouvelles sur la sécurité des constructions. *Mémoire de la Societé de Ingeniers Civils de France*, 103:50–57.

Putz, R. R. (1952). Statistical distribution for ocean waves. *Transactions of the American Geophysical Union*, 33:685–692.

Reiss, R. D. and Thomas, M. (2001). *Statistical Analysis of Extreme Values*. Birkhäuser Verlag, Basel, 2nd ed.

Rencher, A. C. (2002). *Methods of Multivariate Analysis*. John Wiley & Sons, New York, 2nd ed.

Resnick, S. (1987). *Extreme Values, Regular Variation, and Point Processes*. Springer-Verlag, New York.

Roberts, E. M. (1979a). Review of statistics of extreme values with applications to air quality data, Part I: Review. *Journal of the Air Pollution and Control Association*, 29:632–637.

Roberts, E. M. (1979b). Review of statistics of extreme values with applications to air quality data, Part II: Applications. *Journal of the Air Pollution and Control Association*, 29:733–740.

Ross, S. M. (1992). *Applied Probability Models with Optimization Applications*. Dover Publications, New York. (Reprint of the 1970 original).

Rychlik, I. (1996). Fatigue and stochastic loads. *Scandinavian Journal of Statistics*, 23(4):387–404.

Sachs, P. (1972). *Wind Forces in Engineering*. Pergamon Press, Oxford, England.

Schlather, M. (2001). Examples for the coefficient of tail dependence and the domain of attraction of a bivariate extreme value distribution. *Statistics and Probability Letters*, 53(3):325–329.

Sellars, F. (1975). Maximum heights of ocean waves. *Journal of Geophysical Research*, 80:398–404.

Shane, R. and Lynn, W. (1964). Mathematical model for flood risk evaluation. *Journal of the Hydraulics Division, ASCE*, 90:1–20.

Simiu, E., Biétry, J., and Filliben, J. J. (1978). Sampling error in estimation of extreme winds. *Journal of the Structural Division, ASCE*, pp. 491–501.

Simiu, E., Changery, M. J., and Filliben, J. J. (1979). Extreme wind speeds at 129 stations in the contiguous United States. Tech. rep., National Bureau of Standards, Washington, D.C.

Simiu, E. and Filliben, J. J. (1975). Statistical analysis of extreme winds. NBS TR-868. Tech. rep., National Bureau of Standards, Washington, D.C.

Simiu, E. and Filliben, J. J. (1976). Probability distribution of extreme wind speeds. *Journal of the Structural Division, ASCE*, 102, ST9:1861–1877.

Simiu, E., Filliben, J. J., and Shaver, J. R. (1982). Short term records and extreme wind speeds. *Journal of the Structural Division, ASCE*, 108:2571–2577.

Simiu, E. and Scanlan, R. H. (1977). *Wind Effects on Structures: An Introduction to Wind Engineering*. John Wiley & Sons, New York.

Simonoff, J. S. (1996). *Smoothing Methods in Statistics*. Springer-Verlag, New York.

Singpurwalla, N. D. (1972). Extreme values for a lognormal law with applications to air pollution problems. *Technometrics*, 14:703–711.

Sjö, E. (2000). *Crossings and Maxima in Gaussian Fields and Seas*. Doctoral Theses, Lund University, Centre for Mathematical Sciences, Lund, Sweden.

Sjö, E. (2001). Simultaneous distributions of space-time wave characteristics in a Gaussian sea. *Extremes*, 4(3):263–288.

Smith, R. L. (1980). A probabilistic model for fibrous composite local loadsharing. *Proceedings of the Royal Society of London*, A372:539–553.

Smith, R. L. (1981). Asymptotic distribution for the failure of fibrous materials under series-parallel structure and equal load-sharing. *Journal of Applied Mechanics*, 103:75–82.

Smith, R. L. (1985). Maximum likelihood estimation in a class of nonregular cases. *Biometrika*, 72:67–90.

Smith, R. L. (1987). Estimating tails of probability distributions. *Annals of Statistics*, 15:1174–1207.

Smith, R. L. (1990). Extreme value theory. In S. V. W. Ledermann, E. Lloyd and C. Alexander, eds., *Handbook of Applicable Mathematics Supplement*, pp. 437–472. John Wiley & Sons, New York.

Smith, R. L. (1994). Multivariate threshold methods. In J. L. J. Galambos and E. Simiu, eds., *Extreme Value Theory and Applications*, pp. 225–248. Kluwer Academic Publishers, Dordrecht.

Smith, R. L., Tawn, J. A., and Yuen, H. K. (1990). Statistics of multivariate extremes. *International Statistical Review*, 58:47–58.

Smith, R. L. and Weissman, I. (1985). Maximum likelihood estimation of the lower tail of a probability distribution. *Journal of the Royal Statistical Society, B*, 47:285–298.

Sneyers, R. (1984). Extremes in meteorology. In *Statistical Extremes and Applications*. NATO ASI Series. D. Reidel Publishing Company, Dordrecht, The Netherlands.

Spindel, J. E., Board, B. R., and Haibach, E. (1979). *The Statistical Analysis of Fatigue Test Results*. ORE, Utrech.

Takahashi, R. and Sibuya, M. (2002). Metal fatigue, Wicksell transform and extreme values. *Applied Stochastic Models in Business and Industry*, 18(3):301–312.

Tawn, J. A. (1988). Bivariate extreme value theory: models and estimation. *Biometrika*, 75:397–415.

Taylor, H. M. (1994). The Poisson-Weibull flaw model for brittle strength. In *Extreme Value Theory ans Applications: Proceedings of the Conference on Extreme Value Theory and Applications*. Kluwer Academic Publications, Dordrecht, The Netherlands.

Thiruvengadam, A. (1972). Corrosion fatigue at high frequencies and high hydrostatic pressures. stress corrosion cracking of metals. a state of the art. Tech. rep., ASTM, Philadelphia, PA.

Thoft-Christensen, P., ed. (). *Reliability and Optimization of Structural Systems '88*.

Thom, H. C. S. (1967). Asymptotic extreme value distributions applied to wind and waves. In *NATO Seminar on Extreme Value Problems*. Faro, Portugal.

Thom, H. C. S. (1968a). New distributions of extreme winds in the United States. *Journal of the Structural Division, ASCE*, 94:1787–1801.

Thom, H. C. S. (1968b). Toward a universal climatological extreme wind distribution. In *International Research Seminar on Wind effects on Buildings and Structures, National Research Council Proceedings*. Ottawa, Canada.

Thom, H. C. S. (1969). Application of climatological analysis to engineering design data. *Revue Belge Statistique and Recherche Opérationel*, 9:2–13.

Thom, H. C. S. (1971). Asymptotic extreme value distributions of wave height in open ocean. *Journal of Marine Research*, 29:19–27.

Thom, H. C. S. (1973). Extreme wave height distributions over oceans. *Journal of Waterway, Port, Coastal, and Ocean Engineering, ASCE*, 99:355–374.

Thrasher, L. W. and Aagard, P. M. (1970). Measured wave force data on offshore platforms. *Journal of Petroleum Technology*, pp. 339–346.

Tiago de Oliveira, J. (1958). Extremal distributions. *Revista de la Faculdade de Ciencias*, A7:215–227.

Tiago de Oliveira, J. (1979). Extreme values and applications. In *Applications of Statistics and Probability in Soil and Structural Engineering*, vol. III, pp. 127–135. Unisearch, Kensington.

Tide, R. H. R. and van Horn, D. (1966). A statistical study of the static and fatigue properties of high strength prestressing strand. Tech. rep., Fritz Engineering Laboratory, Lehigh University.

Tierney, L. (1982). Asymptotic bounds on the time to fatigue failure of bundles of fibers under local load-sharing. *Advances in Applied Probability*, 14:95–121.

Tilly, G. P. and Moss, D. S. (1982). Long endurance of steel reinforcement. In *IABSE Colloquium on Fatigue of Steel and Concrete Structures*, pp. 229–238. Lausanne.

Todorovic, P. (1978). Stochastic models of floods. *Water Resources Research*, pp. 345–356.

Todorovic, P. (1979). A probabilistic approach to analysis and prediction of floods. In *Proceedings of the 43rd Session of the International Statistical Institute*. Buenos Aires.

Tucker, M. J. (1963). Analysis of records of sea waves. *Proceedings of the Institute of Civil Engineers, London*, 26:305–316.

Wackerly, D. D., Mendenhall, W., and Scheaffer, R. L. (2001). *Mathematical Statistics with Applications*. PWS-Kent Publishing Company, Boston, MA., sixth edition ed.

Walshaw, D. (2000). Modelling extreme wind speeds in regions prone to hurricanes. *Journal of the Royal Statistical Society, C*, 49(1):51–62.

Warner, R. F. and Hulsbos, C. L. (1966). Fatigue properties of prestressing strand. *PCI Journal*, 2:25–46.

Weber, K. H. and Endicott, H. S. (1956). Area effect and its extremal basis for the electric breakdown of transformer oil. *AIEE Transactions, Part III: Power Apparatus and Systems*, 75:371–381.

Weber, K. H. and Endicott, H. S. (1957). Extremal area effect for large area electrodes for the electric breakdown of transformer oil. *AIEE Transactions, Part III: Power Apparatus and Systems*, 76:1091–1098.

Weibull, W. (1951). A statistical distribution function of wide applicability. *Journal of Applied Mechanics*, 18:293–297.

Weibull, W. (1952). Statistical design of fatigue experiments. *Journal of Applied Mechanics*, 19:109–113.

Weibull, W. (1959). Zur abhängigkeit der festigkeit von der probegrösse. *Ingenieur Archiv*, 28:360–362.

Wiegel, R. L. (1964). *Oceanographical Engineering*. Prentice Hall, Englewood Cliffs, N.J.

Wilson, B. W. (1966). Design sea and wind conditions for offshore structures. In *Proceedings of the OECON, Offshore Exploration Conference*, pp. 665–708.

Worms, R. (2001). Vitesse de convergence de l'appoximation de Pareto généralisée de la loi des excès. *Comptes Rendus de l'Académie des Sciences. Série I*, 333(1):65–70.

Wu, J. W. and Li, P. L. (2003). Optimal parameter estimation of the extreme value distribution based on a type II censored sample. *Communications in Statistics, Theory and Methods*, 32(3):533–554.

Yang, C. Y., Tayfun, M. A., and Hsiao, G. C. (1974). Stochastic prediction of extreme waves and sediment transport in coastal waters. In *Stochastic Problems in Mechanics. Proceedings of the Symposium, University of Waterloo, Ontario*, pp. 431–448. Canada Waterloo Press.

Yang, G. L. (1978). Estimation of a biometric function. *Annals of Statistics*, 6:112–116.

Young, D. H. and Bakir, S. T. (1987). Bias correction for a generalized log-gamma regression model. *Technometrics*, 29:183–191.

Yun, S. (2002). On a generalized Pickands estimator of the extreme value index. *Journal of Statistical Planning and Inference*, 102(2):389–409.

Zelenhasic, E. (1970). Theoretical probability distribution for flood peaks. Tech. rep., Hydrologic Paper 42, Colorado State University, Fort Collins, Colorado.

Zidek, J. V., Navin, F. P. D., and Lockhart, R. (1979). Statistics of extremes: an alternate method with application to bridge design codes. *Technometrics*, 231:185–191.

Index

WILEY SERIES IN PROBABILITY AND STATISTICS
ESTABLISHED BY WALTER A. SHEWHART AND SAMUEL S. WILKS

The ***Wiley Series in Probability and Statistics*** is well established and authoritative. It covers many topics of current research interest in both pure and applied statistics and probability theory. Written by leading statisticians and institutions, the titles span both state-of-the-art developments in the field and classical methods.

Reflecting the wide range of current research in statistics, the series encompasses applied, methodological and theoretical statistics, ranging from applications and new techniques made possible by advances in computerized practice to rigorous treatment of theoretical approaches.

This series provides essential and invaluable reading for all statisticians, whether in academia, industry, government, or research.

ABRAHAM and LEDOLTER · Statistical Methods for Forecasting
AGRESTI · Analysis of Ordinal Categorical Data
AGRESTI · An Introduction to Categorical Data Analysis
AGRESTI · Categorical Data Analysis, *Second Edition*
ALTMAN, GILL, and McDONALD · Numerical Issues in Statistical Computing for the Social Scientist
AMARATUNGA and CABRERA · Exploration and Analysis of DNA Microarray and Protein Array Data
ANDĚL · Mathematics of Chance
ANDERSON · An Introduction to Multivariate Statistical Analysis, *Third Edition*
*ANDERSON · The Statistical Analysis of Time Series
ANDERSON, AUQUIER, HAUCK, OAKES, VANDAELE, and WEISBERG · Statistical Methods for Comparative Studies
ANDERSON and LOYNES · The Teaching of Practical Statistics
ARMITAGE and DAVID (editors) · Advances in Biometry
ARNOLD, BALAKRISHNAN, and NAGARAJA · Records
*ARTHANARI and DODGE · Mathematical Programming in Statistics
*BAILEY · The Elements of Stochastic Processes with Applications to the Natural Sciences
BALAKRISHNAN and KOUTRAS · Runs and Scans with Applications
BARNETT · Comparative Statistical Inference, *Third Edition*
BARNETT and LEWIS · Outliers in Statistical Data, *Third Edition*
BARTOSZYNSKI and NIEWIADOMSKA-BUGAJ · Probability and Statistical Inference
BASILEVSKY · Statistical Factor Analysis and Related Methods: Theory and Applications
BASU and RIGDON · Statistical Methods for the Reliability of Repairable Systems
BATES and WATTS · Nonlinear Regression Analysis and Its Applications
BECHHOFER, SANTNER, and GOLDSMAN · Design and Analysis of Experiments for Statistical Selection, Screening, and Multiple Comparisons
BELSLEY · Conditioning Diagnostics: Collinearity and Weak Data in Regression

*Now available in a lower priced paperback edition in the Wiley Classics Library.

* BELSLEY, KUH, and WELSCH · Regression Diagnostics: Identifying Influential Data and Sources of Collinearity
BENDAT and PIERSOL · Random Data: Analysis and Measurement Procedures, *Third Edition*
BERRY, CHALONER, and GEWEKE · Bayesian Analysis in Statistics and Econometrics: Essays in Honor of Arnold Zellner
BERNARDO and SMITH · Bayesian Theory
BHAT and MILLER · Elements of Applied Stochastic Processes, *Third Edition*
BHATTACHARYA and WAYMIRE · Stochastic Processes with Applications
BILLINGSLEY · Convergence of Probability Measures, *Second Edition*
BILLINGSLEY · Probability and Measure, *Third Edition*
BIRKES and DODGE · Alternative Methods of Regression
BLISCHKE AND MURTHY (editors) · Case Studies in Reliability and Maintenance
BLISCHKE AND MURTHY · Reliability: Modeling, Prediction, and Optimization
BLOOMFIELD · Fourier Analysis of Time Series: An Introduction, *Second Edition*
BOLLEN · Structural Equations with Latent Variables
BOROVKOV · Ergodicity and Stability of Stochastic Processes
BOULEAU · Numerical Methods for Stochastic Processes
BOX · Bayesian Inference in Statistical Analysis
BOX · R. A. Fisher, the Life of a Scientist
BOX and DRAPER · Empirical Model-Building and Response Surfaces
*BOX and DRAPER · Evolutionary Operation: A Statistical Method for Process Improvement
BOX, HUNTER, and HUNTER · Statistics for Experimenters: An Introduction to Design, Data Analysis, and Model Building
BOX and LUCEÑO · Statistical Control by Monitoring and Feedback Adjustment
BRANDIMARTE · Numerical Methods in Finance: A MATLAB-Based Introduction
BROWN and HOLLANDER · Statistics: A Biomedical Introduction
BRUNNER, DOMHOF, and LANGER · Nonparametric Analysis of Longitudinal Data in Factorial Experiments
BUCKLEW · Large Deviation Techniques in Decision, Simulation, and Estimation
CAIROLI and DALANG · Sequential Stochastic Optimization
CASTILLO, HADI, BALAKRISHNAN, and SARABIA · Extreme Value and Related Models with Applications in Engineering and Science
CHAN · Time Series: Applications to Finance
CHATTERJEE and HADI · Sensitivity Analysis in Linear Regression
CHATTERJEE and PRICE · Regression Analysis by Example, *Third Edition*
CHERNICK · Bootstrap Methods: A Practitioner's Guide
CHERNICK and FRIIS · Introductory Biostatistics for the Health Sciences
CHILÈS and DELFINER · Geostatistics: Modeling Spatial Uncertainty
CHOW and LIU · Design and Analysis of Clinical Trials: Concepts and Methodologies, *Second Edition*
CLARKE and DISNEY · Probability and Random Processes: A First Course with Applications, *Second Edition*
*COCHRAN and COX · Experimental Designs, *Second Edition*
CONGDON · Applied Bayesian Modelling
CONGDON · Bayesian Statistical Modelling
CONOVER · Practical Nonparametric Statistics, *Third Edition*
COOK · Regression Graphics
COOK and WEISBERG · Applied Regression Including Computing and Graphics
COOK and WEISBERG · An Introduction to Regression Graphics
CORNELL · Experiments with Mixtures, Designs, Models, and the Analysis of Mixture Data, *Third Edition*

*Now available in a lower priced paperback edition in the Wiley Classics Library.

COVER and THOMAS · Elements of Information Theory
COX · A Handbook of Introductory Statistical Methods
*COX · Planning of Experiments
CRESSIE · Statistics for Spatial Data, *Revised Edition*
CSÖRGŐ and HORVÁTH · Limit Theorems in Change Point Analysis
DANIEL · Applications of Statistics to Industrial Experimentation
DANIEL · Biostatistics: A Foundation for Analysis in the Health Sciences, *Eighth Edition*
*DANIEL · Fitting Equations to Data: Computer Analysis of Multifactor Data, *Second Edition*
DASU and JOHNSON · Exploratory Data Mining and Data Cleaning
DAVID and NAGARAJA · Order Statistics, *Third Edition*
*DEGROOT, FIENBERG, and KADANE · Statistics and the Law
DEL CASTILLO · Statistical Process Adjustment for Quality Control
DEMARIS · Regression with Social Data: Modeling Continuous and Limited Response Variables
DEMIDENKO · Mixed Models: Theory and Applications
DENISON, HOLMES, MALLICK and SMITH · Bayesian Methods for Nonlinear Classification and Regression
DETTE and STUDDEN · The Theory of Canonical Moments with Applications in Statistics, Probability, and Analysis
DEY and MUKERJEE · Fractional Factorial Plans
DILLON and GOLDSTEIN · Multivariate Analysis: Methods and Applications
DODGE · Alternative Methods of Regression
*DODGE and ROMIG · Sampling Inspection Tables, *Second Edition*
*DOOB · Stochastic Processes
DOWDY, WEARDEN, and CHILKO · Statistics for Research, *Third Edition*
DRAPER and SMITH · Applied Regression Analysis, *Third Edition*
DRYDEN and MARDIA · Statistical Shape Analysis
DUDEWICZ and MISHRA · Modern Mathematical Statistics
DUNN and CLARK · Basic Statistics: A Primer for the Biomedical Sciences, *Third Edition*
DUPUIS and ELLIS · A Weak Convergence Approach to the Theory of Large Deviations
*ELANDT-JOHNSON and JOHNSON · Survival Models and Data Analysis
ENDERS · Applied Econometric Time Series
ETHIER and KURTZ · Markov Processes: Characterization and Convergence
EVANS, HASTINGS, and PEACOCK · Statistical Distributions, *Third Edition*
FELLER · An Introduction to Probability Theory and Its Applications, Volume I, *Third Edition,* Revised; Volume II, *Second Edition*
FISHER and VAN BELLE · Biostatistics: A Methodology for the Health Sciences
FITZMAURICE, LAIRD, and WARE · Applied Longitudinal Analysis
*FLEISS · The Design and Analysis of Clinical Experiments
FLEISS · Statistical Methods for Rates and Proportions, *Third Edition*
FLEMING and HARRINGTON · Counting Processes and Survival Analysis
FULLER · Introduction to Statistical Time Series, *Second Edition*
FULLER · Measurement Error Models
GALLANT · Nonlinear Statistical Models
GHOSH, MUKHOPADHYAY, and SEN · Sequential Estimation
GIESBRECHT and GUMPERTZ · Planning, Construction, and Statistical Analysis of Comparative Experiments
GIFI · Nonlinear Multivariate Analysis
GLASSERMAN and YAO · Monotone Structure in Discrete-Event Systems
GNANADESIKAN · Methods for Statistical Data Analysis of Multivariate Observations, *Second Edition*

*Now available in a lower priced paperback edition in the Wiley Classics Library.

GOLDSTEIN and LEWIS · Assessment: Problems, Development, and Statistical Issues
GREENWOOD and NIKULIN · A Guide to Chi-Squared Testing
GROSS and HARRIS · Fundamentals of Queueing Theory, *Third Edition*
*HAHN and SHAPIRO · Statistical Models in Engineering
HAHN and MEEKER · Statistical Intervals: A Guide for Practitioners
HALD · A History of Probability and Statistics and their Applications Before 1750
HALD · A History of Mathematical Statistics from 1750 to 1930
HAMPEL · Robust Statistics: The Approach Based on Influence Functions
HANNAN and DEISTLER · The Statistical Theory of Linear Systems
HEIBERGER · Computation for the Analysis of Designed Experiments
HEDAYAT and SINHA · Design and Inference in Finite Population Sampling
HELLER · MACSYMA for Statisticians
HINKELMAN and KEMPTHORNE: · Design and Analysis of Experiments, Volume 1: Introduction to Experimental Design
HOAGLIN, MOSTELLER, and TUKEY · Exploratory Approach to Analysis of Variance
HOAGLIN, MOSTELLER, and TUKEY · Exploring Data Tables, Trends and Shapes
*HOAGLIN, MOSTELLER, and TUKEY · Understanding Robust and Exploratory Data Analysis
HOCHBERG and TAMHANE · Multiple Comparison Procedures
HOCKING · Methods and Applications of Linear Models: Regression and the Analysis of Variance, *Second Edition*
HOEL · Introduction to Mathematical Statistics, *Fifth Edition*
HOGG and KLUGMAN · Loss Distributions
HOLLANDER and WOLFE · Nonparametric Statistical Methods, *Second Edition*
HOSMER and LEMESHOW · Applied Logistic Regression, *Second Edition*
HOSMER and LEMESHOW · Applied Survival Analysis: Regression Modeling of Time to Event Data
HUBER · Robust Statistics
HUBERTY · Applied Discriminant Analysis
HUNT and KENNEDY · Financial Derivatives in Theory and Practice
HUSKOVA, BERAN, and DUPAC · Collected Works of Jaroslav Hajek—with Commentary
HUZURBAZAR · Flowgraph Models for Multistate Time-to-Event Data
IMAN and CONOVER · A Modern Approach to Statistics
JACKSON · A User's Guide to Principle Components
JOHN · Statistical Methods in Engineering and Quality Assurance
JOHNSON · Multivariate Statistical Simulation
JOHNSON and BALAKRISHNAN · Advances in the Theory and Practice of Statistics: A Volume in Honor of Samuel Kotz
JOHNSON and BHATTACHARYYA · Statistics: Principles and Methods, *Fifth Edition*
JOHNSON and KOTZ · Distributions in Statistics
JOHNSON and KOTZ (editors) · Leading Personalities in Statistical Sciences: From the Seventeenth Century to the Present
JOHNSON, KOTZ, and BALAKRISHNAN · Continuous Univariate Distributions, Volume 1, *Second Edition*
JOHNSON, KOTZ, and BALAKRISHNAN · Continuous Univariate Distributions, Volume 2, *Second Edition*
JOHNSON, KOTZ, and BALAKRISHNAN · Discrete Multivariate Distributions
JOHNSON, KOTZ, and KEMP · Univariate Discrete Distributions, *Second Edition*
JUDGE, GRIFFITHS, HILL, LÜTKEPOHL, and LEE · The Theory and Practice of Econometrics, *Second Edition*
JUREČKOVÁ and SEN · Robust Statistical Procedures: Aymptotics and Interrelations

*Now available in a lower priced paperback edition in the Wiley Classics Library.

JUREK and MASON · Operator-Limit Distributions in Probability Theory
KADANE · Bayesian Methods and Ethics in a Clinical Trial Design
KADANE AND SCHUM · A Probabilistic Analysis of the Sacco and Vanzetti Evidence
KALBFLEISCH and PRENTICE · The Statistical Analysis of Failure Time Data, *Second Edition*
KASS and VOS · Geometrical Foundations of Asymptotic Inference
KAUFMAN and ROUSSEEUW · Finding Groups in Data: An Introduction to Cluster Analysis
KEDEM and FOKIANOS · Regression Models for Time Series Analysis
KENDALL, BARDEN, CARNE, and LE · Shape and Shape Theory
KHURI · Advanced Calculus with Applications in Statistics, *Second Edition*
KHURI, MATHEW, and SINHA · Statistical Tests for Mixed Linear Models
*KISH · Statistical Design for Research
KLEIBER and KOTZ · Statistical Size Distributions in Economics and Actuarial Sciences
KLUGMAN, PANJER, and WILLMOT · Loss Models: From Data to Decisions
KLUGMAN, PANJER, and WILLMOT · Solutions Manual to Accompany Loss Models: From Data to Decisions
KOTZ, BALAKRISHNAN, and JOHNSON · Continuous Multivariate Distributions, Volume 1, *Second Edition*
KOTZ and JOHNSON (editors) · Encyclopedia of Statistical Sciences: Volumes 1 to 9 with Index
KOTZ and JOHNSON (editors) · Encyclopedia of Statistical Sciences: Supplement Volume
KOTZ, READ, and BANKS (editors) · Encyclopedia of Statistical Sciences: Update Volume 1
KOTZ, READ, and BANKS (editors) · Encyclopedia of Statistical Sciences: Update Volume 2
KOVALENKO, KUZNETZOV, and PEGG · Mathematical Theory of Reliability of Time-Dependent Systems with Practical Applications
LACHIN · Biostatistical Methods: The Assessment of Relative Risks
LAD · Operational Subjective Statistical Methods: A Mathematical, Philosophical, and Historical Introduction
LAMPERTI · Probability: A Survey of the Mathematical Theory, *Second Edition*
LANGE, RYAN, BILLARD, BRILLINGER, CONQUEST, and GREENHOUSE · Case Studies in Biometry
LARSON · Introduction to Probability Theory and Statistical Inference, *Third Edition*
LAWLESS · Statistical Models and Methods for Lifetime Data, *Second Edition*
LAWSON · Statistical Methods in Spatial Epidemiology
LE · Applied Categorical Data Analysis
LE · Applied Survival Analysis
LEE and WANG · Statistical Methods for Survival Data Analysis, *Third Edition*
LEPAGE and BILLARD · Exploring the Limits of Bootstrap
LEYLAND and GOLDSTEIN (editors) · Multilevel Modelling of Health Statistics
LIAO · Statistical Group Comparison
LINDVALL · Lectures on the Coupling Method
LINHART and ZUCCHINI · Model Selection
LITTLE and RUBIN · Statistical Analysis with Missing Data, *Second Edition*
LLOYD · The Statistical Analysis of Categorical Data
MAGNUS and NEUDECKER · Matrix Differential Calculus with Applications in Statistics and Econometrics, *Revised Edition*
MALLER and ZHOU · Survival Analysis with Long Term Survivors
MALLOWS · Design, Data, and Analysis by Some Friends of Cuthbert Daniel
MANN, SCHAFER, and SINGPURWALLA · Methods for Statistical Analysis of Reliability and Life Data

*Now available in a lower priced paperback edition in the Wiley Classics Library.

MANTON, WOODBURY, and TOLLEY · Statistical Applications Using Fuzzy Sets
MARCHETTE · Random Graphs for Statistical Pattern Recognition
MARDIA and JUPP · Directional Statistics
MASON, GUNST, and HESS · Statistical Design and Analysis of Experiments with Applications to Engineering and Science, *Second Edition*
McCULLOCH and SEARLE · Generalized, Linear, and Mixed Models
McFADDEN · Management of Data in Clinical Trials
* McLACHLAN · Discriminant Analysis and Statistical Pattern Recognition
McLACHLAN, DO, and AMBROISE · Analyzing Microarray Gene Expression Data
McLACHLAN and KRISHNAN · The EM Algorithm and Extensions
McLACHLAN and PEEL · Finite Mixture Models
McNEIL · Epidemiological Research Methods
MEEKER and ESCOBAR · Statistical Methods for Reliability Data
MEERSCHAERT and SCHEFFLER · Limit Distributions for Sums of Independent Random Vectors: Heavy Tails in Theory and Practice
MICKEY, DUNN, and CLARK · Applied Statistics: Analysis of Variance and Regression, *Third Edition*
*MILLER · Survival Analysis, *Second Edition*
MONTGOMERY, PECK, and VINING · Introduction to Linear Regression Analysis, *Third Edition*
MORGENTHALER and TUKEY · Configural Polysampling: A Route to Practical Robustness
MUIRHEAD · Aspects of Multivariate Statistical Theory
MULLER and STOYAN · Comparison Methods for Stochastic Models and Risks
MURRAY · X-STAT 2.0 Statistical Experimentation, Design Data Analysis, and Nonlinear Optimization
MURTHY, XIE, and JIANG · Weibull Models
MYERS and MONTGOMERY · Response Surface Methodology: Process and Product Optimization Using Designed Experiments, *Second Edition*
MYERS, MONTGOMERY, and VINING · Generalized Linear Models. With Applications in Engineering and the Sciences
*NELSON · Accelerated Testing, Statistical Models, Test Plans, and Data Analyses
NELSON · Applied Life Data Analysis
NEWMAN · Biostatistical Methods in Epidemiology
OCHI · Applied Probability and Stochastic Processes in Engineering and Physical Sciences
OKABE, BOOTS, SUGIHARA, and CHIU · Spatial Tesselations: Concepts and Applications of Voronoi Diagrams, *Second Edition*
OLIVER and SMITH · Influence Diagrams, Belief Nets and Decision Analysis
PALTA · Quantitative Methods in Population Health: Extensions of Ordinary Regressions
PANKRATZ · Forecasting with Dynamic Regression Models
PANKRATZ · Forecasting with Univariate Box-Jenkins Models: Concepts and Cases
*PARZEN · Modern Probability Theory and Its Applications
PEÑA, TIAO, and TSAY · A Course in Time Series Analysis
PIANTADOSI · Clinical Trials: A Methodologic Perspective
PORT · Theoretical Probability for Applications
POURAHMADI · Foundations of Time Series Analysis and Prediction Theory
PRESS · Bayesian Statistics: Principles, Models, and Applications
PRESS · Subjective and Objective Bayesian Statistics, *Second Edition*
PRESS and TANUR · The Subjectivity of Scientists and the Bayesian Approach
PUKELSHEIM · Optimal Experimental Design
PURI, VILAPLANA, and WERTZ · New Perspectives in Theoretical and Applied Statistics
PUTERMAN · Markov Decision Processes: Discrete Stochastic Dynamic Programming
*RAO · Linear Statistical Inference and Its Applications, *Second Edition*

*Now available in a lower priced paperback edition in the Wiley Classics Library.

RAUSAND and HØYLAND · System Reliability Theory: Models, Statistical Methods, and Applications, *Second Edition*
RENCHER · Linear Models in Statistics
RENCHER · Methods of Multivariate Analysis, *Second Edition*
RENCHER · Multivariate Statistical Inference with Applications
* RIPLEY · Spatial Statistics
RIPLEY · Stochastic Simulation
ROBINSON · Practical Strategies for Experimenting
ROHATGI and SALEH · An Introduction to Probability and Statistics, *Second Edition*
ROLSKI, SCHMIDLI, SCHMIDT, and TEUGELS · Stochastic Processes for Insurance and Finance
ROSENBERGER and LACHIN · Randomization in Clinical Trials: Theory and Practice
ROSS · Introduction to Probability and Statistics for Engineers and Scientists
ROUSSEEUW and LEROY · Robust Regression and Outlier Detection
RUBIN · Multiple Imputation for Nonresponse in Surveys
RUBINSTEIN · Simulation and the Monte Carlo Method
RUBINSTEIN and MELAMED · Modern Simulation and Modeling
RYAN · Modern Regression Methods
RYAN · Statistical Methods for Quality Improvement, *Second Edition*
SALTELLI, CHAN, and SCOTT (editors) · Sensitivity Analysis
*SCHEFFE · The Analysis of Variance
SCHIMEK · Smoothing and Regression: Approaches, Computation, and Application
SCHOTT · Matrix Analysis for Statistics
SCHOUTENS · Levy Processes in Finance: Pricing Financial Derivatives
SCHUSS · Theory and Applications of Stochastic Differential Equations
SCOTT · Multivariate Density Estimation: Theory, Practice, and Visualization
*SEARLE · Linear Models
SEARLE · Linear Models for Unbalanced Data
SEARLE · Matrix Algebra Useful for Statistics
SEARLE, CASELLA, and McCULLOCH · Variance Components
SEARLE and WILLETT · Matrix Algebra for Applied Economics
SEBER and LEE · Linear Regression Analysis, *Second Edition*
*SEBER · Multivariate Observations
SEBER and WILD · Nonlinear Regression
SENNOTT · Stochastic Dynamic Programming and the Control of Queueing Systems
*SERFLING · Approximation Theorems of Mathematical Statistics
SHAFER and VOVK · Probability and Finance: It's Only a Game!
SILVAPULLE and SEN · Constrained Statistical Inference: Order, Inequality and Shape Constraints
SMALL and McLEISH · Hilbert Space Methods in Probability and Statistical Inference
SRIVASTAVA · Methods of Multivariate Statistics
STAPLETON · Linear Statistical Models
STAUDTE and SHEATHER · Robust Estimation and Testing
STOYAN, KENDALL, and MECKE · Stochastic Geometry and Its Applications, *Second Edition*
STOYAN and STOYAN · Fractals, Random Shapes and Point Fields: Methods of Geometrical Statistics
STYAN · The Collected Papers of T. W. Anderson: 1943–1985
SUTTON, ABRAMS, JONES, SHELDON, and SONG · Methods for Meta-Analysis in Medical Research
TANAKA · Time Series Analysis: Nonstationary and Noninvertible Distribution Theory
THOMPSON · Empirical Model Building
THOMPSON · Sampling, *Second Edition*
THOMPSON · Simulation: A Modeler's Approach

*Now available in a lower priced paperback edition in the Wiley Classics Library.

THOMPSON and SEBER · Adaptive Sampling
THOMPSON, WILLIAMS, and FINDLAY · Models for Investors in Real World Markets
TIAO, BISGAARD, HILL, PEÑA, and STIGLER (editors) · Box on Quality and Discovery: with Design, Control, and Robustness
TIERNEY · LISP-STAT: An Object-Oriented Environment for Statistical Computing and Dynamic Graphics
TSAY · Analysis of Financial Time Series
UPTON and FINGLETON · Spatial Data Analysis by Example, Volume II: Categorical and Directional Data
VAN BELLE · Statistical Rules of Thumb
VAN BELLE, FISHER, HEAGERTY, and LUMLEY · Biostatistics: A Methodology for the Health Sciences, *Second Edition*
VESTRUP · The Theory of Measures and Integration
VIDAKOVIC · Statistical Modeling by Wavelets
VINOD and REAGLE · Preparing for the Worst: Incorporating Downside Risk in Stock Market Investments
WALLER and GOTWAY · Applied Spatial Statistics for Public Health Data
WEERAHANDI · Generalized Inference in Repeated Measures: Exact MANOVA and Mixed Models
WEISBERG · Applied Linear Regression, *Third Edition*
WELSH · Aspects of Statistical Inference
WESTFALL and YOUNG · Resampling-Based Multiple Testing: Examples and Methods for *p*-Value Adjustment
WHITTAKER · Graphical Models in Applied Multivariate Statistics
WINKER · Optimization Heuristics in Economics: Applications of Threshold Accepting
WONNACOTT and WONNACOTT · Econometrics, *Second Edition*
WOODING · Planning Pharmaceutical Clinical Trials: Basic Statistical Principles
WOODWORTH · Biostatistics
WOOLSON and CLARKE · Statistical Methods for the Analysis of Biomedical Data, *Second Edition*
WU and HAMADA · Experiments: Planning, Analysis, and Parameter Design Optimization
YANG · The Construction Theory of Denumerable Markov Processes
*ZELLNER · An Introduction to Bayesian Inference in Econometrics
ZHOU, OBUCHOWSKI, and McCLISH · Statistical Methods in Diagnostic Medicine

*Now available in a lower priced paperback edition in the Wiley Classics Library.

Printed and bound by CPI Group (UK) Ltd, Croydon, CR0 4YY
16/09/2026
14963443-0002